BIG IDEAS MATH®

Algebra 1

Student Journal

- Maintaining Mathematical Proficiency
- Exploration Journal
- Notetaking with Vocabulary
- Extra Practice

Erie, Pennsylvania

Photo Credits

Cover Image Allies Interactive/Shutterstock.com

198 Eric Isselee/Shutterstock.com,
Vasyl Helevachuk/Shutterstock.com

Printed in the United States

ISBN 13: 978-1-60840-852-8
ISBN 10: 1-60840-852-3

789-VLP-18 17 16 15

Contents

Contents

Contents

Contents

Contents

Contents

Contents

About the Student Journal

Maintaining Mathematical Proficiency

The Maintaining Mathematical Proficiency corresponds to the Pupil Edition Chapter Opener. Here you have the opportunity to practice prior skills necessary to move forward.

Exploration Journal

The Exploration pages correspond to the Explorations and accompanying exercises in the Pupil Edition. Here you have room to show your work and record your answers.

Notetaking with Vocabulary

This student-friendly notetaking component is designed to be a reference for key vocabulary, properties, and core concepts from the lesson. There is room to add definitions in your words and take notes about the core concepts.

Extra Practice

Each section of the Pupil Edition has an additional Practice with room for you to show your work and record your answers.

Name________________________________ Date__________

Chapter 1 Maintaining Mathematical Proficiency

Add or subtract.

1. $-1 + (-3)$ **2.** $0 + (-12)$ **3.** $5 - (-2)$ **4.** $-4 - 7$

5. Find two pairs of integers whose sum is -6.

6. In a city, the record monthly high temperature for March is 56°F. The record monthly low temperature for March is −4°F. What is the range of temperatures for the month of March?

Multiply or divide.

7. $-2(13)$ **8.** $-8 \bullet (-5)$ **9.** $-14 \div 2$ **10.** $-30 \div (-3)$

11. Find two pairs of integers whose product is -20.

12. A football team loses 3 yards in 3 consecutive plays. What is the total yardage gained?

Name ______________________________ Date __________

1.1 Solving Simple Equations
For use with Exploration 1.1

Essential Question How can you use simple equations to solve real-life problems?

1 EXPLORATION: Measuring Angles

Go to *BigIdeasMath.com* for an interactive tool to investigate this exploration.

Work with a partner. Use a protractor to measure the angles of each quadrilateral. Complete the table to organize your results. (The notation $m\angle A$ denotes the measure of angle A.) How precise are your measurements?

a.

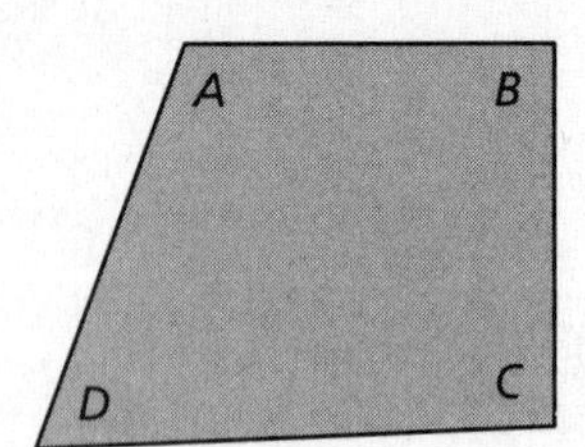

b.

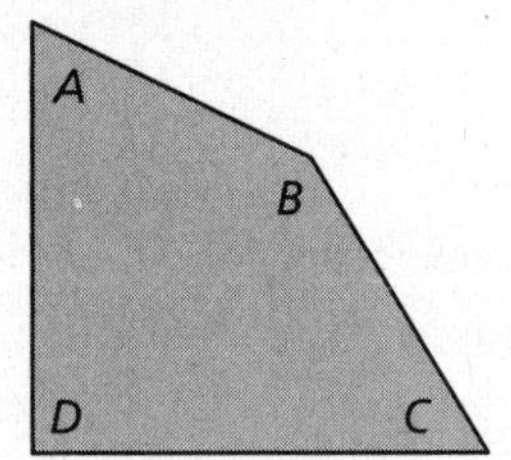

c.

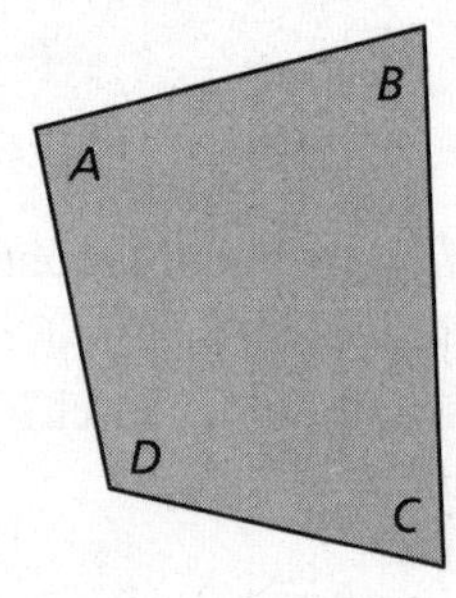

Quadrilateral	$m\angle A$ (degrees)	$m\angle B$ (degrees)	$m\angle C$ (degrees)	$m\angle D$ (degrees)	$m\angle A + m\angle B + m\angle C + m\angle D$
a.					
b.					
c.					

2 EXPLORATION: Making a Conjecture

Go to *BigIdeasMath.com* for an interactive tool to investigate this exploration.

Work with a partner. Use the completed table in Exploration 1 to write a conjecture about the sum of the angle measures of a quadrilateral. Draw three quadrilaterals that are different from those in Exploration 1 and use them to justify your conjecture.

3 EXPLORATION: Applying Your Conjecture

Go to *BigIdeasMath.com* for an interactive tool to investigate this exploration.

Work with a partner. Use the conjecture you wrote in Exploration 2 to write an equation for each quadrilateral. Then solve the equation to find the value of x. Use a protractor to check the reasonableness of your answer.

a.

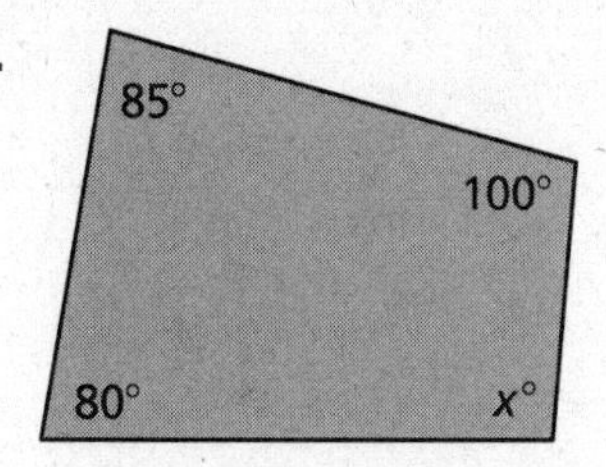

b.

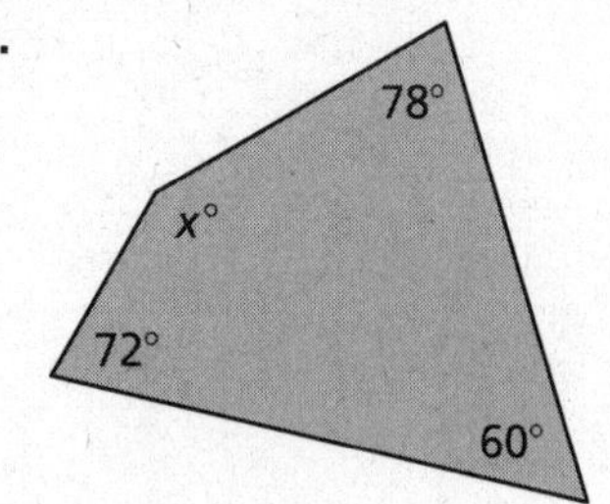

c.

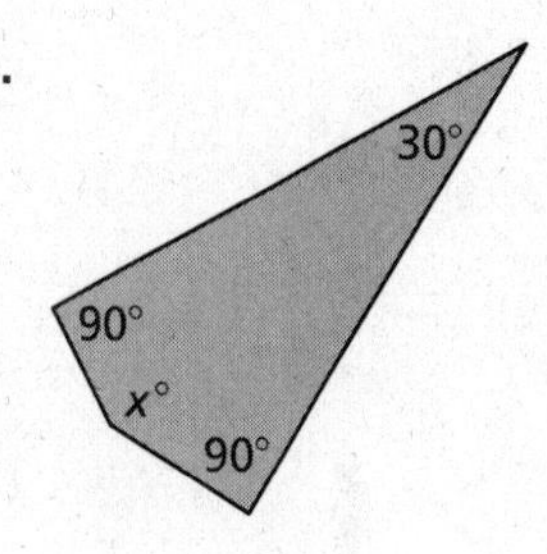

Communicate Your Answer

4. How can you use simple equations to solve real-life problems?

5. Draw your own quadrilateral and cut it out. Tear off the four corners of the quadrilateral and rearrange them to affirm the conjecture you wrote in Exploration 2. Explain how this affirms the conjecture.

Name ______________________________ Date __________

1.1 Notetaking with Vocabulary

For use after Lesson 1.1

In your own words, write the meaning of each vocabulary term.

conjecture

rule

theorem

equation

linear equation in one variable

solution

inverse operations

equivalent equations

Core Concepts

Addition Property of Equality

Words Adding the same number to each side of an equation produces an equivalent equation.

Algebra If $a = b$, then $a + c = b + c$.

Notes:

Name ______________________________ Date __________

1.1 Notetaking with Vocabulary (continued)

Subtraction Property of Equality

Words Subtracting the same number from each side of an equation produces an equivalent equation.

Algebra If $a = b$, then $a - c = b - c$.

Notes:

Multiplication Property of Equality

Words Multiplying each side of an equation by the same nonzero number produces an equivalent equation.

Algebra If $a = b$, then $a \bullet c = b \bullet c, c \neq 0$.

Notes:

Division Property of Equality

Words Dividing each side of an equation by the same nonzero number produces an equivalent equation.

Algebra If $a = b$, then $a \div c = b \div c, c \neq 0$.

Notes:

Four Step Approach to Problem Solving

1. **Understand the Problem** What is the unknown? What information is being given? What is being asked?

2. **Make a Plan** This plan might involve one or more of the problem-solving strategies shown on the following page.

3. **Solve the Problem** Carry out your plan. Check that each step is correct.

4. **Look Back** Examine your solution. Check that your solution makes sense in the original statement of the problem.

Notes:

Name ______________________________ Date __________

1.1 Notetaking with Vocabulary (continued)

Common Problem-Solving Strategies

Use a verbal model.	Guess, check, and revise.
Draw a diagram.	Sketch a graph or number line.
Write an equation.	Make a table.
Look for a pattern.	Make a list.
Work backward.	Break the problem into parts.

Notes:

Extra Practice

In Exercises 1–9, solve the equation. Justify each step. Check your solution.

1. $w + 4 = 16$

2. $x + 7 = -12$

3. $-15 + w = 6$

4. $z - 5 = 8$

5. $-2 = y - 9$

6. $7q = 35$

7. $4b = -52$

8. $3 = \frac{q}{11}$

9. $\frac{n}{-2} = -15$

10. A coupon subtracts $17.95 from the price p of a pair of headphones. You pay $71.80 for the headphones after using the coupon. Write and solve an equation to find the original price of the headphones.

11. After a party, you have $\frac{2}{5}$ of the brownies you made left over. There are 16 brownies left. How many brownies did you make for the party?

Name____ Date__________

1.2 Solving Multi-Step Equations

For use with Exploration 1.2

Essential Question How can you use multi-step equations to solve real-life problems?

1 EXPLORATION: Solving for the Angle Measures of a Polygon

Go to *BigIdeasMath.com* for an interactive tool to investigate this exploration.

Work with a partner. The sum S of the angle measures of a polygon with n sides can be found using the formula $S = 180(n - 2)$. Write and solve an equation to find each value of x. Justify the steps in your solution. Then find the angle measures of each polygon. How can you check the reasonableness of your answers?

a.

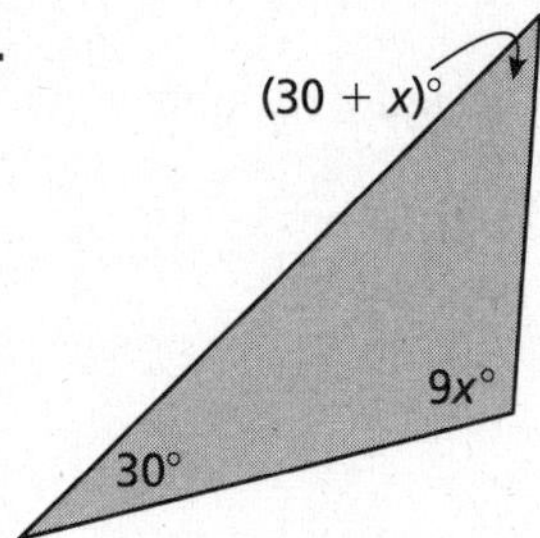

b.

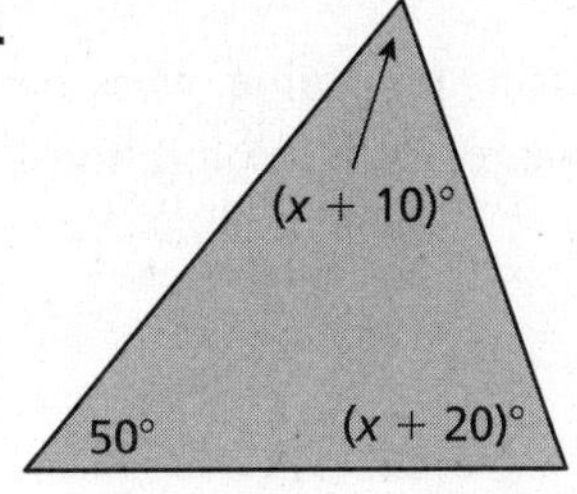

c.

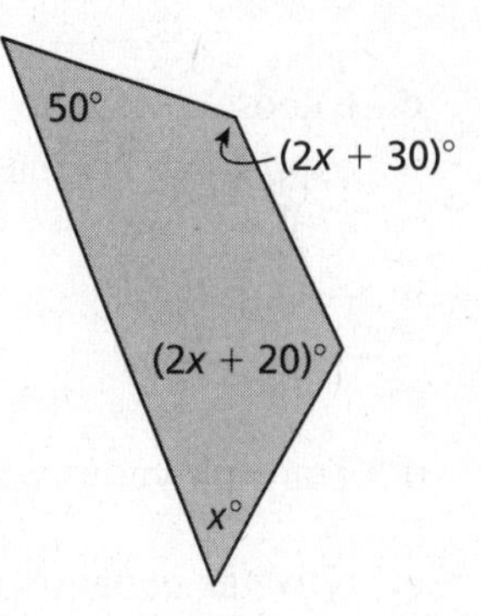

d.

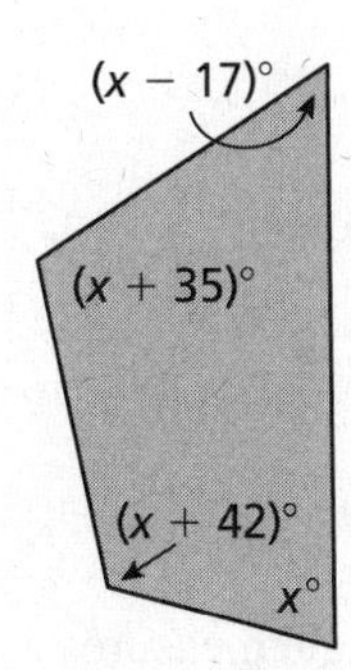

e.

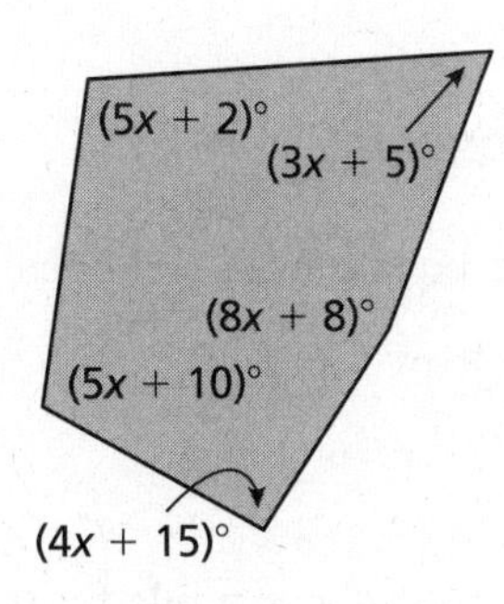

f.

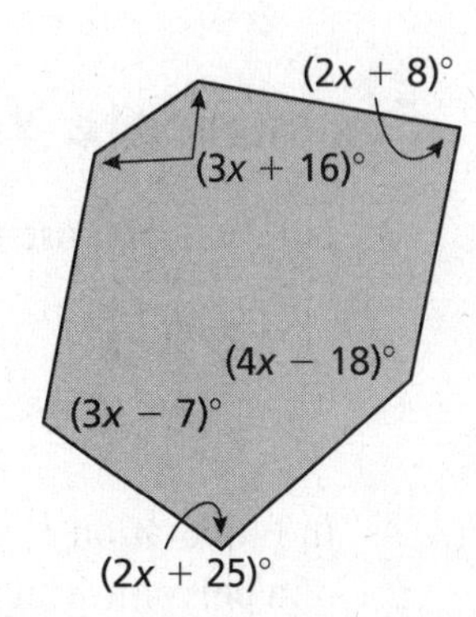

2 EXPLORATION: Writing a Multi-Step Equation

Go to *BigIdeasMath.com* for an interactive tool to investigate this exploration.

Work with a partner.

a. Draw an irregular polygon.

b. Measure the angles of the polygon. Record the measurements on a separate sheet of paper.

c. Choose a value for x. Then, using this value, work backward to assign a variable expression to each angle measure, as in Exploration 1.

d. Trade polygons with your partner.

e. Solve an equation to find the angle measures of the polygon your partner drew. Do your answers seem reasonable? Explain.

Communicate Your Answer

3. How can you use multi-step equations to solve real-life problems?

4. In Exploration 1, you were given the formula for the sum S of the angle measures of a polygon with n sides. Explain why this formula works.

5. The sum of the angle measures of a polygon is $1080°$. How many sides does the polygon have? Explain how you found your answer.

Name____ Date__________

1.2 Notetaking with Vocabulary
For use after Lesson 1.2

In your own words, write the meaning of each vocabulary term.

inverse operations

mean

Core Concepts

Solving Multi-Step Equations

To solve a multi-step equation, simplify each side of the equation, if necessary. Then use inverse operations to isolate the variable.

Notes:

Extra Practice

In Exercises 1–14, solve the equation. Check your solution.

1. $3x + 4 = 19$

2. $5z - 13 = -3$

3. $17 = z - (-9)$

4. $15 = 2 + 4 - d$

5. $\dfrac{f}{4} - 5 = -9$

6. $\dfrac{q + (-5)}{3} = 8$

7. $5x + 3x = 28$

8. $5z - 2z - 4 = -7$

9. $12x + 4 + 2x = 39$

10. $9z - 5 - 4z = -5$

1.2 Notetaking with Vocabulary (continued)

11. $3(z + 7) = 21$

12. $-4(z - 12) = 42$

13. $33 = 12r - 3(9 - r)$

14. $7 + 3(2g - 6) = -29$

15. You can represent an odd integer with the expression $2n + 1$, where n is any integer. Write and solve an equation to find three consecutive odd integers that have a sum of 63.

16. One angle of a triangle has a measure of $66°$. The measure of the third angle is $57°$ more than $\frac{1}{2}$ the measure of the second angle. The sum of the angle measures of a triangle is $180°$. What is the measure of the second angle? What is the measure of the third angle?

17. Your cousin is 8 years older than your brother. Three years ago, your cousin was twice as old as your brother. How old is your cousin now? How old is your brother now?

Name ______________________________ Date __________

1.3 Solving Equations with Variables on Both Sides

For use with Exploration 1.3

Essential Question How can you solve an equation that has variables on both sides?

1 EXPLORATION: Perimeter

Work with a partner. The two polygons have the same perimeter. Use this information to write and solve an equation involving x. Explain the process you used to find the solution. Then find the perimeter of each polygon.

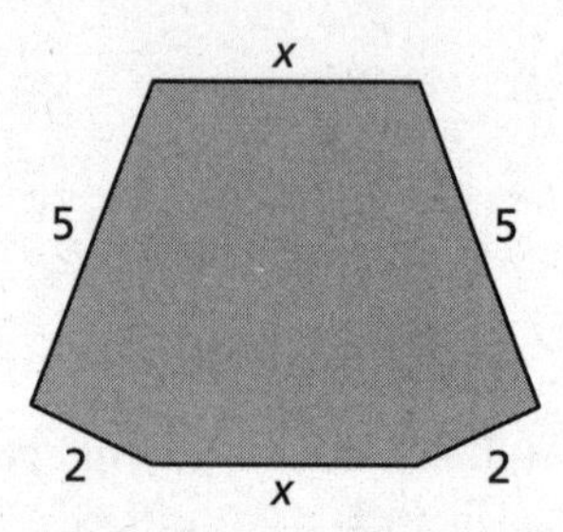

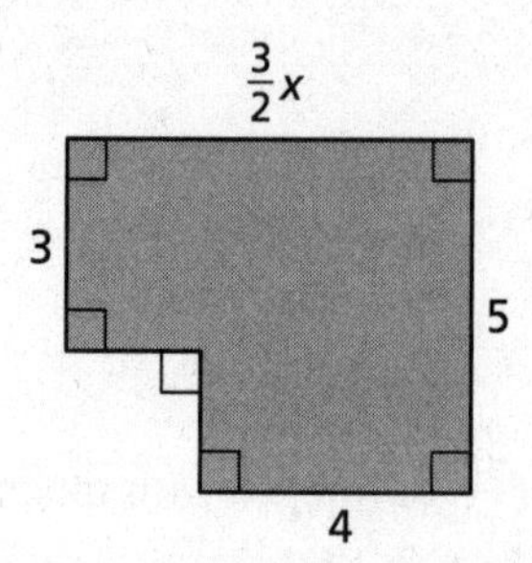

2 EXPLORATION: Perimeter and Area

Work with a partner.

- Each figure has the unusual property that the value of its perimeter (in feet) is equal to the value of its area (in square feet). Use this information to write an equation for each figure.
- Solve each equation for x. Explain the process you used to find the solution.
- Find the perimeter and area of each figure.

Name_______________________________________ Date__________

2 EXPLORATION: Perimeter and Area (continued)

a.

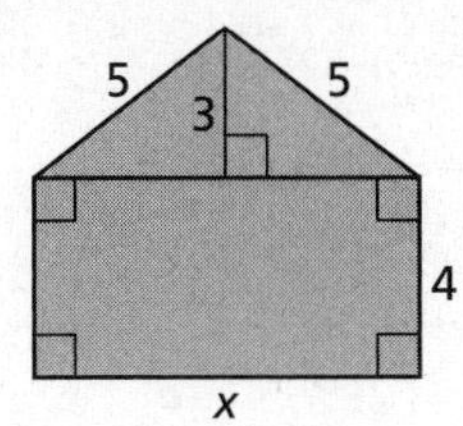

b.

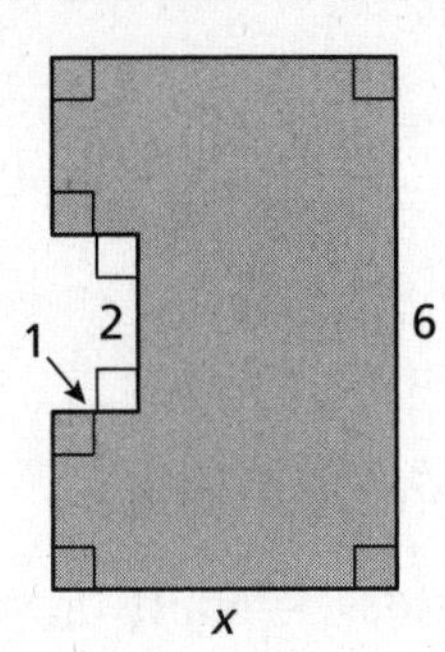

c.

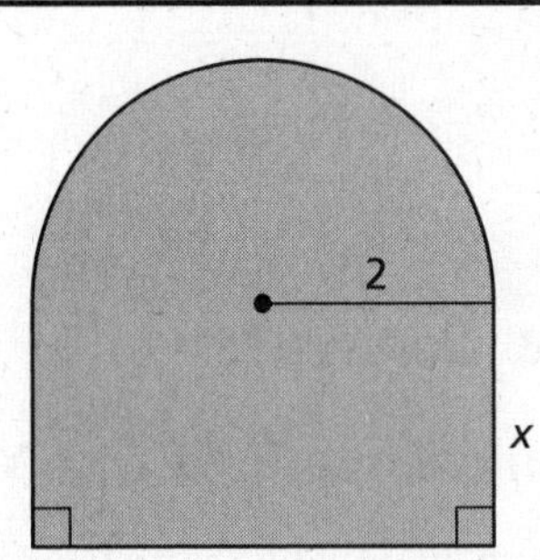

Communicate Your Answer

3. How can you solve an equation that has variables on both sides?

4. Write three equations that have the variable x on both sides. The equations should be different from those you wrote in Explorations 1 and 2. Have your partner solve the equations.

Name ____________________ Date __________

1.3 Notetaking with Vocabulary

For use after Lesson 1.3

In your own words, write the meaning of each vocabulary term.

identity

Core Concepts

Solving Equations with Variables on Both Sides

To solve an equation with variables on both sides, simplify one or both sides of the equation, if necessary. Then use inverse operations to collect the variable terms on one side, collect the constant terms on the other side, and isolate the variable.

Notes:

Special Solutions of Linear Equations

Equations do not always have one solution. An equation that is true for all values of the variable is an **identity** and has *infinitely many solutions*. An equation that is not true for any value of the variable has *no solution*.

Notes:

Name____ Date__________

1.3 Notetaking with Vocabulary (continued)

Steps for Solving Linear Equations

Here are several steps you can use to solve a linear equation. Depending on the equation, you may not need to use some steps.

Step 1 Use the Distributive Property to remove any grouping symbols.

Step 2 Simplify the expression on each side of the equation.

Step 3 Collect the variable terms on one side of the equation and the constant terms on the other side.

Step 4 Isolate the variable.

Step 5 Check your solution.

Notes:

Extra Practice

In Exercises 1–10, solve the equation. Check your solution.

1. $12 - 3x = -6x$

2. $7 - 5z = 17 + 5z$

3. $3k + 45 = 8k + 25$

4. $\frac{3}{4}(48 - 16x) = 4(4 + 2x)$

5. $5q + 6 = 2q - 2 + q$

6. $8 + 6x - 10x = 16 - 8x$

1.3 Notetaking with Vocabulary (continued)

7. $6a - 4 = 3a + 5$

8. $2(4b - 6) = 4(3b - 7)$

9. $8(2r - 3) - r = 3(3r + 2)$

10. $3x - 8(2x + 3) = -6(2x + 5)$

In Exercises 11–14, solve the equation. Determine whether the equation has *one solution*, *no solution*, or *infinitely many solutions*.

11. $6(4s + 12) = 8(3s - 14)$

12. $16f + 24 = 8(2f + 3)$

13. $\frac{1}{2}(10 + 12n) = \frac{1}{3}(15n + 15)$

14. $\frac{2}{3}(6j + 9) = 3j + 7$

15. The value of the surface area of a rectangular prism is equal to the value of the volume of the rectangular prism. Write and solve an equation to find the value of x.

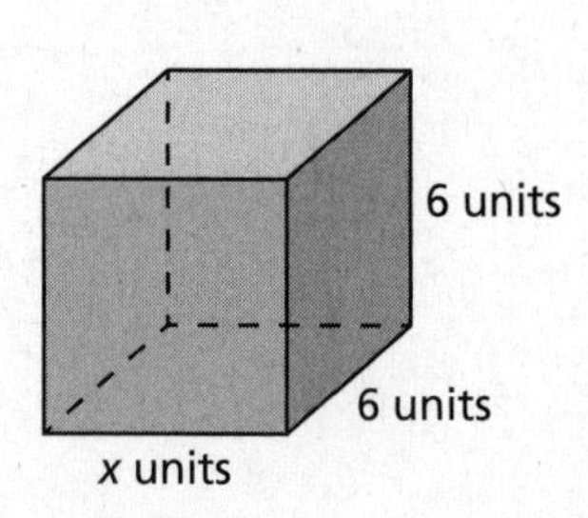

Name____ Date__________

1.4 Solving Absolute Value Equations

For use with Exploration 1.4

Essential Question How can you solve an absolute value equation?

1 EXPLORATION: Solving an Absolute Value Equation Algebraically

Work with a partner. Consider the absolute value equation $|x + 2| = 3$.

a. Describe the values of $x + 2$ that make the equation true. Use your description to write two linear equations that represent the solutions of the absolute value equation.

b. Use the linear equations you wrote in part (a) to find the solutions of the absolute value equation.

c. How can you use linear equations to solve an absolute value equation?

2 EXPLORATION: Solving an Absolute Value Equation Graphically

Go to *BigIdeasMath.com* for an interactive tool to investigate this exploration.

Work with a partner. Consider the absolute value equation $|x + 2| = 3$.

a. On a real number line, locate the point for which $x + 2 = 0$.

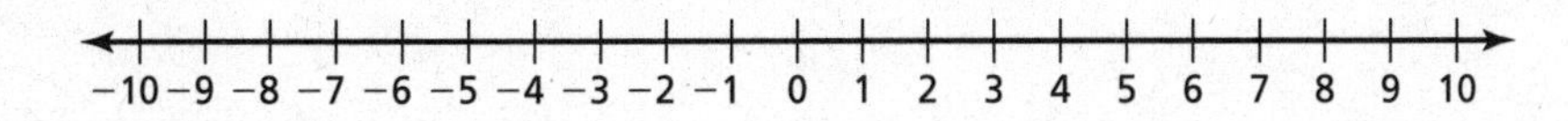

b. Locate the points that are 3 units from the point you found in part (a). What do you notice about those points?

c. How can you use a number line to solve an absolute value equation?

Name ______________________________ Date ________

3 EXPLORATION: Solving an Absolute Value Equation Numerically

Go to *BigIdeasMath.com* for an interactive tool to investigate this exploration.

Work with a partner. Consider the absolute value equation $|x + 2| = 3$.

a. Use a spreadsheet, as shown, to solve the absolute value equation.

	A	B
1	*x*	|x + 2|
2	-6	4
3	-5	
4	-4	
5	-3	
6	-2	
7	-1	
8	0	
9	1	
10	2	
11		

abs(A2 + 2)

b. Compare the solutions you found using the spreadsheet with those you found in Explorations 1 and 2. What do you notice?

c. How can you use a spreadsheet to solve an absolute value equation?

Communicate Your Answer

4. How can you solve an absolute value equation?

5. What do you like or dislike about the algebraic, graphical, and numerical methods for solving an absolute value equation? Give reasons for your answers.

Name____ Date__________

Notetaking with Vocabulary

For use after Lesson 1.4

In your own words, write the meaning of each vocabulary term.

absolute value equation

extraneous solution

Core Concepts

Properties of Absolute Value

Let a and b be real numbers. Then the following properties are true.

1. $|a| \geq 0$

2. $|-a| = |a|$

3. $|ab| = |a||b|$

4. $\left|\frac{a}{b}\right| = \frac{|a|}{|b|}, b \neq 0$

Notes:

Name ____________________ Date ________

1.4 Notetaking with Vocabulary (continued)

Solving Absolute Value Equations

To solve $|ax + b| = c$ when $c \geq 0$, solve the related linear equations

$$ax + b = c \quad or \quad ax + b = -c.$$

When $c < 0$, the absolute value equation $|ax + b| = c$ has no solution because absolute value always indicates a number that is not negative.

Notes:

Solving Equations with Two Absolute Values

To solve $|ax + b| = |cx + d|$, solve the related linear equations

$$ax + b = cx + d \quad or \quad ax + b = -(cx + d).$$

Notes:

Extra Practice

In Exercises 1–5, solve the equation. Graph the solution(s), if possible.

1. $|3x + 12| = 0$

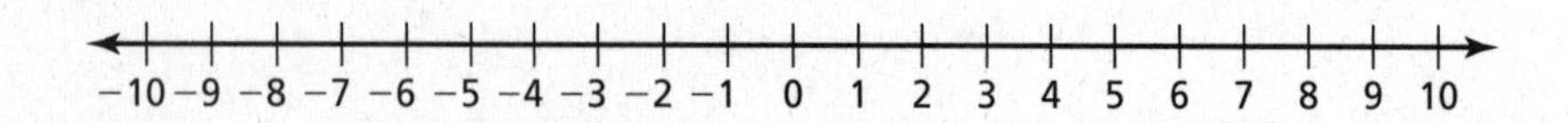

Name____ Date__________

1.4 Notetaking with Vocabulary (continued)

2. $|y + 2| = 8$

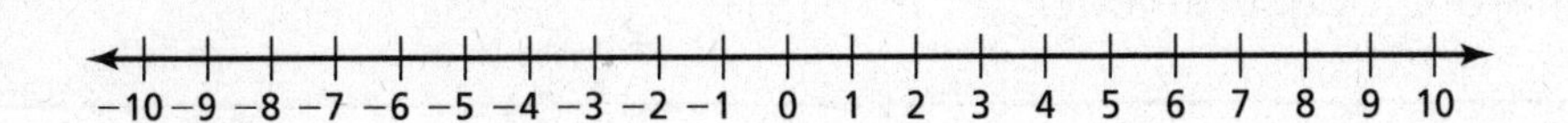

3. $-4|7 - 6k| = 14$

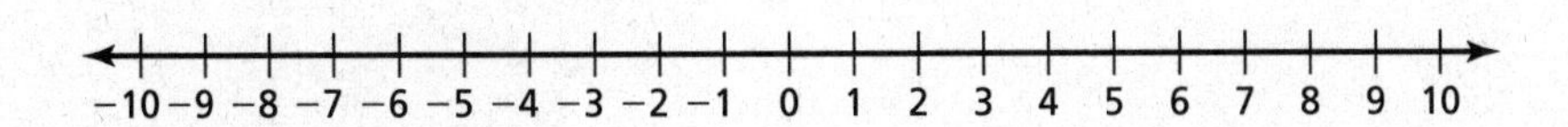

4. $\left|\dfrac{d}{3}\right| = 3$

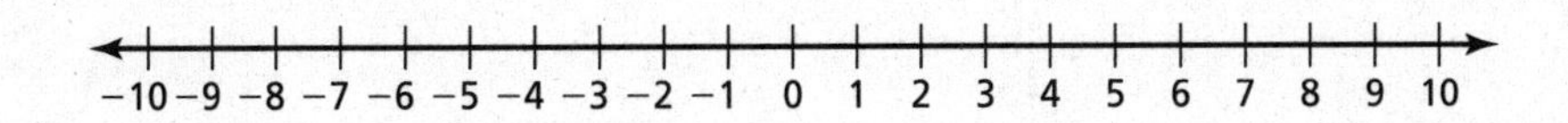

5. $3|2x + 5| + 10 = 37$

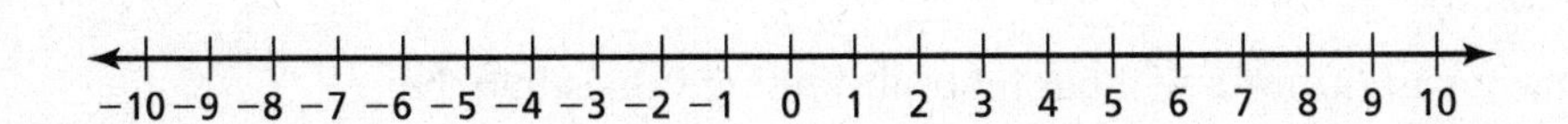

In Exercises 6–9, solve the equation. Check your solutions.

6. $|20x| = |4x + 16|$

7. $|p + 4| = |p - 2|$

8. $|4q + 9| = |2q - 1|$

9. $|2x - 7| = |2x + 9|$

Name ______________________________ Date __________

1.5 Rewriting Equations and Formulas

For use with Exploration 1.5

Essential Question How can you use a formula for one measurement to write a formula for a different measurement?

1 EXPLORATION: Using an Area Formula

Work with a partner.

a. Write a formula for the area A of a parallelogram.

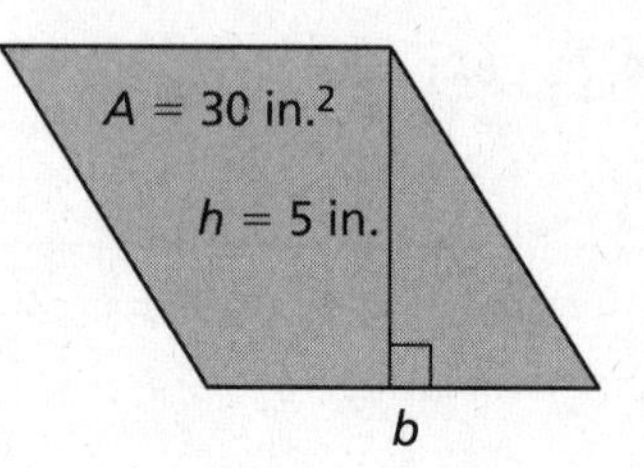

b. Substitute the given values into the formula. Then solve the equation for b. Justify each step.

c. Solve the formula in part (a) for b without first substituting values into the formula. Justify each step.

d. Compare how you solved the equations in parts (b) and (c). How are the processes similar? How are they different?

Name____ Date__________

2 EXPLORATION: Using Area, Circumference, and Volume Formulas

Work with a partner. Write the indicated formula for each figure. Then write a new formula by solving for the variable whose value is not given. Use the new formula to find the value of the variable.

a. Area A of a trapezoid

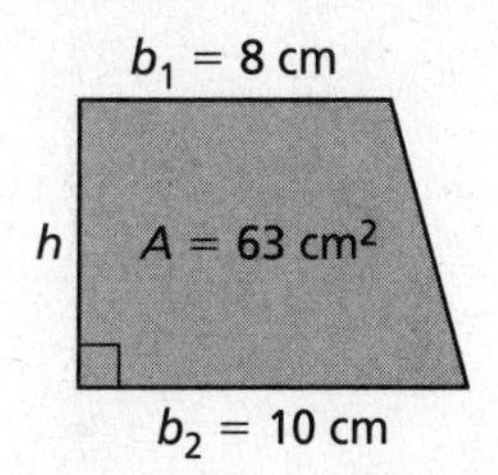

b. Circumference C of a circle

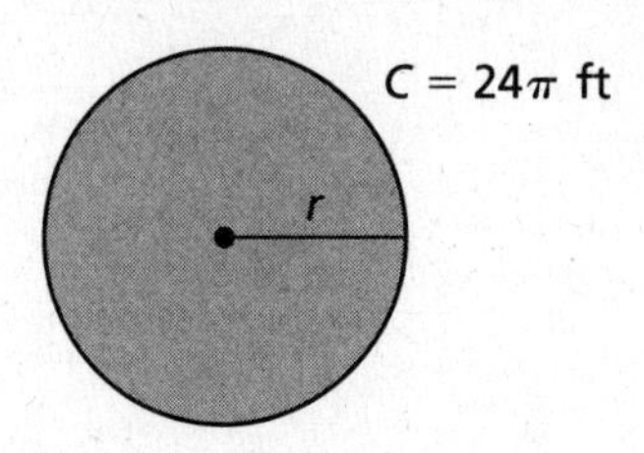

c. Volume V of a rectangular prism

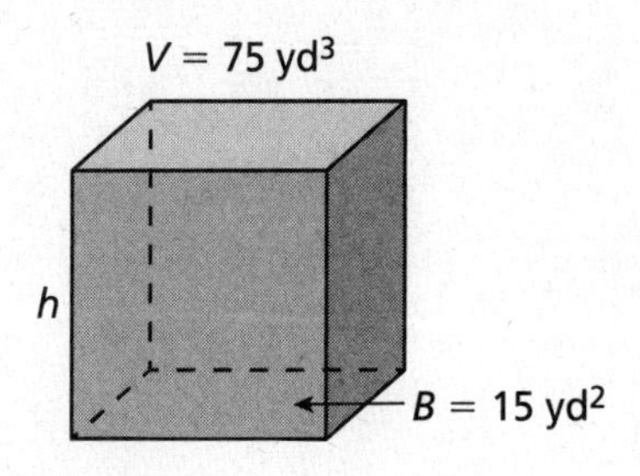

d. Volume V of a cone

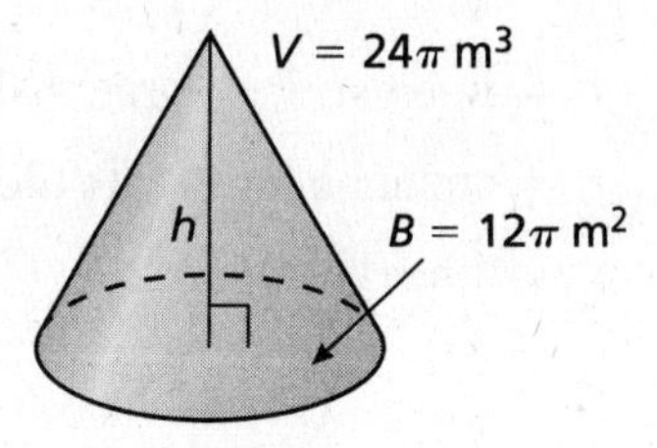

Communicate Your Answer

3. How can you use a formula for one measurement to write a formula for a different measurement? Give an example that is different from those given in Explorations 1 and 2.

Name ____________________ Date ________

1.5 Notetaking with Vocabulary

For use after Lesson 1.5

In your own words, write the meaning of each vocabulary term.

literal equation

formula

Core Concepts

Common Formulas

Temperature F = degrees Fahrenheit, C = degrees Celsius

$C = \frac{5}{9}(F - 32)$

Simple Interest I = interest, P = principal,

r = annual interest rate (decimal form),

t = time (years)

$I = Prt$

Distance d = distance traveled, r = rate, t = time

$d = rt$

Notes:

Name____ Date__________

Extra Practice

In Exercises 1–6, solve the literal equation for *y*.

1. $y - 2x = 15$

2. $4x + y = 2$

3. $5x - 2 = 8 + 5y$

4. $y + x = 11$

5. $3x - y = -4$

6. $3x + 1 = 7 - 4y$

In Exercises 7–12, solve the literal equation for *x*.

7. $y = 10x - 4x$

8. $q = 3x + 9xz$

9. $r = 4 + 7x - sx$

10. $y + 4x = 10x - 6$

11. $4g + r = 2r - 2x$

12. $3z + 8 = 12 + 3x - z$

In Exercises 13–16, solve the formula for the indicated variable.

13. Area of a triangle: $A = \frac{1}{2}bh$; Solve for b.

14. Volume of a cone: $V = \frac{1}{3}\pi r^2 h$; Solve for h.

15. Ohm's Law: $I = \frac{V}{R}$; Solve for R.

16. Ideal Gas Law: $PV = nRT$; Solve for R.

17. The amount A of money in an account after simple interest has been earned is given by the formula $A = P + Prt$ where P is the principal, r is the annual interest rate in decimal form, and t is the time in years.

a. Solve the formula for r.

b. The amount of money in an account after interest has been earned is \$1080, the principal is \$1000, and the time is 2 years. What is the annual interest rate?

c. Solve the formula for P.

Name____ Date__________

Chapter 2 Maintaining Mathematical Proficiency

Graph the number.

1. $|-2|$

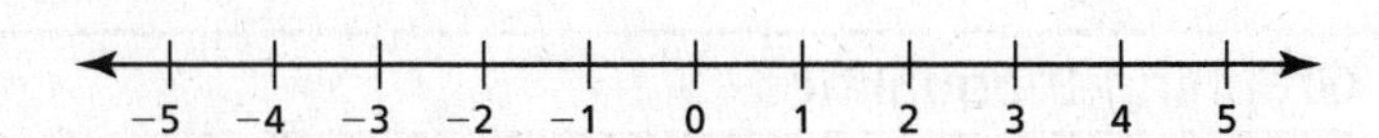

2. $-3 + |-3|$

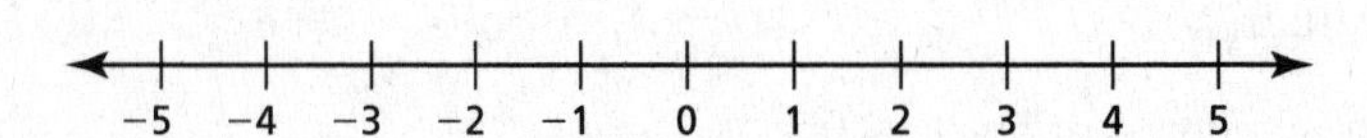

3. $-1 - |-4|$

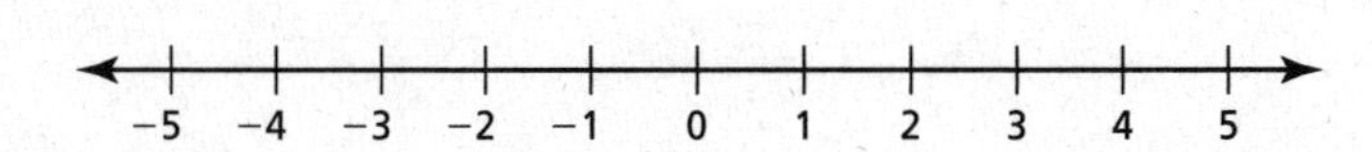

4. $2 + |2|$

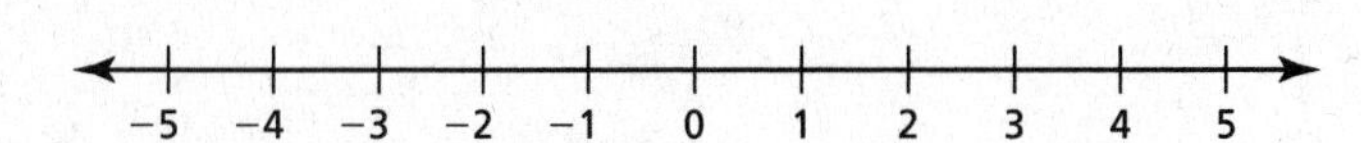

Complete the statement with <, >, or = .

5. 6 _____ 5 **6.** -2 _____ 3 **7.** -4 _____ -7

8. -8 _____ -5 **9.** $|-5|$ _____ 5 **10.** -7 _____ $|-6|$

11. A number a is to the right of a number b on the number line. Which is greater, $-a$ or $-b$?

12. A number a is to the left of a number b on the number line. Which is greater, $|-a|$ or $|-b|$?

Name ____ Date __________

2.1 Writing and Graphing Inequalities

For use with Exploration 2.1

Essential Question How can you use an inequality to describe a real-life statement?

1 EXPLORATION: Writing and Graphing Inequalities

Go to *BigIdeasMath.com* for an interactive tool to investigate this exploration.

Work with a partner. Write an inequality for each statement. Then sketch the graph of the numbers that make each inequality true.

a. Statement The temperature t in Sweden is at least $-10°C$.

Inequality

Graph

−40 −30 −20 −10 0 10 20 30 40

b. Statement The elevation e of Alabama is at most 2407 feet.

Inequality

−3000 −2000 −1000 0 1000 2000 3000

Graph

2 EXPLORATION: Writing Inequalities

Work with a partner. Write an inequality for each graph. Then, in words, describe all the values of x that make each inequality true.

a.

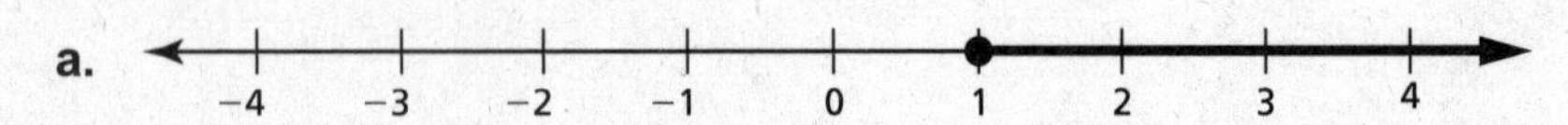

Name____ Date__________

2.1 Writing and Graphing Inequalities (continued)

2 EXPLORATION: Writing Inequalities (continued)

b. 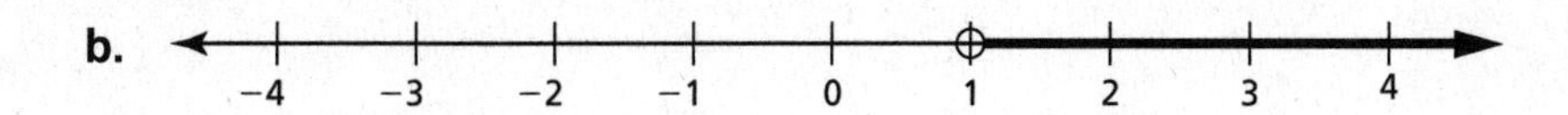

c. 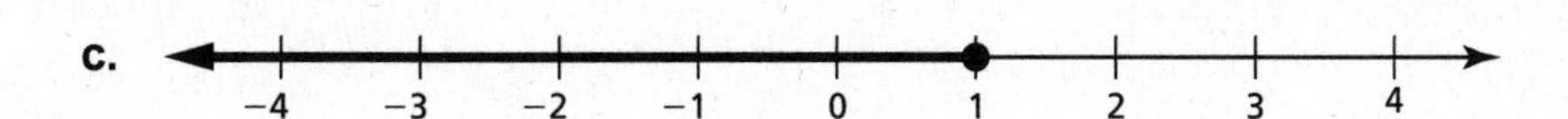

d.

−4 −3 −2 −1 0 1 2 3 4

Communicate Your Answer

3. How can you use an inequality to describe a real-life statement?

4. Write a real-life statement that involves each inequality.

a. $x < 3.5$

b. $x \leq 6$

c. $x > -2$

d. $x \geq 10$

Name ______________________________ Date __________

2.1 Notetaking with Vocabulary

For use after Lesson 2.1

In your own words, write the meaning of each vocabulary term.

inequality

solution of an inequality

solution set

graph of an inequality

Core Concepts

Representing Linear Inequalities

Words	Algebra	Graph
x is less than 2	$x < 2$	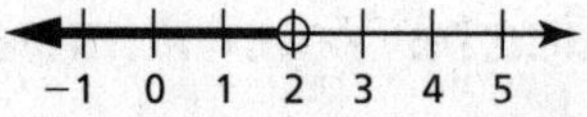
x is greater than 2	$x > 2$	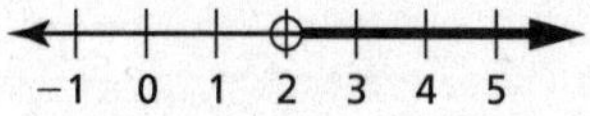
x is less than or equal to 2	$x \leq 2$	
x is greater than or equal to 2	$x \geq 2$	

Notes:

Extra Practice

In Exercises 1–4, write the sentence as an inequality.

1. Twelve is greater than or equal to five times a number n.

2. One-third of a number h is less than 15.

3. Seven is less than or equal to the difference of a number q and 6.

4. The sum of a number u and 14 is more than 6.

In Exercises 5 and 6, tell whether the value is a solution of the inequality.

5. $d - 7 < 12;\ d = 19$

6. $9 \geq 3n + 6;\ n = 1$

In Exercises 7–10, graph the inequality.

7. $x \geq 3$

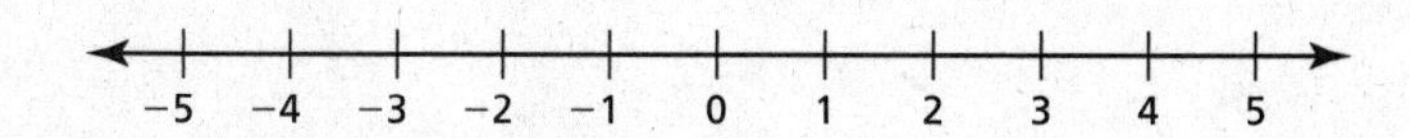

8. $x \leq 4$

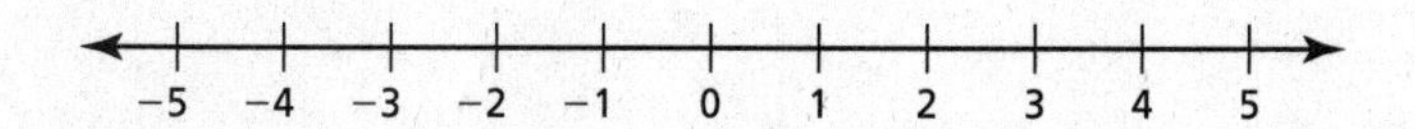

Name ____ Date __________

9. $x > -1$

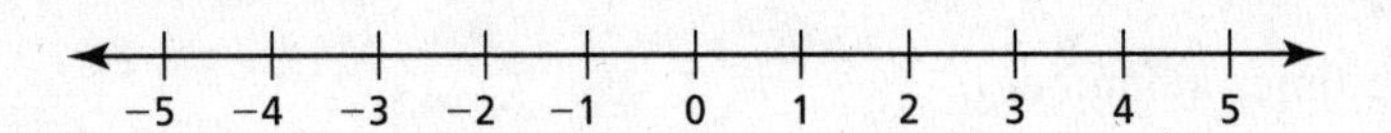

10. $x < 1$

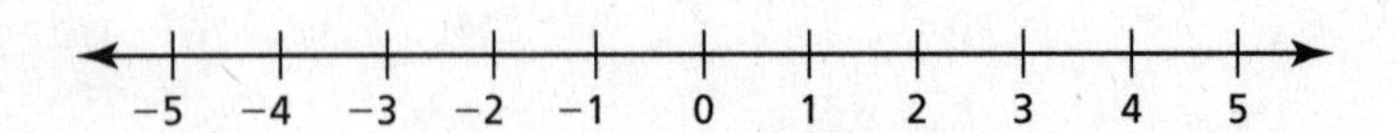

In Exercises 11–14, write an inequality that represents the graph.

11.

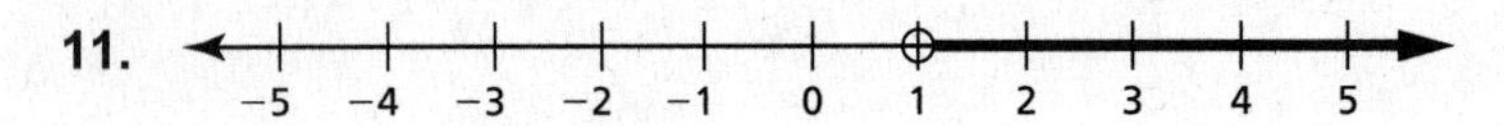

12.

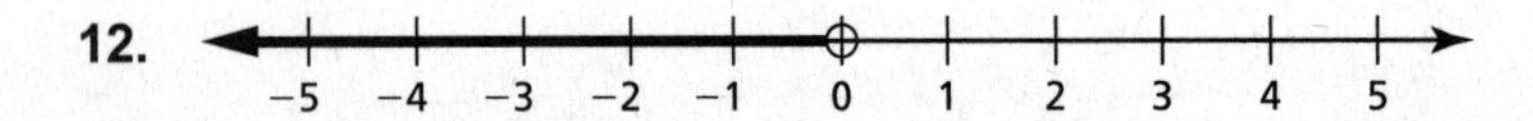

13.

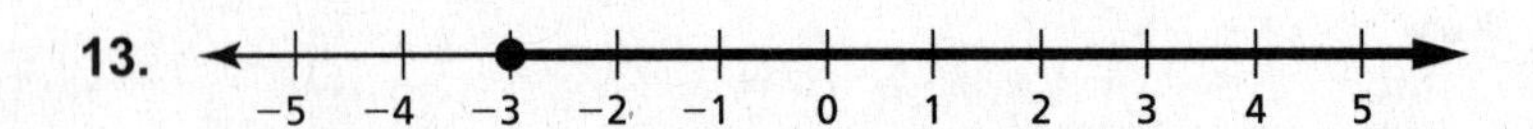

14.

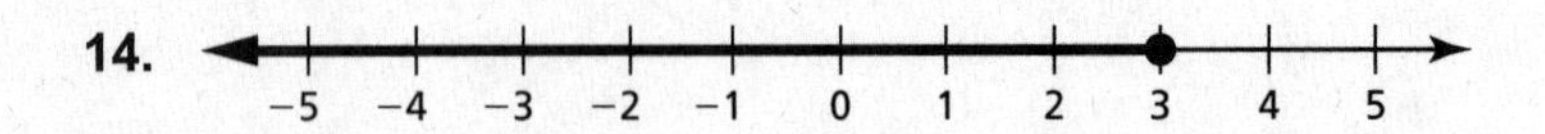

Name____ Date__________

2.2 Solving Inequalities Using Addition or Subtraction
For use with Exploration 2.2

Essential Question How can you use addition or subtraction to solve an inequality?

1 EXPLORATION: Quarterback Passing Efficiency

Work with a partner. The National Collegiate Athletic Association (NCAA) uses the following formula to rank the passing efficiencies P of quarterbacks.

$$P = \frac{8.4Y + 100C + 330T - 200N}{A}$$

Y = total length of all completed passes (in **Y**ards)

C = **C**ompleted passes

T = passes resulting in a **T**ouchdown

N = i**N**tercepted passes

A = **A**ttempted passes

M = inco**M**plete passes

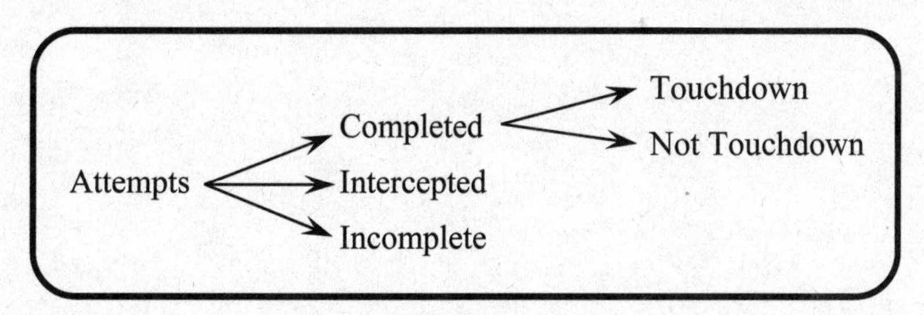

Determine whether each inequality must be true. Explain your reasoning.

a. $T < C$

b. $C + N \leq A$

c. $N < A$

d. $A - C \geq M$

Name ____________________ Date __________

2 EXPLORATION: Finding Solutions of Inequalities

Work with a partner. Use the passing efficiency formula to create a passing record that makes each inequality true. Record your results in the table. Then describe the values of P that make each inequality true.

	Attempts	Completions	Yards	Touchdowns	Interceptions
a.					
b.					
c.					

a. $P < 0$

b. $P + 100 \geq 250$

c. $P - 250 > -80$

Communicate Your Answer

3. How can you use addition or subtraction to solve an inequality?

4. Solve each inequality.

a. $x + 3 < 4$

b. $x - 3 \geq 5$

c. $4 > x - 2$

d. $-2 \leq x + 1$

Name____________________ Date__________

2.2 Notetaking with Vocabulary
For use after Lesson 2.2

In your own words, write the meaning of each vocabulary term.

equivalent inequalities

Notes:

Core Concepts

Addition Property of Inequality

Words Adding the same number to each side of an inequality produces an equivalent inequality.

Numbers

$$\begin{array}{rcr} -3 & < & 2 \\ \underline{+4} & & \underline{+4} \\ 1 & < & 6 \end{array} \qquad \begin{array}{rcr} -3 & \geq & -10 \\ \underline{+3} & & \underline{+3} \\ 0 & \geq & -7 \end{array}$$

Algebra If $a > b$, then $a + c > b + c$. If $a \geq b$, then $a + c \geq b + c$.

If $a < b$, then $a + c < b + c$. If $a \leq b$, then $a + c \leq b + c$.

Notes:

Subtraction Property of Inequality

Words Subtracting the same number from each side of an inequality produces an equivalent inequality.

Numbers

$$\begin{array}{rcr} -3 & \le & 1 \\ \underline{-5} & & \underline{-5} \\ -8 & \le & -4 \end{array} \qquad \begin{array}{rcr} 7 & > & -20 \\ \underline{-7} & & \underline{-7} \\ 0 & > & -27 \end{array}$$

Algebra If $a > b$, then $a - c > b - c$. If $a \ge b$, then $a - c \ge b - c$.

If $a < b$, then $a - c < b - c$. If $a \le b$, then $a - c \le b - c$.

Notes:

Extra Practice

In Exercises 1–6, solve the inequality. Graph the solution.

1. $x - 3 < -4$

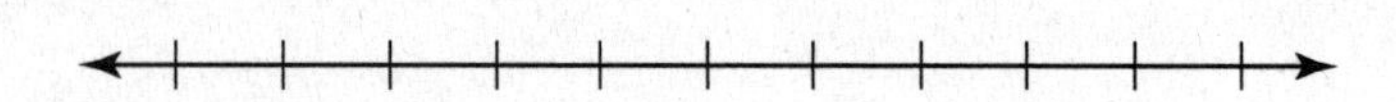

2. $-3 > -3 + h$

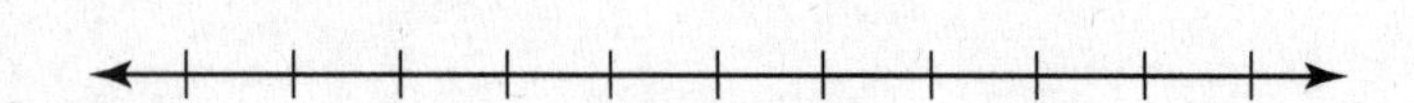

2.2 Notetaking with Vocabulary (continued)

3. $s - (-1) \geq 2$

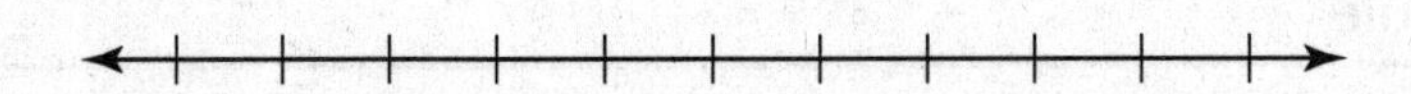

4. $6 - 9 + u < -2$

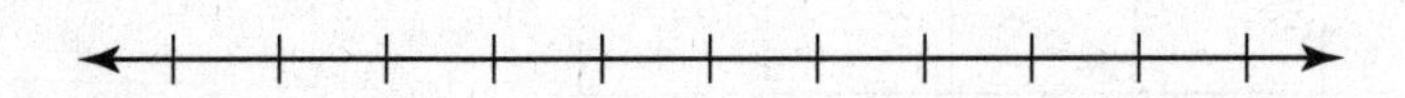

5. $12 \leq 4c - 3c + 10$

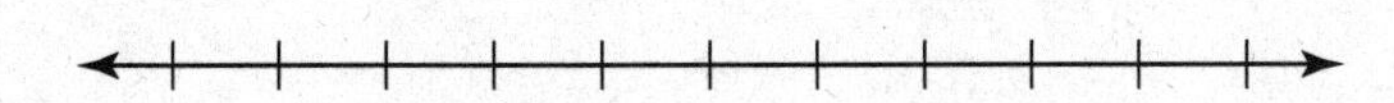

6. $15 - 7p + 8p > 15 - 2$

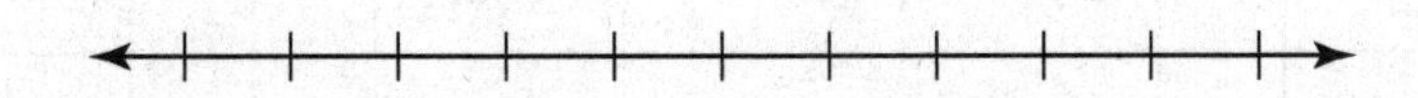

7. You have \$15 to spend on groceries. You have \$12.25 worth of groceries already in your cart.

a. Write an inequality that represents how much more money *m* you can spend on groceries.

b. Solve the inequality.

Name ______________________________ Date __________

2.3 Solving Inequalities Using Multiplication or Division

For use with Exploration 2.3

Essential Question How can you use division to solve an inequality?

1 EXPLORATION: Writing a Rule

Work with a partner.

a. Complete the table. Decide which graph represents the solution of the inequality $6 < 3x$. Write the solution of the inequality.

x	-1	0	1	2	3	4	5
$3x$	-3						
$6 \overset{?}{<} 3x$	No						

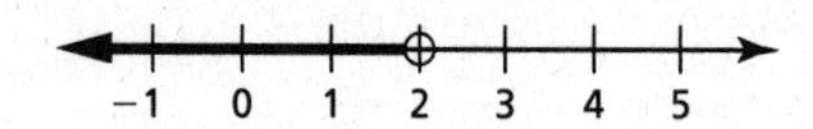

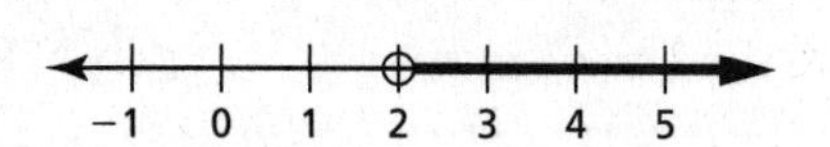

b. Use a table to solve each inequality. Then write a rule that describes how to use division to solve the inequalities.

i. $2x < 4$

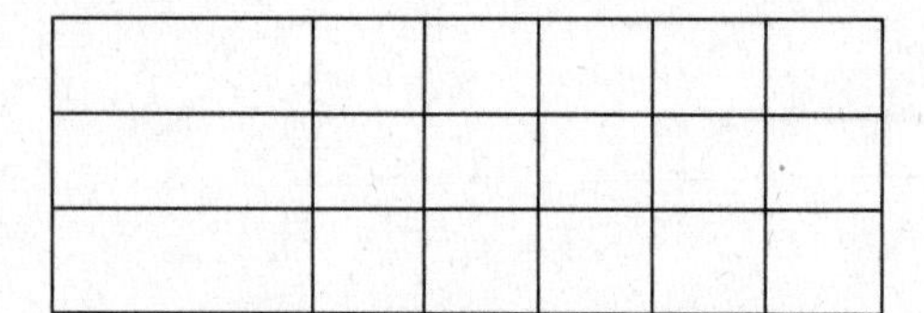

ii. $3 \geq 3x$

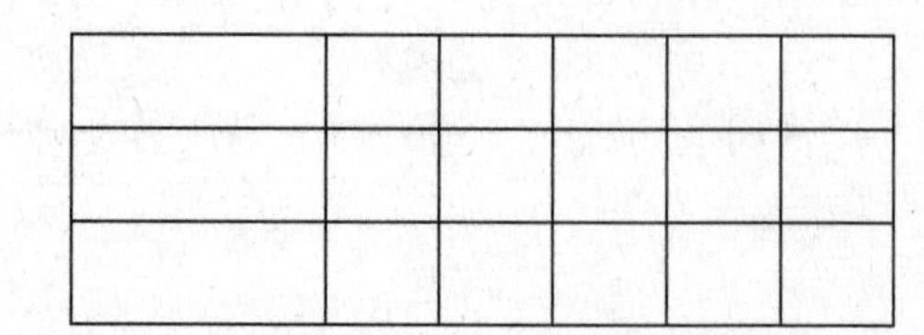

iii. $2x < 8$

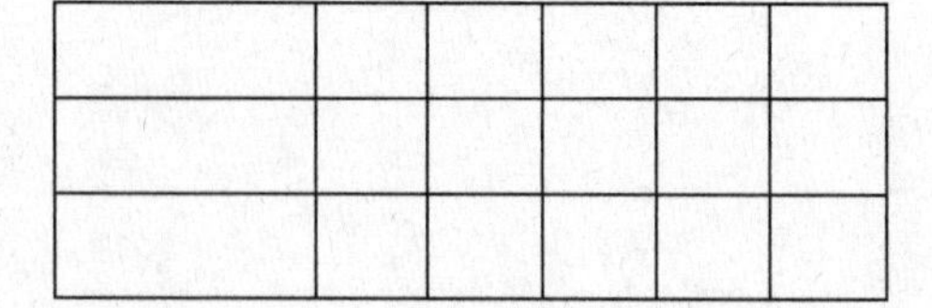

iv. $6 \geq 3x$

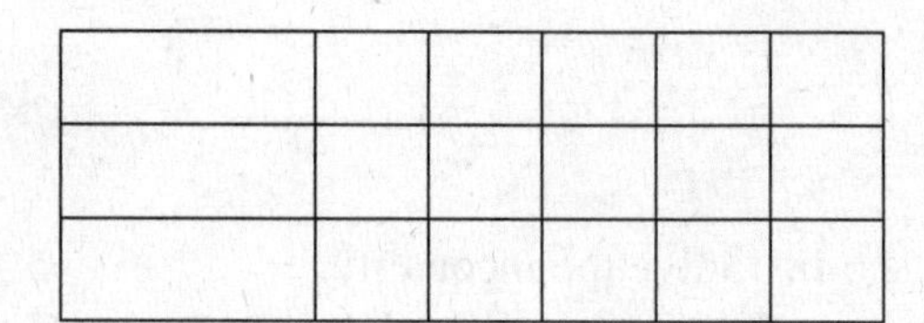

Name_________________________________ Date__________

2 EXPLORATION: Writing a Rule

Work with a partner.

a. Complete the table. Decide which graph represents the solution of the inequality $6 < -3x$. Write the solution of the inequality.

x	−5	−4	−3	−2	−1	0	1
$-3x$							
$6 \overset{?}{<} -3x$							

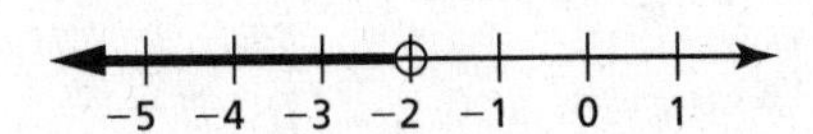

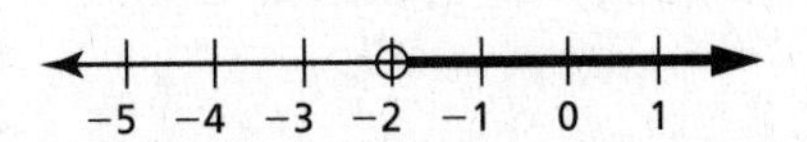

b. Use a table to solve each inequality. Then write a rule that describes how to use division to solve the inequalities.

i. $-2x < 4$

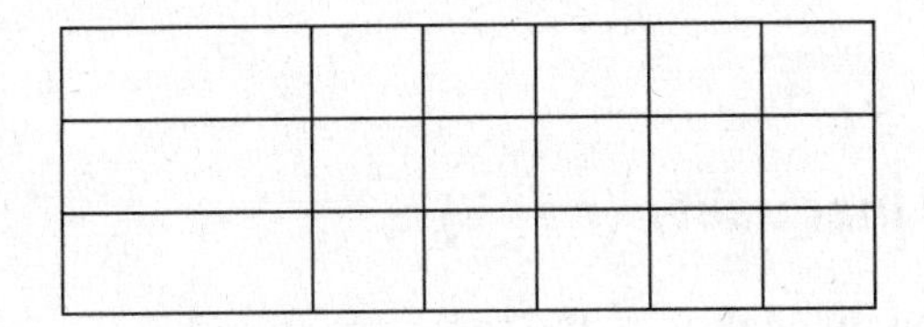

ii. $3 \geq -3x$

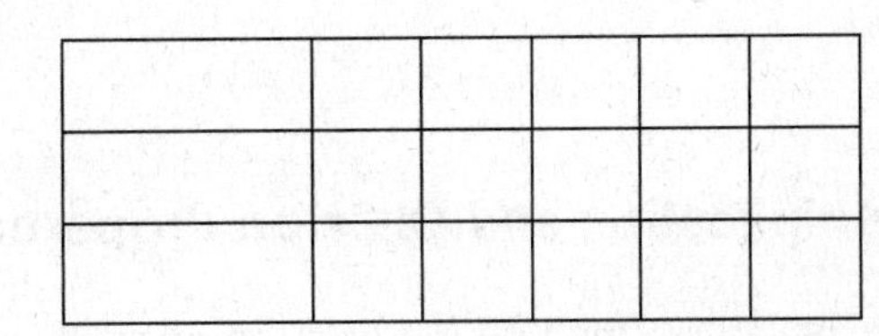

iii. $-2x < 8$

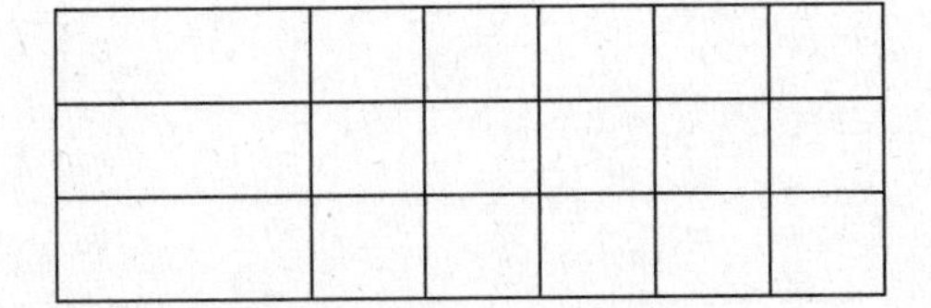

iv. $6 \geq -3x$

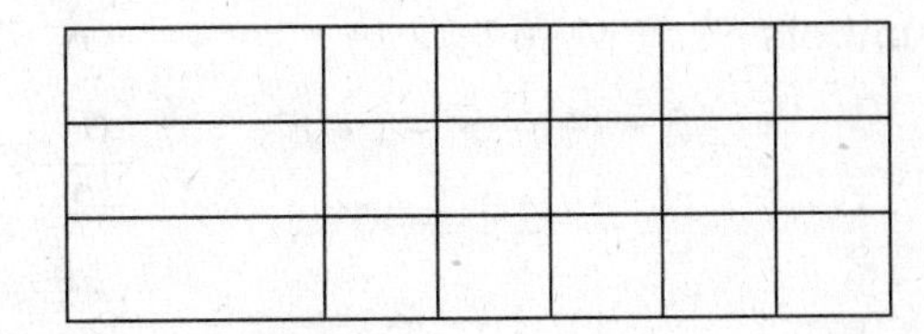

Communicate Your Answer

3. How can you use division to solve an inequality?

4. Use the rules you wrote in Explorations 1(b) and 2(b) to solve each inequality.

a. $7x < -21$ **b.** $12 \leq 4x$ **c.** $10 < -5x$ **d.** $-3x \leq 0$

Name ______________________________ Date __________

2.3 Notetaking with Vocabulary
For use after Lesson 2.3

Core Concepts

Multiplication and Division Properties of Inequality ($c > 0$)

Words Multiplying or dividing each side of an inequality by the same *positive* number produces an equivalent inequality.

Numbers

$$-6 < 8 \qquad\qquad 6 > -8$$

$$\mathbf{2} \bullet (-6) < \mathbf{2} \bullet 8 \qquad\qquad \frac{6}{\mathbf{2}} > \frac{-8}{\mathbf{2}}$$

$$-12 < 16 \qquad\qquad 3 > -4$$

Algebra If $a > b$ and $c > 0$, then $ac > bc$. If $a > b$ and $c > 0$, then $\frac{a}{c} > \frac{b}{c}$.

If $a < b$ and $c > 0$, then $ac < bc$. If $a < b$ and $c > 0$, then $\frac{a}{c} < \frac{b}{c}$.

These properties are also true for $\leq$ and $\geq$.

Notes:

Multiplication and Division Properties of Inequality ($c < 0$)

Words When multiplying or dividing each side of an inequality by the same *negative* number, the direction of the inequality symbol must be reversed to produce an equivalent inequality.

Numbers

$$-6 < 8 \qquad\qquad 6 > -8$$

$$\mathbf{-2} \bullet (-6) > \mathbf{-2} \bullet 8 \qquad\qquad \frac{6}{\mathbf{-2}} < \frac{-8}{\mathbf{-2}}$$

$$12 > -16 \qquad\qquad -3 < 4$$

Algebra If $a > b$ and $c < 0$, then $ac < bc$. If $a > b$ and $c < 0$, then $\frac{a}{c} < \frac{b}{c}$.

If $a < b$ and $c < 0$, then $ac > bc$. If $a < b$ and $c < 0$, then $\frac{a}{c} > \frac{b}{c}$.

These properties are also true for $\leq$ and $\geq$.

Notes:

Extra Practice

In Exercises 1–8, solve the inequality. Graph the solution.

1. $6x < -30$

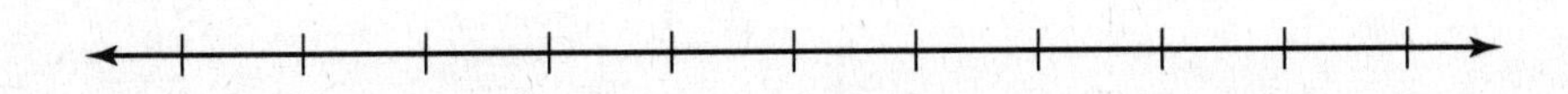

2. $48 \le 16f$

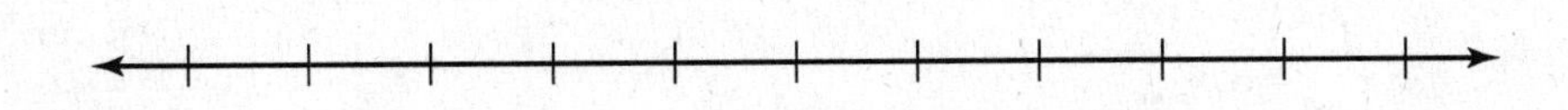

3. $-\frac{6}{7} \le \frac{3}{7}f$

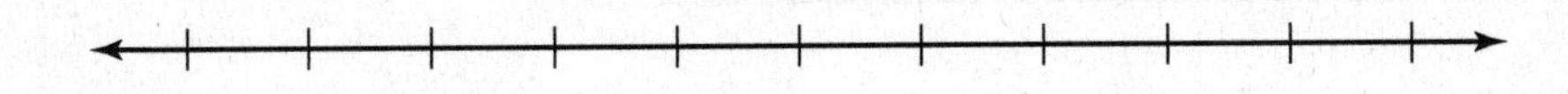

4. $-4m \ge -16$

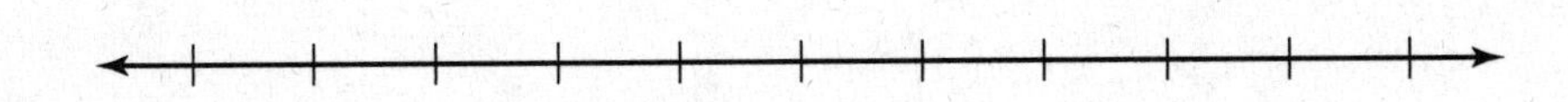

5. $\dfrac{x}{-6} > \dfrac{1}{3}$

Name ____ Date __________

2.3 Notetaking with Vocabulary (continued)

6. $1 \le -\frac{1}{4}y$

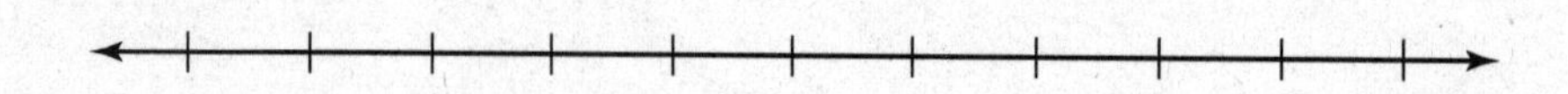

7. $-\frac{2}{3} < -4x$

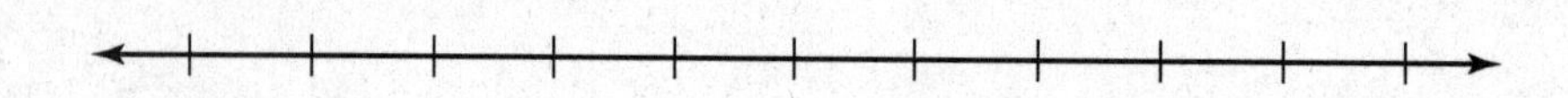

8. $-\frac{4}{5}x \ge -2$

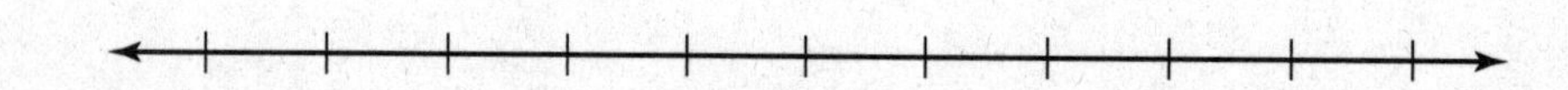

9. There are at most 36 red and blue marbles in a bag. The number of red marbles is twice the number of blue marbles. Write and solve an inequality that represents the greatest number of red marbles r in the bag.

Name_______________________________ Date__________

2.4 Solving Multi-Step Inequalities

For use with Exploration 2.4

Essential Question How can you solve a multi-step inequality?

1 EXPLORATION: Solving a Multi-Step Inequality

Go to *BigIdeasMath.com* for an interactive tool to investigate this exploration.

Work with a partner.

- Use what you already know about solving equations and inequalities to solve each multi-step inequality. Justify each step.

a. $2x + 3 \leq x + 5$

b. $-2x + 3 > x + 9$

c. $27 \geq 5x + 4x$

d. $-8x + 2x - 16 < -5x + 7x$

e. $3(x - 3) - 5x > -3x - 6$

f. $-5x - 6x \leq 8 - 8x - x$

2.4 Solving Multi-Step Inequalities (continued)

1 EXPLORATION: Solving a Multi-Step Inequality (continued)

- Match each inequality with its graph. Use a graphing calculator to check your answer.

a. $2x + 3 \le x + 5$

b. $-2x + 3 > x + 9$

c. $27 \ge 5x + 4x$

d. $-8x + 2x - 16 < -5x + 7x$

e. $3(x - 3) - 5x > -3x - 6$

f. $-5x - 6x \le 8 - 8x - x$

A.

4

−6 6

−4

B.

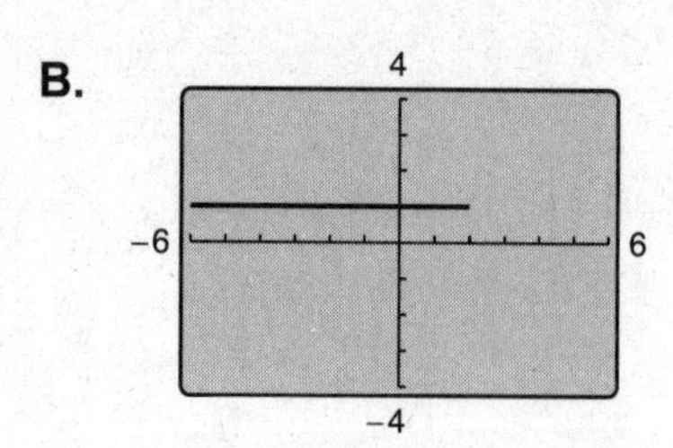

C.

4

−6 6

−4

D.

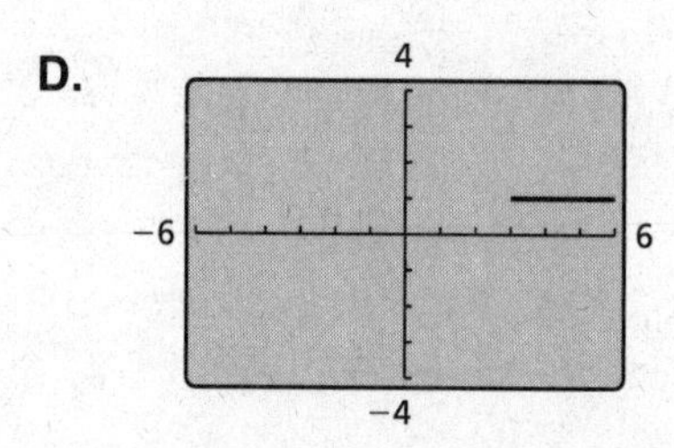

E.

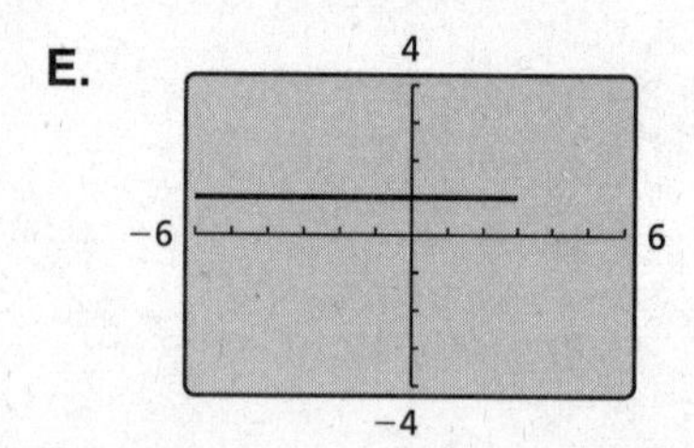

F.

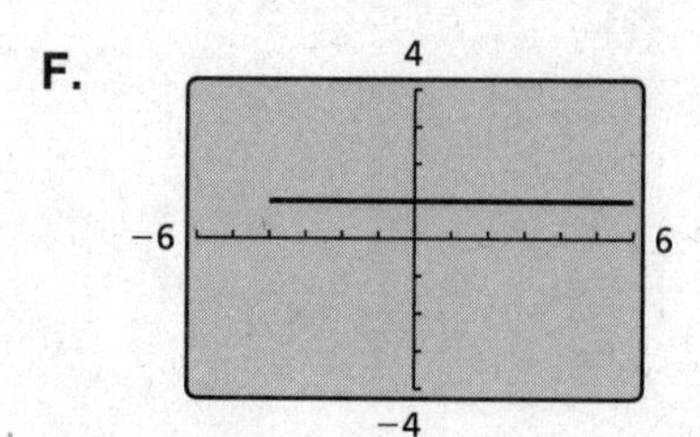

Communicate Your Answer

2. How can you solve a multi-step inequality?

3. Write two different multi-step inequalities whose solutions are represented by the graph.

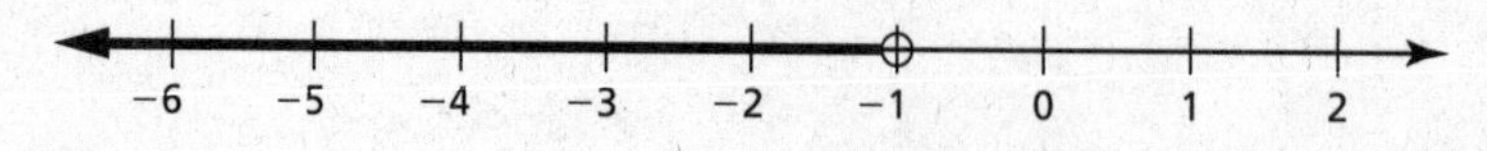

Name____ Date__________

2.4 Notetaking with Vocabulary

For use after Lesson 2.4

Notes:

Extra Practice

In Exercises 1–5, solve the inequality. Graph the solution.

1. $3x - 2 < 10$

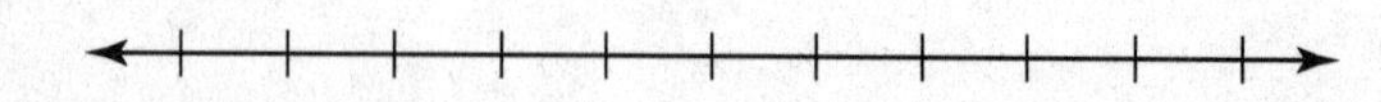

2. $4a + 8 \geq 0$

3. $2 + \dfrac{b}{-3} \leq 3$

4. $-\dfrac{c}{2} - 6 > -8$

5. $8 \leq -4(d + 1)$

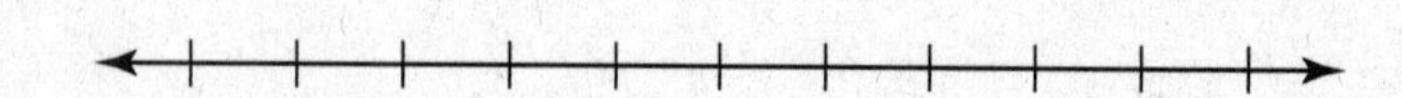

Name______________________________ Date__________

2.4 Notetaking with Vocabulary (continued)

In Exercises 6–10, solve the inequality.

6. $5 - 2n > 8 - 4n$

7. $6h - 18 < 6h + 1$

8. $3p + 4 \geq -4p + 25$

9. $7j - 4j + 6 < -2 + 3j$

10. $12\left(\frac{1}{4}w + 3\right) \leq 3(w - 4)$

11. Find the value of k for which the solution of the inequality $k(4r - 5) \geq -12r - 9$ is all real numbers.

12. Find the value of k that makes the inequality $2kx - 3k < 2x + 4 + 3kx$ have no solution.

Name ____ Date __________

2.5 Solving Compound Inequalities

For use with Exploration 2.5

Essential Question How can you use inequalities to describe intervals on the real number line?

1 EXPLORATION: Describing Intervals on the Real Number Line

Work with a partner. In parts (a)–(d), use two inequalities to describe the interval.

a. Half-Open Interval

b. Half-Open Interval

c. Closed Interval

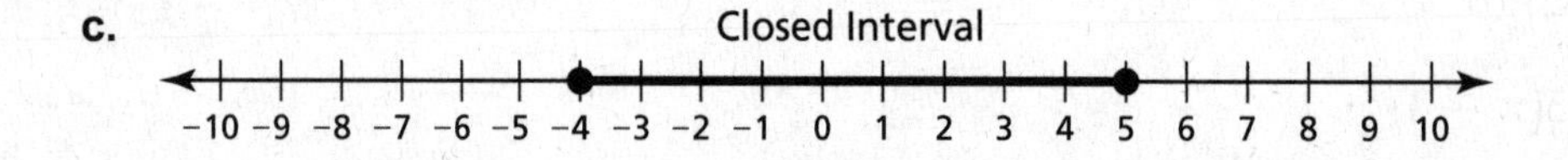

d. Open Interval

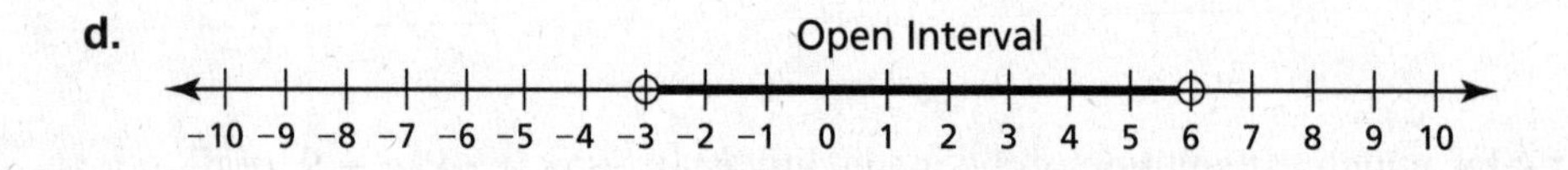

e. Do you use "and" or "or" to connect the two inequalities in parts (a)–(d)? Explain.

2 EXPLORATION: Describing Two Infinite Intervals

Work with a partner. In parts (a)–(d), use two inequalities to describe the interval.

a. 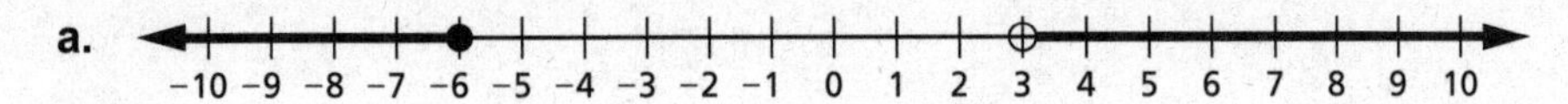

b. 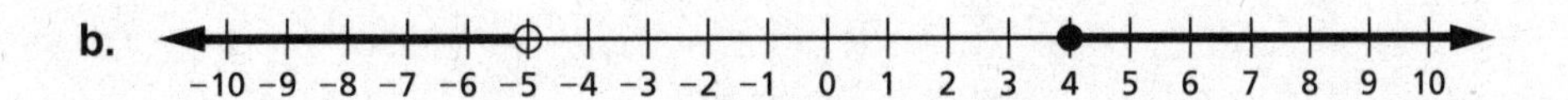

c. 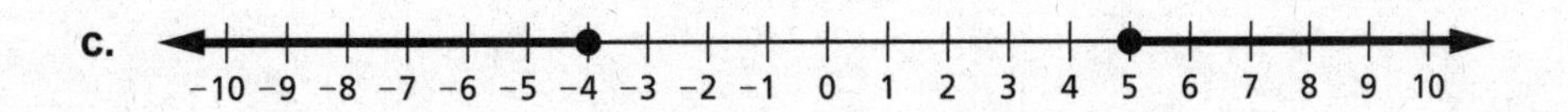

d. 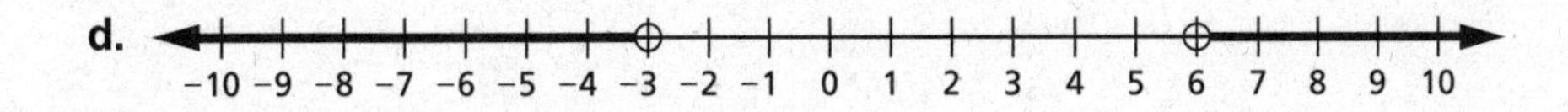

e. Do you use "and" or "or" to connect the two inequalities in parts (a)–(d)? Explain.

Communicate Your Answer

3. How can you use inequalities to describe intervals on the real number line?

Name ______________________________ Date __________

2.5 Notetaking with Vocabulary
For use after Lesson 2.5

In your own words, write the meaning of each vocabulary term.

compound inequality

Notes:

Name____ Date__________

2.5 Notetaking with Vocabulary (continued)

Extra Practice

In Exercises 1–5, write the sentence as an inequality. Graph the inequality.

1. A number u is less than 7 and greater than 3.

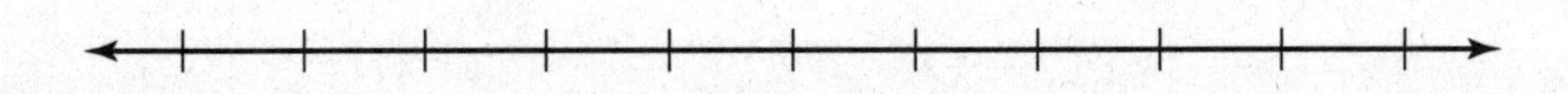

2. A number d is less than -2 or greater than or equal to 2.

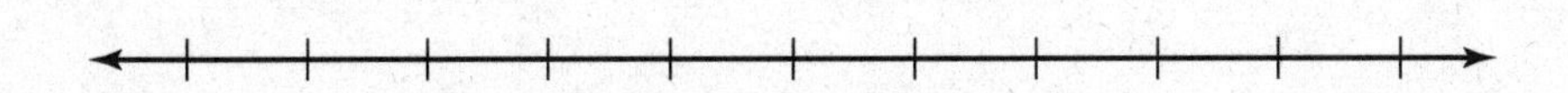

3. A number s is no less than -2.4 and fewer than 4.2.

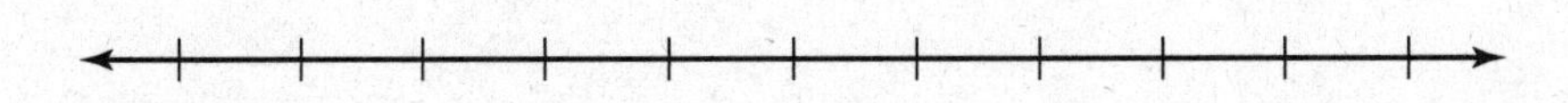

4. A number c is more than -4 or at most $-6\frac{1}{2}$.

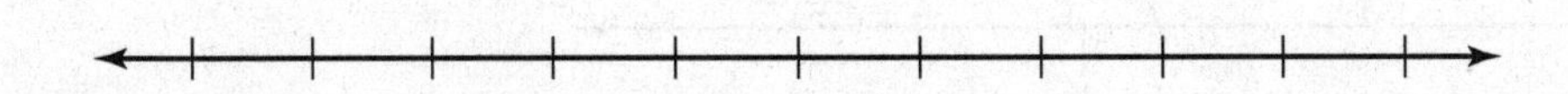

5. A number c is no less than -1.5 and less than 5.3.

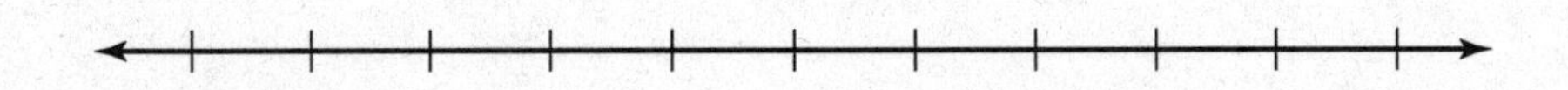

Name ____ Date __________

2.5 Notetaking with Vocabulary (continued)

In Exercises 6–10, solve the inequality. Graph the solution.

6. $4 < x - 3 \leq 7$

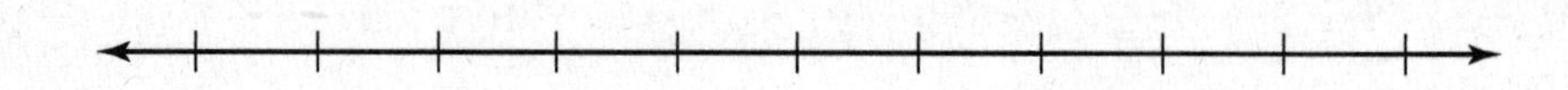

7. $15 \geq -5g \geq -10$

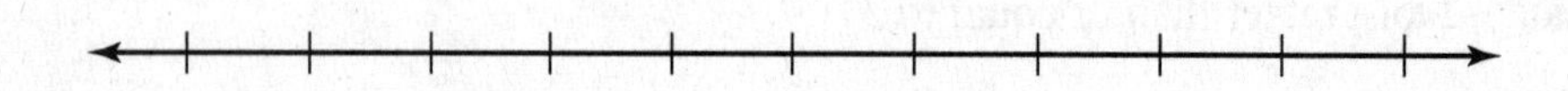

8. $z + 4 < 2$ *or* $-3z < -27$

9. $2t + 6 < 10$ *or* $-t + 7 \leq 2$

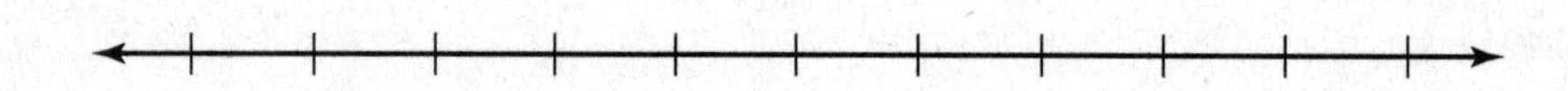

10. $-8 \leq \frac{1}{3}(6x + 24) \leq 12$

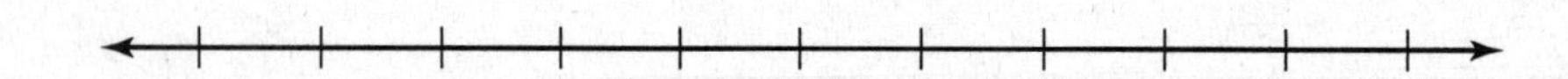

11. A certain machine operates properly when the relative humidity h satisfies the inequality $-60 \leq 2(h - 50) \leq 60$. Solve for h to find the range of values for which the machine operates properly.

Name____ Date__________

2.6 Solving Absolute Value Inequalities
For use with Exploration 2.6

Essential Question How can you solve an absolute value inequality?

1 EXPLORATION: Solving an Absolute Value Inequality Algebraically

Work with a partner. Consider the absolute value inequality $|x + 2| \leq 3$.

a. Describe the values of $x + 2$ that make the inequality true. Use your description to write two linear inequalities that represent the solutions of the absolute value inequality.

b. Use the linear inequalities you wrote in part (a) to find the solutions of the absolute value inequality.

c. How can you use linear inequalities to solve an absolute value inequality?

2 EXPLORATION: Solving an Absolute Value Inequality Graphically

Go to *BigIdeasMath.com* for an interactive tool to investigate this exploration.

Work with a partner. Consider the absolute value inequality $|x + 2| \leq 3$.

a. On a real number line, locate the point for which $x + 2 = 0$.

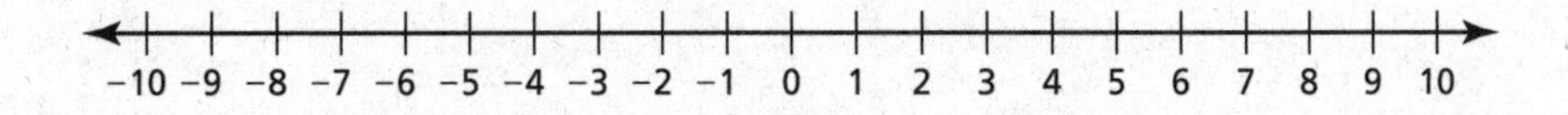

b. Locate the points that are within 3 units from the point you found in part (a). What do you notice about these points?

c. How can you use a number line to solve an absolute value inequality?

3 EXPLORATION: Solving an Absolute Value Inequality Numerically

Go to *BigIdeasMath.com* for an interactive tool to investigate this exploration.

Work with a partner. Consider the absolute value inequality $|x + 2| \le 3$.

a. Use a spreadsheet, as shown, to solve the absolute value inequality.

	A	B
1	x	$\|x + 2\|$
2	-6	4
3	-5	
4	-4	
5	-3	
6	-2	
7	-1	
8	0	
9	1	
10	2	
11		

abs(A2 + 2)

b. Compare the solutions you found using the spreadsheet with those you found in Explorations 1 and 2. What do you notice?

c. How can you use a spreadsheet to solve an absolute value inequality?

Communicate Your Answer

4. How can you solve an absolute value inequality?

5. What do you like or dislike about the algebraic, graphical, and numerical methods for solving an absolute value inequality? Give reasons for your answers.

Name____ Date__________

2.6 Notetaking with Vocabulary
For use after Lesson 2.6

In your own words, write the meaning of each vocabulary term.

absolute value inequality

absolute deviation

Notes:

Core Concepts

Solving Absolute Value Inequalities

To solve $|ax + b| < c$ for $c > 0$, solve the compound inequality

$$ax + b > -c \quad \textit{and} \quad ax + b < c.$$

To solve $|ax + b| > c$ for $c > 0$, solve the compound inequality

$$ax + b < -c \quad \textit{or} \quad ax + b > c.$$

In the inequalities above, you can replace $<$ with $\leq$ and $>$ with $\geq$.

Notes:

Name ______________________ Date __________

2.6 Notetaking with Vocabulary (continued)

Extra Practice

In Exercises 1–9, solve the inequality. Graph the solution, if possible.

1. $|y + 2| < 8$

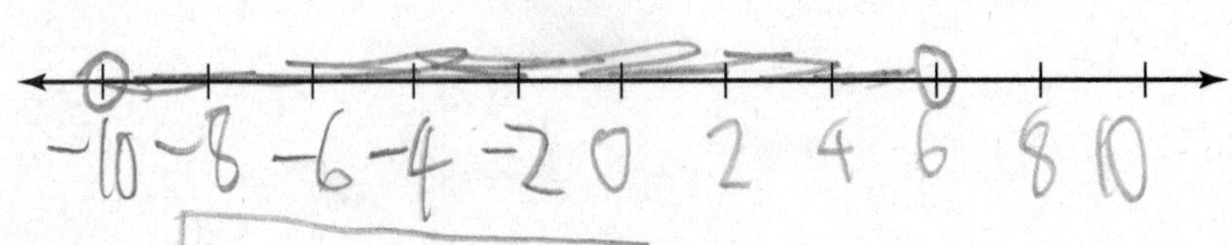

2. $\left|\frac{q}{3}\right| > 2$

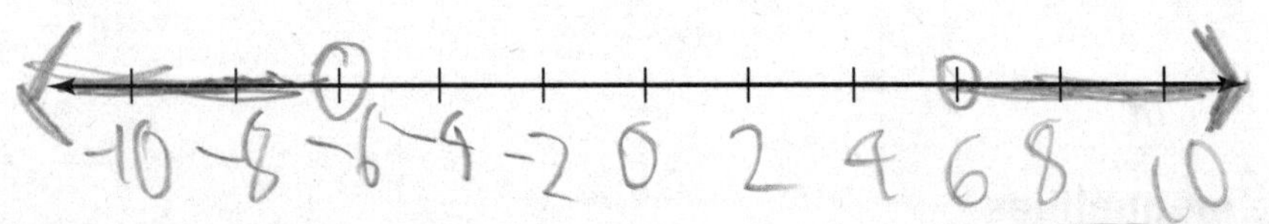

3. $3|2a + 5| + 10 \leq 37$

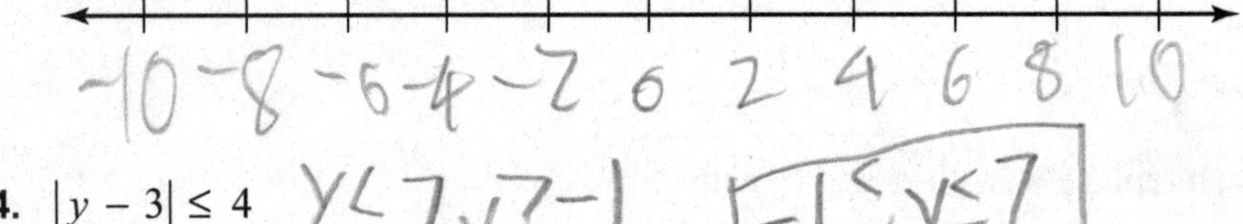

4. $|y - 3| \leq 4$

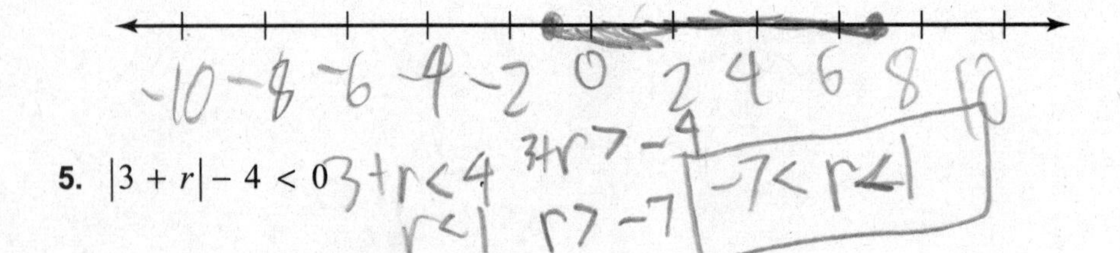

5. $|3 + r| - 4 < 0$

Name Anit V. Date

2.6 Notetaking with Vocabulary (continued)

6. $|f + 12| > -4$

All real numbers

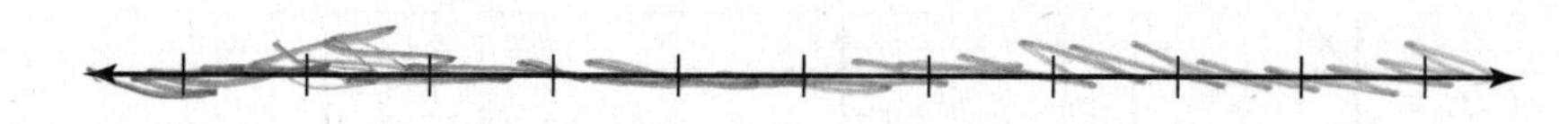

7. $\left|\frac{x}{4} - 7\right| < -2$

No solution

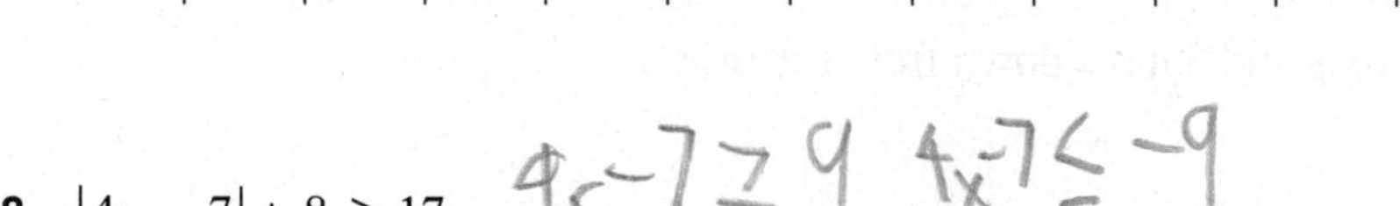

8. $|4x - 7| + 8 \geq 17$

$4x - 7 \geq 9$ $4x - 7 \leq -9$

$4x \geq 16$ $4x \leq -2$

$x \geq 4$ $x \leq -\frac{1}{2}$

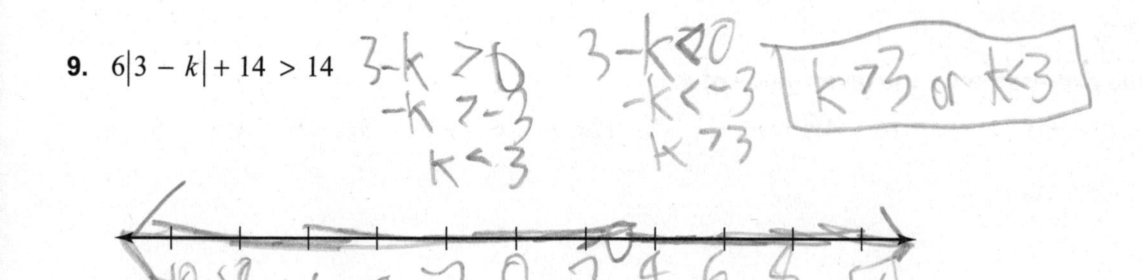

9. $6|3 - k| + 14 > 14$

$3 - k > 0$ $3 - k < 0$

$-k > -3$ $-k < -3$

$k < 3$ $k > 3$

$k > 3$ or $k < 3$

10. At a certain company, the average starting salary s for a new worker is \$25,000. The actual salary has an absolute deviation of at most \$1800. Write and solve an inequality to find the range of the starting salaries.

$1800 \geq |s - 25000|$ $|s - 25000| \geq -1800$

$26800 \geq s$ $s \geq 23200$

$23200 \leq s \leq 26{,}800$

Name ____ Date __________

Chapter 3 Maintaining Mathematical Proficiency

Plot the point in a coordinate plane. Describe the location of the point.

1. $A(-3, 1)$ **2.** $B(2, 2)$ **3.** $C(1, 0)$ **4.** $D(5, 2)$

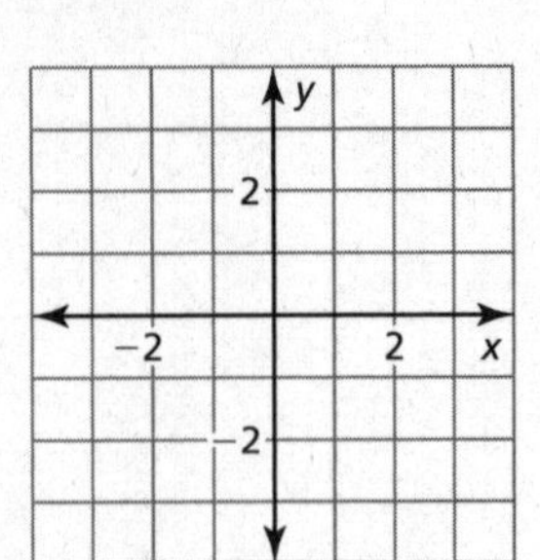

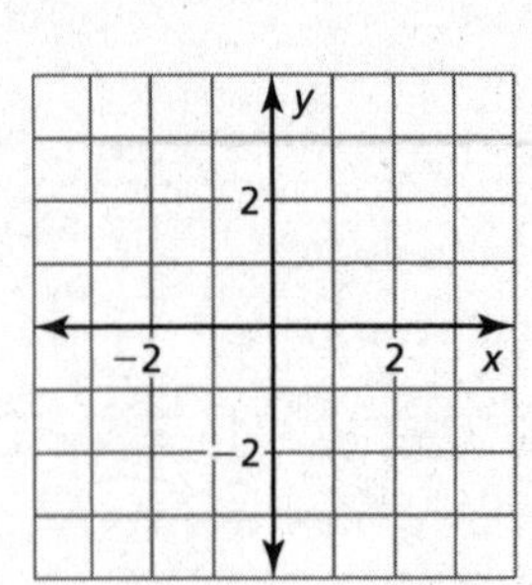

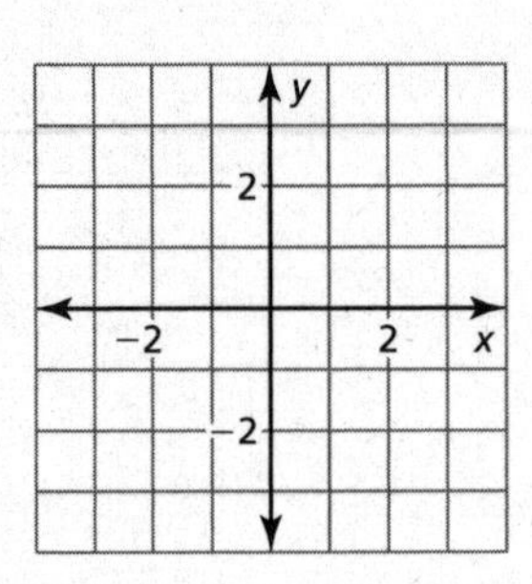

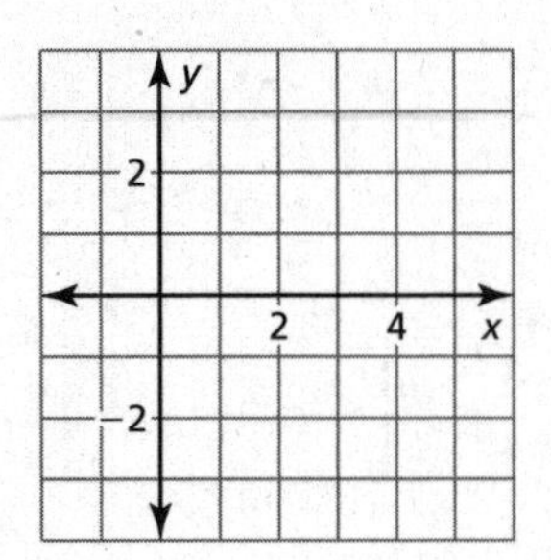

5. Plot the point that is on the y-axis and 5 units down from the origin.

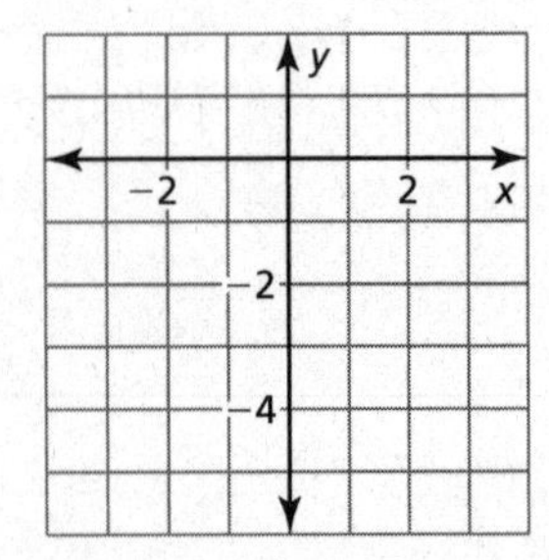

Evaluate the expression for the given value of *x*.

6. $2x + 1; x = 3$ **7.** $16 - 4x; x = -4$ **8.** $12x + 7; x = -2$ **9.** $-9 - 3x; x = 5$

10. The length of a side of a square is represented by $(24 - 3x)$ feet. What is the length of the side of the square when $x = 6$?

Name_______________________ Date__________

3.1 Functions
For use with Exploration 3.1

Essential Question What is a function?

1 EXPLORATION: Describing a Function

Work with a partner. Functions can be described in many ways.

- by an equation
- by an input-output table
- using words
- by a graph
- as a set of ordered pairs

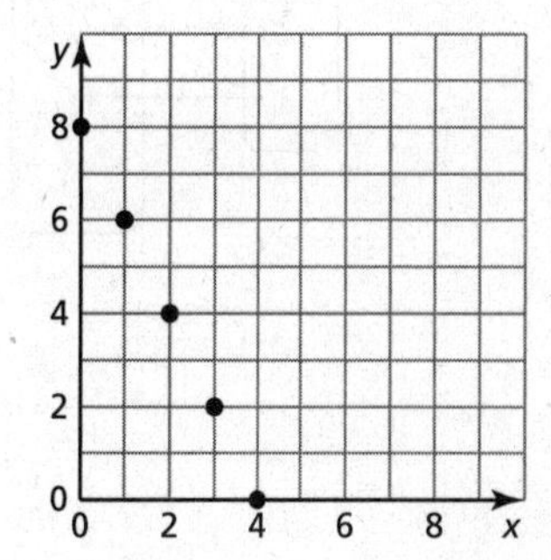

a. Explain why the graph shown represents a function.

b. Describe the function in two other ways.

2 EXPLORATION: Identifying Functions

Work with a partner. Determine whether each relation represents a function. Explain your reasoning.

a.

Input, x	0	1	2	3	4
Output, y	8	8	8	8	8

b.

Input, x	8	8	8	8	8
Output, y	0	1	2	3	4

Name ______________________________ Date __________

2 EXPLORATION: Identifying Functions (continued)

c.

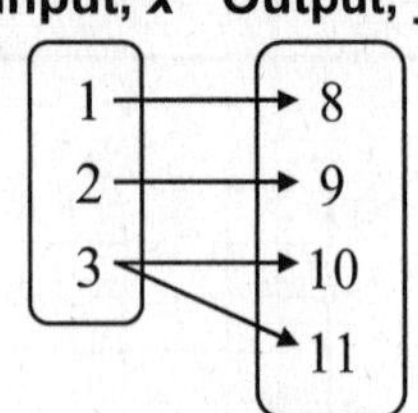

d.

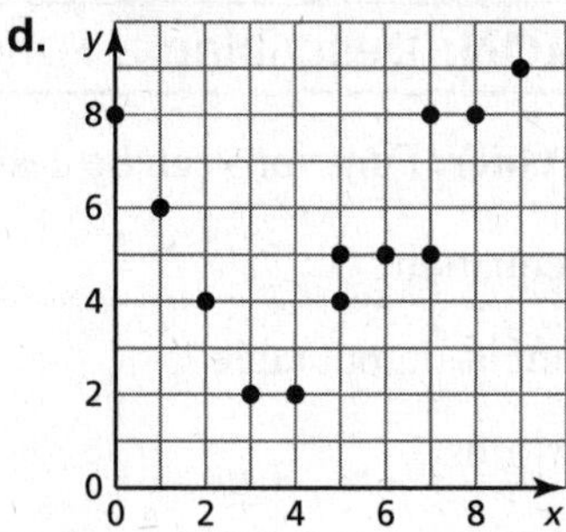

e. $(-2, 5), (-1, 8), (0, 6), (1, 6), (2, 7)$

f. $(-2, 0), (-1, 0), (-1, 1), (0, 1), (1, 2), (2, 2)$

g. Each radio frequency x in a listening area has exactly one radio station y.

h. The same television station x can be found on more than one channel y.

i. $x = 2$

j. $y = 2x + 3$

Communicate Your Answer

3. What is a function? Give examples of relations, other than those in Explorations 1 and 2, that (a) are functions and (b) are not functions.

Name____ Date__________

3.1 Notetaking with Vocabulary
For use after Lesson 3.1

In your own words, write the meaning of each vocabulary term.

relation

function

domain

range

independent variable

dependent variable

Notes:

Core Concepts

Vertical Line Test

Words A graph represent a function when no vertical line passes through more than one point on the graph.

Examples Function

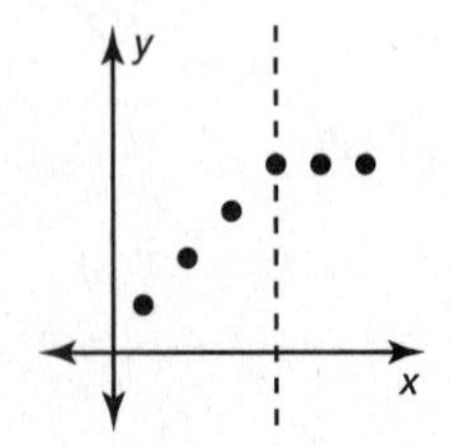

Not a function

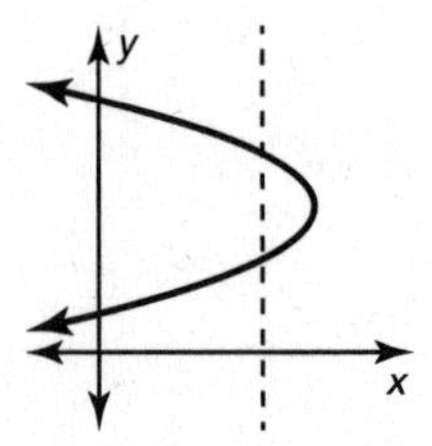

Notes:

The Domain and Range of a Function

The **domain** of a function is the set of all possible input values.

The **range** of a function is the set of all possible output values.

Notes:

Name____ Date__________

Extra Practice

In Exercises 1 and 2, determine whether the relation is a function. Explain.

1.

Input, x	-2	0	1	-2
Output, y	4	5	4	5

2. $(0, 3), (1, 1), (2, 1), (3, 0)$

In Exercises 3 and 4, determine whether the graph represents a function. Explain.

3.

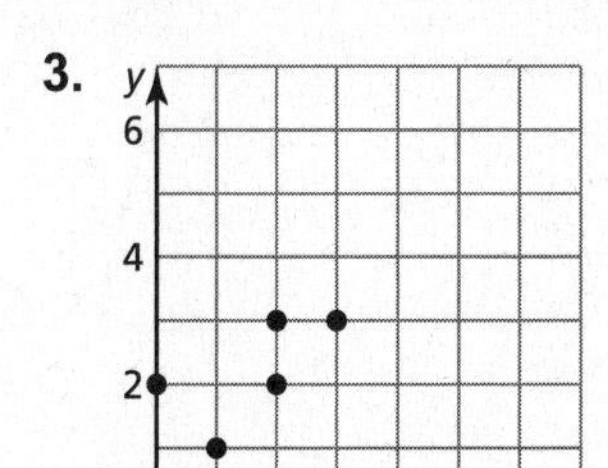

4.

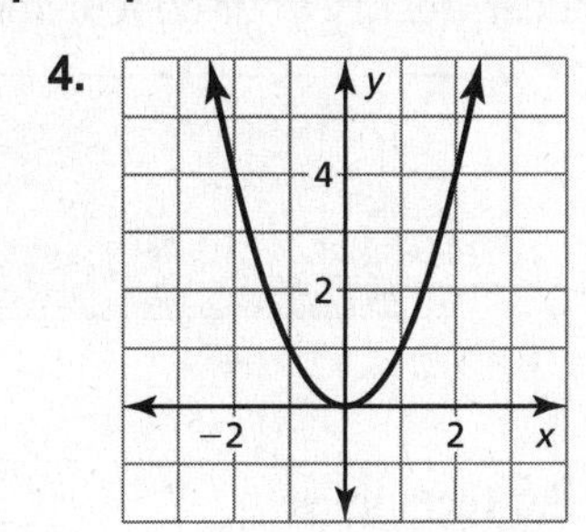

In Exercises 5 and 6, find the domain and range of the function represented by the graph.

5.

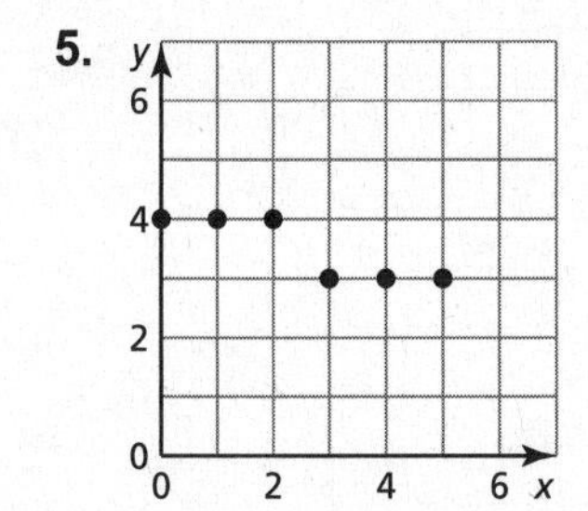

6.

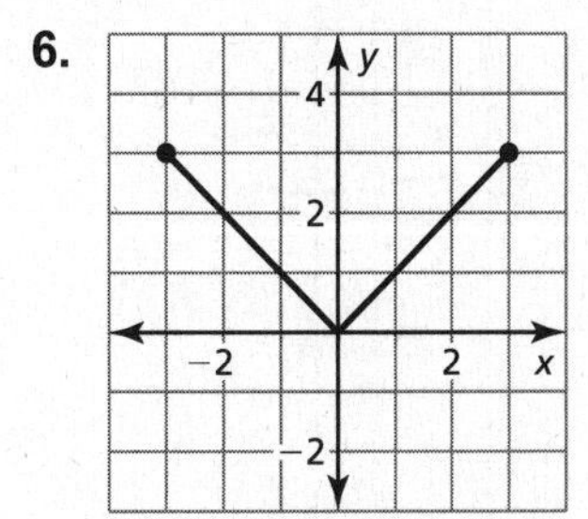

7. The function $y = 12x$ represents the number y of pages of text a computer printer can print in x minutes.

   a. Identify the independent and dependent variables.

   b. The domain is 1, 2, 3, and 4. What is the range?

Name ____ Date __________

3.2 Linear Functions
For use with Exploration 3.2

Essential Question How can you determine whether a function is linear or nonlinear?

1 EXPLORATION: Finding Patterns for Similar Figures

Go to *BigIdeasMath.com* for an interactive tool to investigate this exploration.

Work with a partner. Complete each table for the sequence of similar figures. (In parts (a) and (b), use the rectangle shown.) Graph the data in each table. Decide whether each pattern is linear or nonlinear. Justify your conclusion.

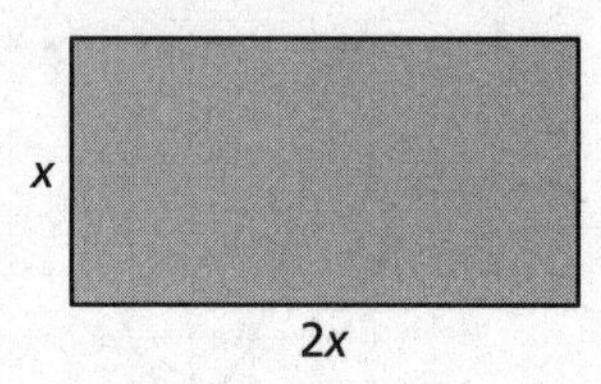

a. perimeters of similar rectangles

x	1	2	3	4	5
P					

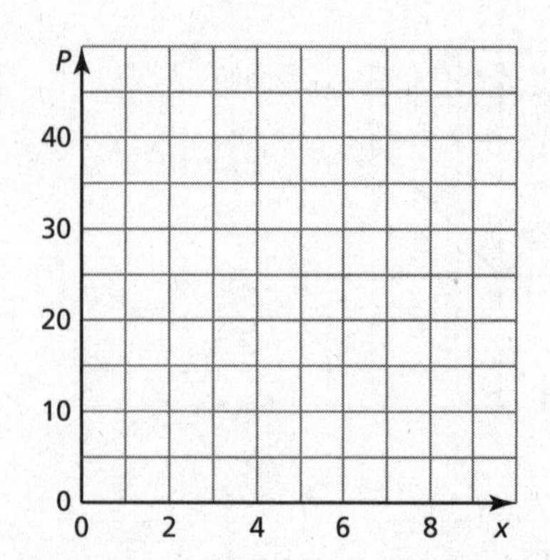

b. areas of similar rectangles

x	1	2	3	4	5
A					

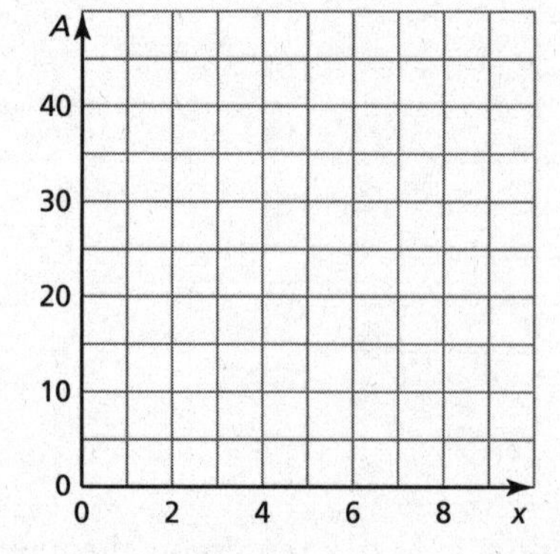

Name______________________________ Date__________

3.2 Linear Functions (continued)

1 EXPLORATION: Finding Patterns for Similar Figures (continued)

c. circumferences of circles of radius r

r	1	2	3	4	5
C					

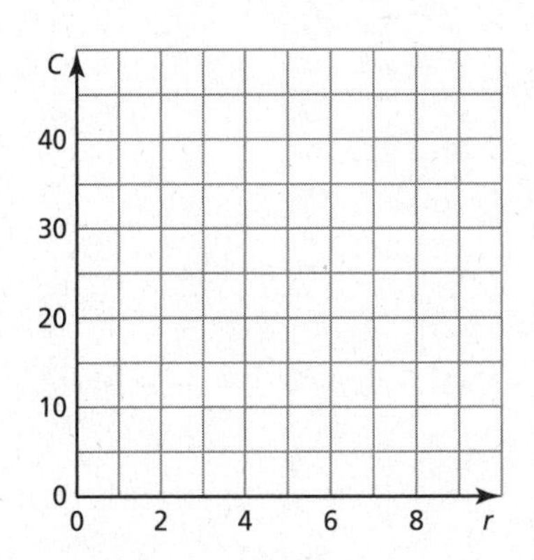

d. areas of circles of radius r

r	1	2	3	4	5
A					

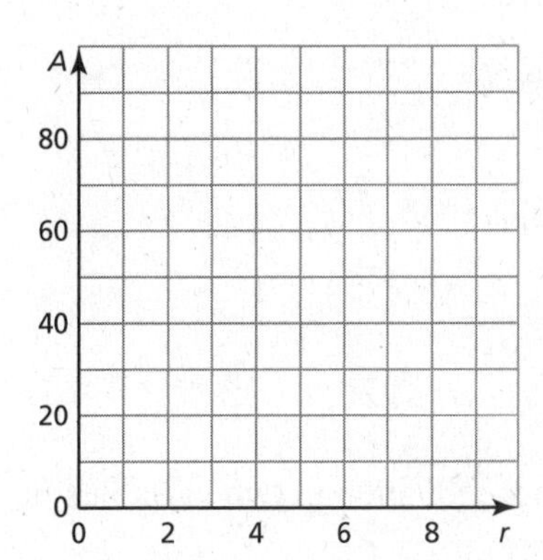

Communicate Your Answer

2. How do you know that the patterns you found in Exploration 1 represent functions?

3. How can you determine whether a function is linear or nonlinear?

4. Describe two real-life patterns: one that is linear and one that is nonlinear. Use patterns that are different from those described in Exploration 1.

Name ______________________________ Date __________

3.2 Notetaking with Vocabulary

For use after Lesson 3.2

In your own words, write the meaning of each vocabulary term.

linear equation in two variables

linear function

nonlinear function

solution of a linear equation in two variables

discrete domain

continuous domain

Notes:

Name____ Date__________

Core Concepts

Representations of Functions

Words An output is 3 more than the input.

Equation $y = x + 3$

Input-Output Table

Input, x	Output, y
−1	2
0	3
1	4
2	5

Mapping Diagram

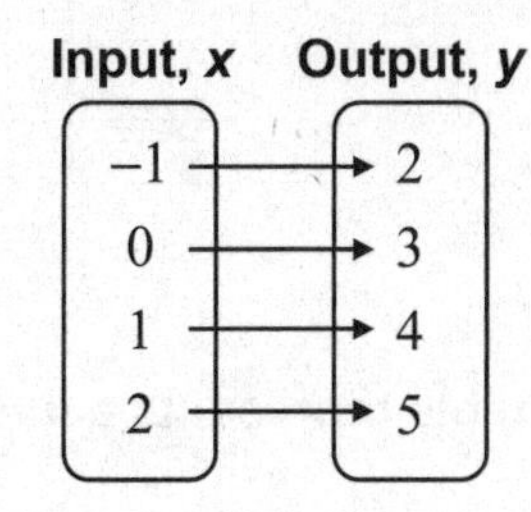

Graph

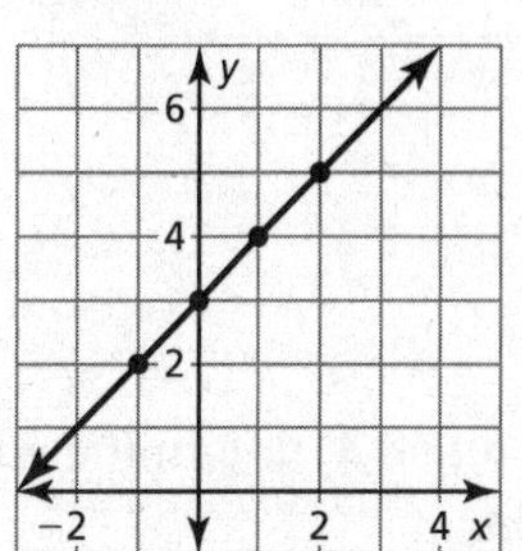

Notes:

Discrete and Continuous Domains

A **discrete domain** is a set of input values that consists of only certain numbers in an interval.

Example: Integers from 1 to 5

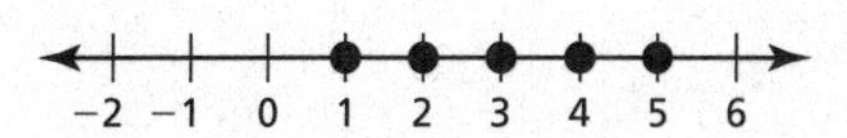

A **continuous domain** is a set of input values that consists of all numbers in an interval.

Example: All numbers from 1 to 5

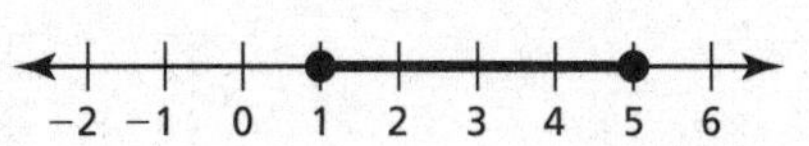

Notes:

Name ______________________________ Date __________

3.2 Notetaking with Vocabulary (continued)

Extra Practice

In Exercises 1 and 2, determine whether the graph represents a *linear* or *nonlinear* function. Explain.

1. 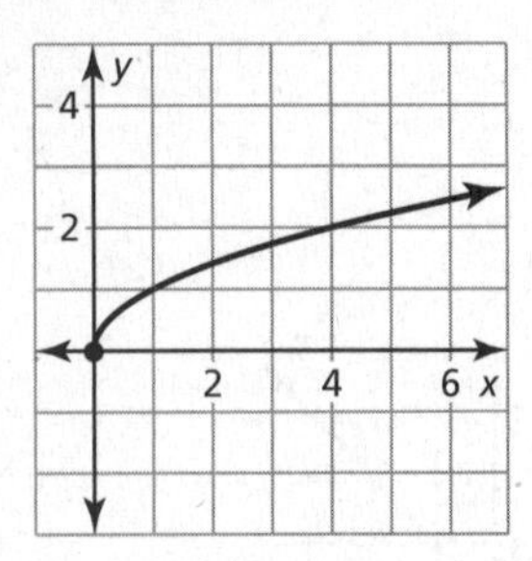

2. 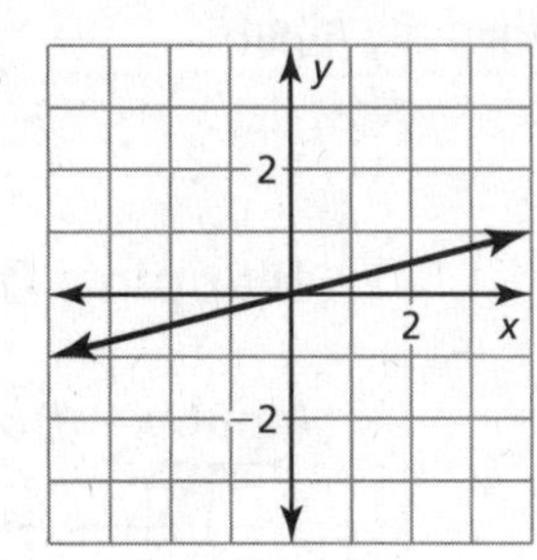

In Exercises 3 and 4, determine whether the table represents a *linear* or *nonlinear* function. Explain.

3.

x	1	2	3	4
y	−1	2	5	8

4.

x	−1	0	1	2
y	0	−1	0	3

In Exercises 5 and 6, determine whether the equation represents a *linear* or *nonlinear* function. Explain.

5. $y = 3 - 2x$

6. $y = -\frac{3}{4}x^3$

In Exercises 7 and 8, find the domain of the function represented by the graph. Determine whether the domain is *discrete* or *continuous*. Explain.

7. 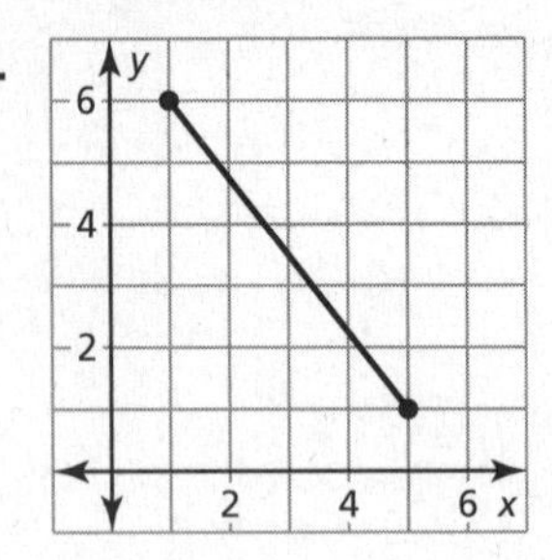

8. 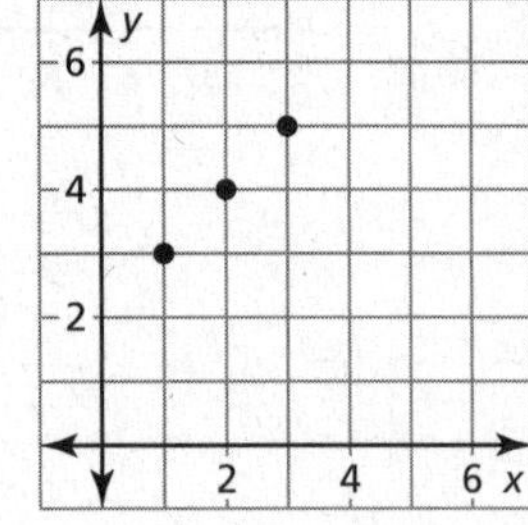

Name____ Date__________

3.3 Function Notation

For use with Exploration 3.3

Essential Question How can you use function notation to represent a function?

1 EXPLORATION: Matching Functions with Their Graphs

Work with a partner. Match each function with its graph.

a. $f(x) = 2x - 3$

b. $g(x) = -x + 2$

c. $h(x) = x^2 - 1$

d. $j(x) = 2x^2 - 3$

A.

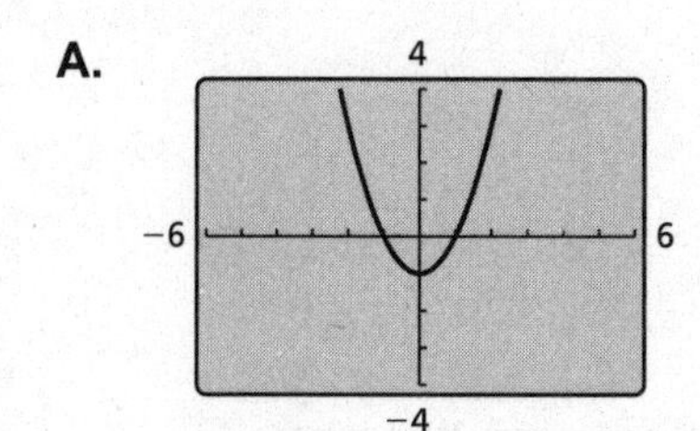

B.

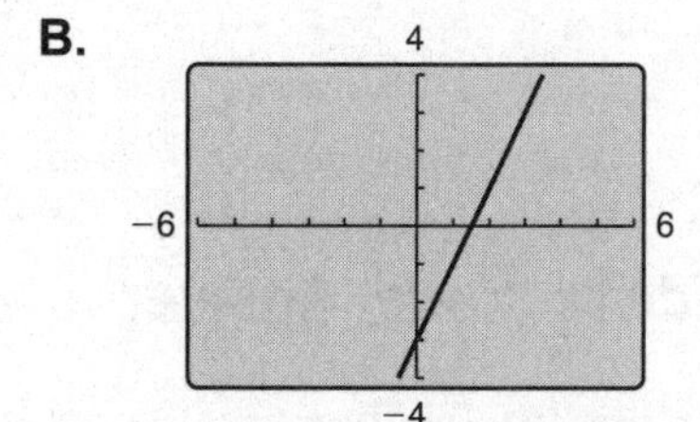

C.

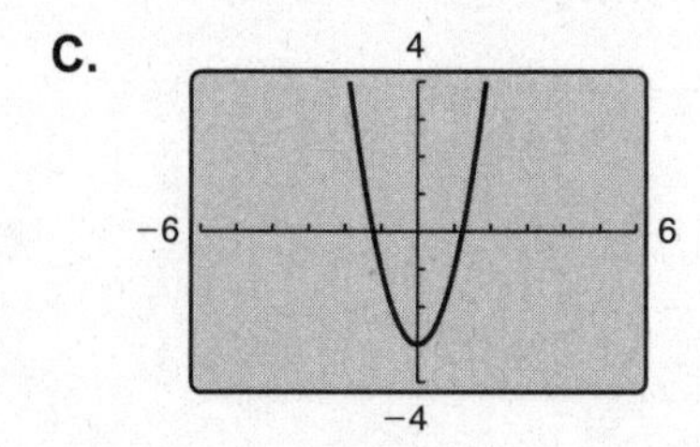

D.

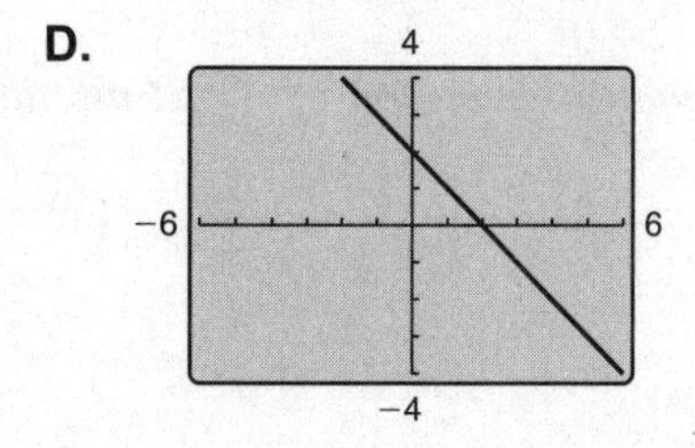

2 EXPLORATION: Evaluating a Function

Go to *BigIdeasMath.com* for an interactive tool to investigate this exploration.

Work with a partner. Consider the function

$f(x) = -x + 3.$

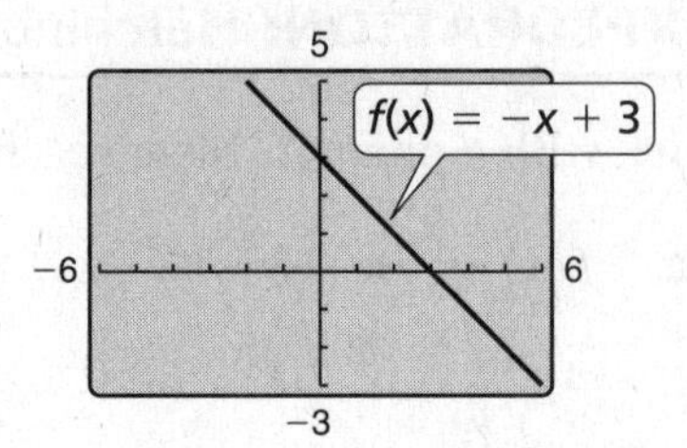

Locate the points $(x, f(x))$ on the graph.

Explain how you found each point.

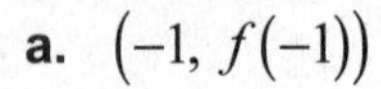

a. $(-1, f(-1))$

b. $(0, f(0))$

c. $(1, f(1))$

d. $(2, f(2))$

Communicate Your Answer

3. How can you use function notation to represent a function? How are standard notation and function notation similar? How are they different?

Standard Notation	***Function Notation***
$y = 2x + 5$	$f(x) = 2x + 5$

Name____ Date__________

3.3 Notetaking with Vocabulary

For use after Lesson 3.3

In your own words, write the meaning of each vocabulary term.

function notation

Notes:

Name ______________________________ Date __________

3.3 Notetaking with Vocabulary (continued)

Extra Practice

In Exercises 1–6, evaluate the function when $x = -4$, 0, and 2.

1. $f(x) = -x + 4$

2. $g(x) = 5x$

3. $h(x) = 7 - 2x$

4. $s(x) = 12 - 0.25x$

5. $t(x) = 6 + 3x - 2$

6. $u(x) = -2 - 2x + 7$

7. Let $n(t)$ be the number of DVDs you have in your collection after t trips to the video store. Explain the meaning of each statement.

a. $n(0) = 8$

b. $n(3) = 14$

c. $n(5) > n(3)$

d. $n(7) - n(2) = 10$

In Exercises 8–11, find the value of *x* so that the function has the given value.

8. $b(x) = -3x + 1;\ b(x) = -20$

9. $r(x) = 4x - 3;\ r(x) = 33$

10. $m(x) = -\frac{3}{5}x - 4;\ m(x) = 2$

11. $w(x) = \frac{5}{6}x - 3;\ w(x) = -18$

Name____ Date__________

3.3 Notetaking with Vocabulary (continued)

In Exercises 12 and 13, graph the linear function.

12. $s(x) = \frac{1}{2}x - 2$

x	-4	-2	0	2	4
s(x)					

13. $t(x) = 1 - 2x$

x	-2	-1	0	1	2
t(x)					

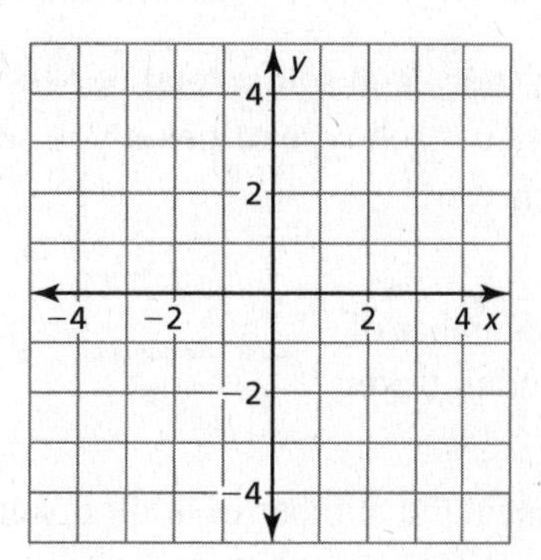

14. The function $B(m) = 50m + 150$ represents the balance (in dollars) in your savings account after m months. The table shows the balance in your friend's savings account. Who has the better savings plan? Explain.

Month	Balance
2	\$330
4	\$410
6	\$490

Name ______________________________ Date __________

3.4 Graphing Linear Equations in Standard Form

For use with Exploration 3.4

Essential Question How can you describe the graph of the equation $Ax + By = C$?

1 EXPLORATION: Using a Table to Plot Points

Go to *BigIdeasMath.com* for an interactive tool to investigate this exploration.

Work with a partner. You sold a total of $16 worth of tickets to a fundraiser. You lost track of how many of each type of ticket you sold. Adult tickets are $4 each. Child tickets are $2 each.

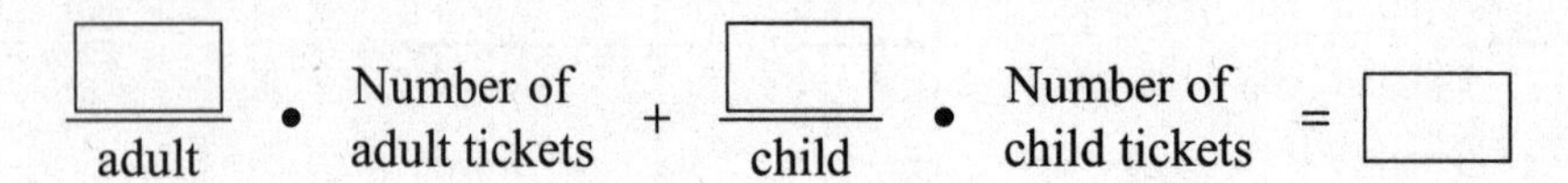

a. Let x represent the number of adult tickets. Let y represent the number of child tickets. Use the verbal model to write an equation that relates x and y.

b. Complete the table to show the different combinations of tickets you might have sold.

x					
y					

c. Plot the points from the table. Describe the pattern formed by the points.

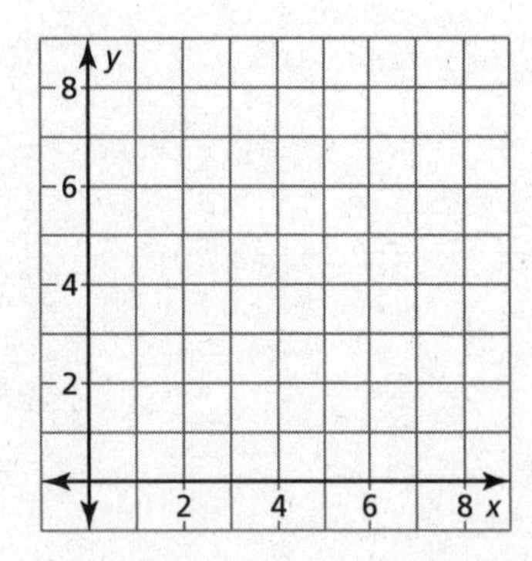

d. If you remember how many adult tickets you sold, can you determine how many child tickets you sold? Explain your reasoning.

3.4 Graphing Linear Equations in Standard Form (continued)

2 EXPLORATION: Rewriting and Graphing an Equation

Go to *BigIdeasMath.com* for an interactive tool to investigate this exploration.

Work with a partner. You sold a total of \$48 worth of cheese. You forgot how many pounds of each type of cheese you sold. Swiss cheese costs \$8 per pound. Cheddar cheese costs \$6 per pound.

$$\frac{\square}{\text{pound}} \bullet \text{Pounds of Swiss} + \frac{\square}{\text{pound}} \bullet \text{Pounds of cheddar} = \square$$

a. Let x represent the number of pounds of Swiss cheese. Let y represent the number of pounds of cheddar cheese. Use the verbal model to write an equation that relates x and y.

b. Solve the equation for y. Then use a graphing calculator to graph the equation. Given the real-life context of the problem, find the domain and range of the function.

c. The ***x*-intercept** of a graph is the x-coordinate of a point where the graph crosses the x-axis. The ***y*-intercept** of a graph is the y-coordinate of a point where the graph crosses the y-axis. Use the graph to determine the x- and y-intercepts.

d. How could you use the equation you found in part (a) to determine the x- and y-intercepts? Explain your reasoning.

e. Explain the meaning of the intercepts in the context of the problem.

Communicate Your Answer

3. How can you describe the graph of the equation $Ax + By = C$?

4. Write a real-life problem that is similar to those shown in Explorations 1 and 2.

Name ______________________________ Date __________

3.4 Notetaking with Vocabulary
For use after Lesson 3.4

In your own words, write the meaning of each vocabulary term.

standard form

x-intercept

y-intercept

Core Concepts

Horizontal and Vertical Lines

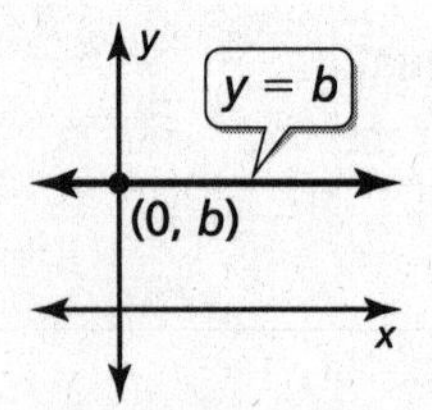

The graph of $y = b$ is a horizontal line. The line passes through the point $(0, b)$.

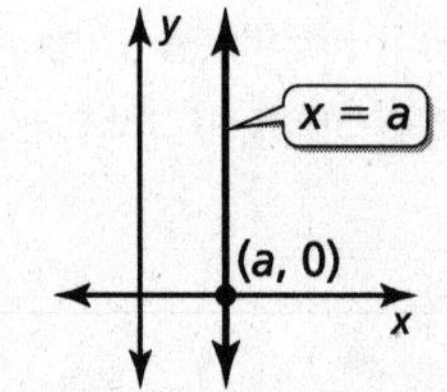

The graph of $x = a$ is a vertical line. The line passes through the point $(a, 0)$.

Notes:

Name______________________________ Date__________

3.4 Notetaking with Vocabulary (continued)

Using Intercepts to Graph Equations

The ***x*-intercept** of a graph is the *x*-coordinate of a point where the graph crosses the *x*-axis. It occurs when $y = 0$.

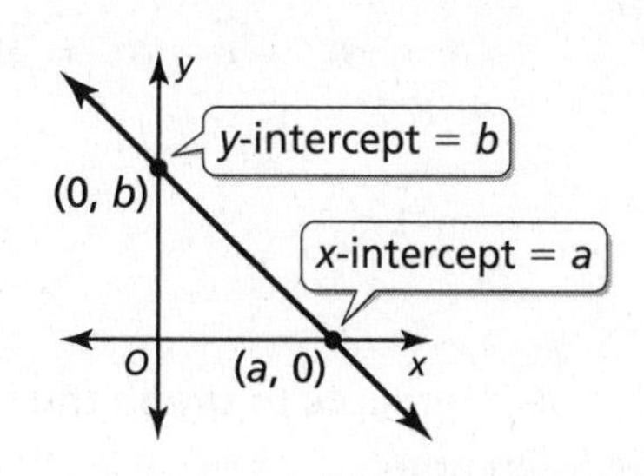

The ***y*-intercept** of a graph is the *y*-coordinate of a point where the graph crosses the *y*-axis. It occurs when $x = 0$.

To graph the linear equation $Ax + By = C$, find the intercepts and draw the line that passes through the two intercepts.

- To find the *x*-intercept, let $y = 0$ and solve for *x*.
- To find the *y*-intercept, let $x = 0$ and solve for *y*.

Notes:

Extra Practice

In Exercises 1 and 2, graph the linear equation.

1. $y = -3$

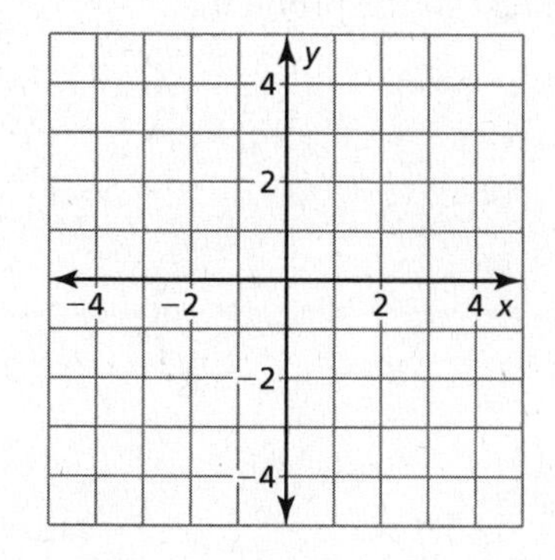

2. $x = 2$

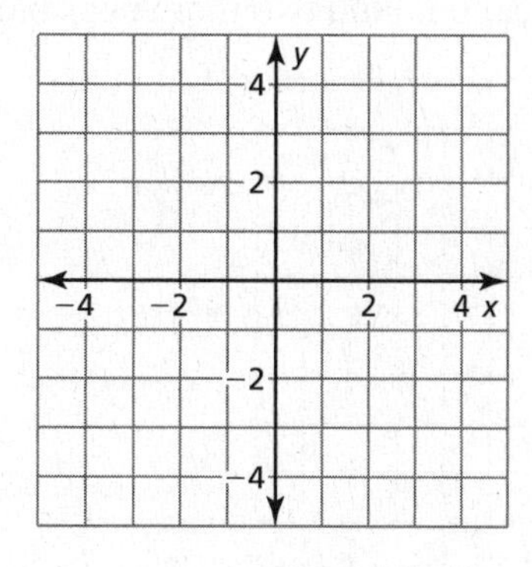

3.4 Notetaking with Vocabulary (continued)

In Exercises 3–5, find the *x*- and *y*-intercepts of the graph of the linear equation.

3. $3x + 4y = 12$ **4.** $-x - 4y = 16$ **5.** $5x - 2y = -30$

In Exercises 6 and 7, use intercepts to graph the linear equation. Label the points corresponding to the intercepts.

6. $-8x + 12y = 24$

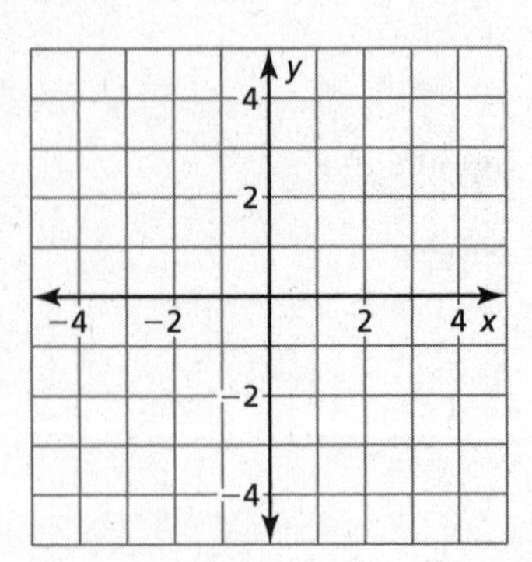

7. $2x + y = 4$

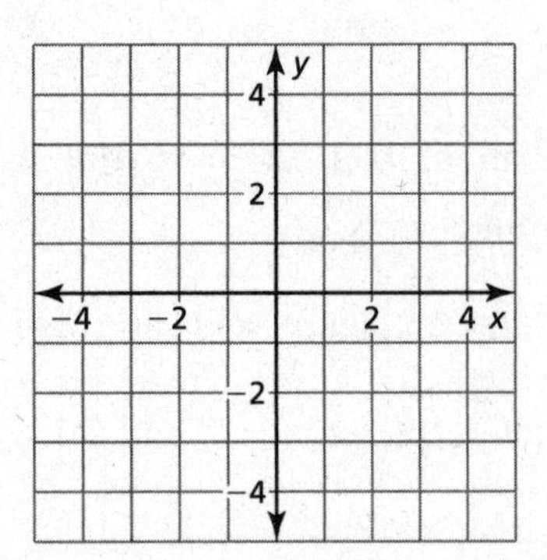

8. The school band is selling sweatshirts and baseball caps to raise \$9000 to attend a band competition. Sweatshirts cost \$25 each and baseball caps cost \$10 each. The equation $25x + 10y = 9000$ models this situation, where x is the number of sweatshirts sold and y is the number of baseball caps sold.

a. Find and interpret the intercepts.

b. If 258 sweatshirts are sold, how many baseball caps are sold?

c. Graph the equation. Find two more possible solutions in the context of the problem.

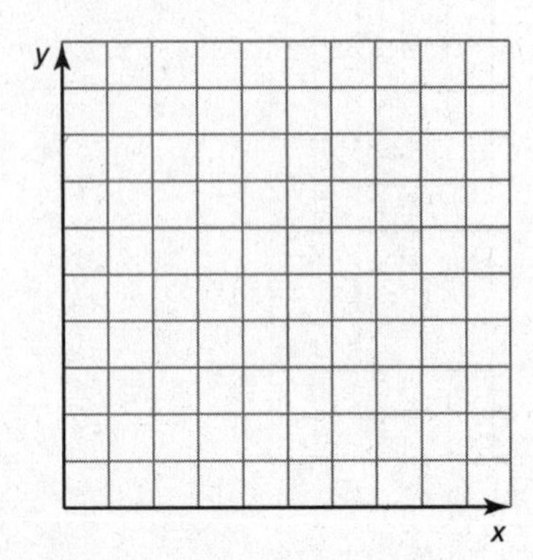

Name______________________________ Date__________

3.5 Graphing Linear Equations in Slope-Intercept Form

For use with Exploration 3.5

Essential Question How can you describe the graph of the equation $y = mx + b$?

1 EXPLORATION: Finding Slopes and y-Intercepts

Work with a partner. Find the slope and y-intercept of each line.

a.

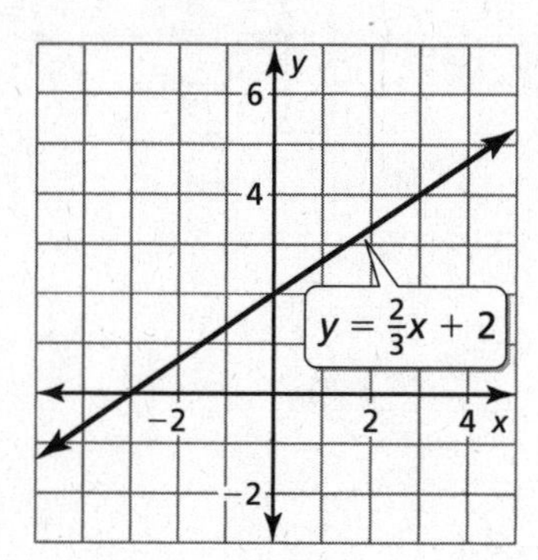

b.

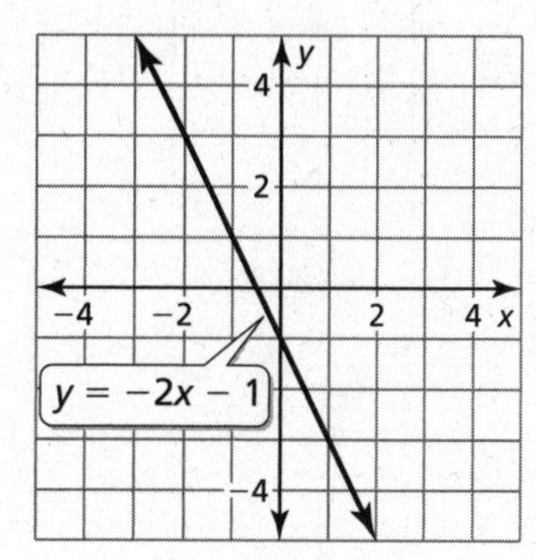

2 EXPLORATION: Writing a Conjecture

Go to *BigIdeasMath.com* for an interactive tool to investigate this exploration.

Work with a partner. Graph each equation. Then complete the table. Use the completed table to write a conjecture about the relationship between the graph of $y = mx + b$ and the values of m and b.

Equation	Description of graph	Slope of graph	y-Intercept
a. $y = -\frac{2}{3}x + 3$	Line	$-\frac{2}{3}$	3
b. $y = 2x - 2$			
c. $y = -x + 1$			
d. $y = x - 4$			

a.

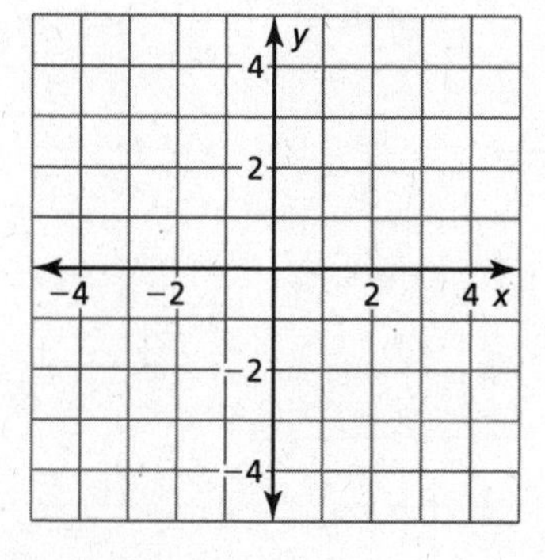

b.

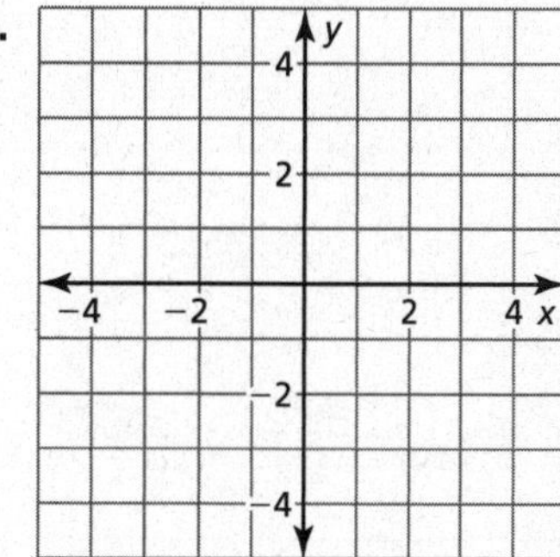

Name ______________________________ Date

2 EXPLORATION: Writing a Conjecture (continued)

c. 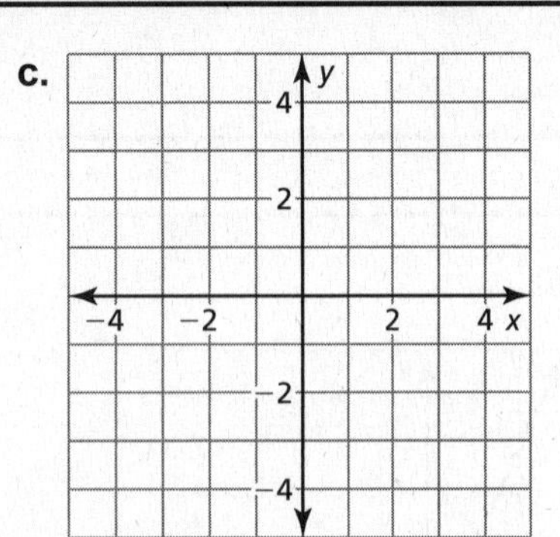

d. 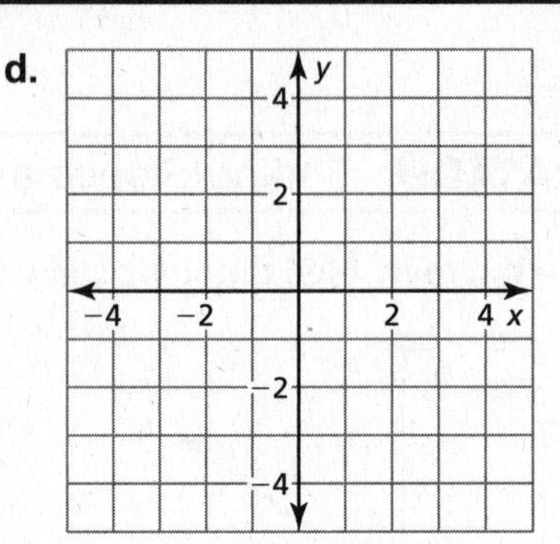

Communicate Your Answer

3. How can you describe the graph of the equation $y = mx + b$?

   a. How does the value of m affect the graph of the equation?

   b. How does the value of b affect the graph of the equation?

   c. Check your answers to parts (a) and (b) by choosing one equation from Exploration 2 and (1) varying only m and (2) varying only b.

Name____ Date__________

3.5 Notetaking with Vocabulary
For use after Lesson 3.5

In your own words, write the meaning of each vocabulary term.

slope

rise

run

slope-intercept form

constant function

Core Concepts

Slope

The **slope** m of a nonvertical line passing through two points (x_1, y_1) and (x_2, y_2) is the ratio of the **rise** (change in y) to the **run** (change in x).

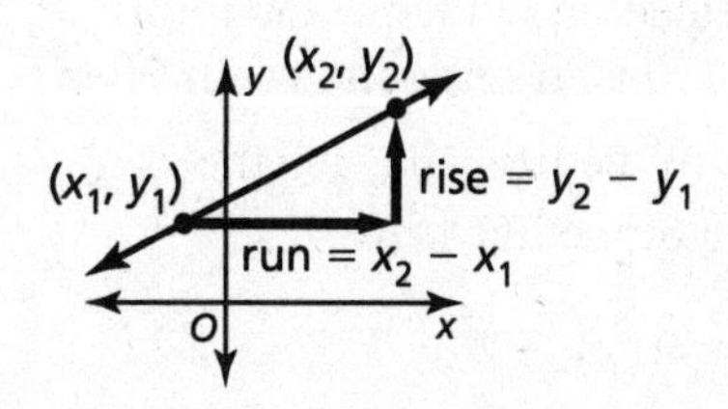

$$\text{slope} = m = \frac{\text{rise}}{\text{run}} = \frac{\text{change in } y}{\text{change in } x} = \frac{y_2 - y_1}{x_2 - x_1}$$

When the line rises from left to right, the slope is positive. When the line falls from left to right, the slope is negative.

Notes:

3.5 Notetaking with Vocabulary (continued)

Slope

Positive slope

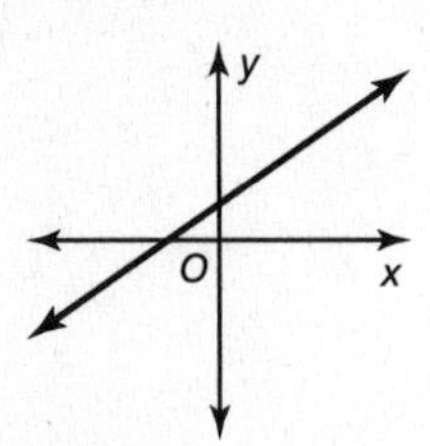

The line rises from left to right.

Negative slope

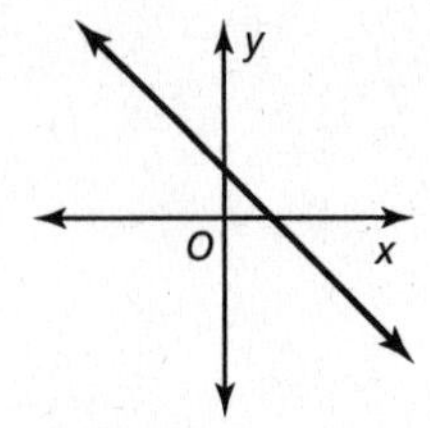

The line falls from left to right.

Slope of 0

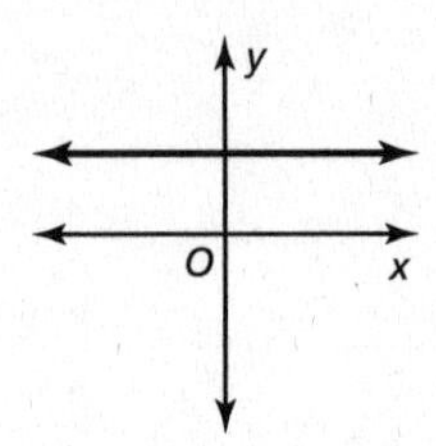

The line is horizontal.

Undefined slope

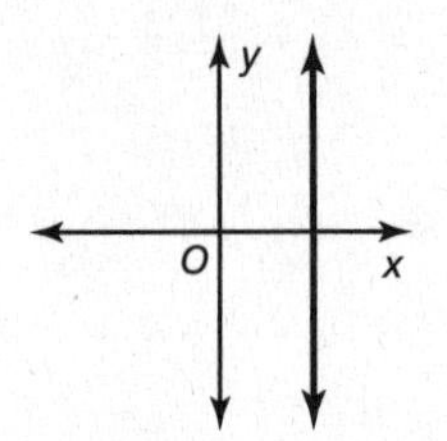

The line is vertical.

Notes:

Slope-Intercept Form

Words A linear equation written in the form $y = mx + b$ is in **slope-intercept form.** The slope of the line is m, and the y-intercept of the line is b.

Algebra $y = mx + b$

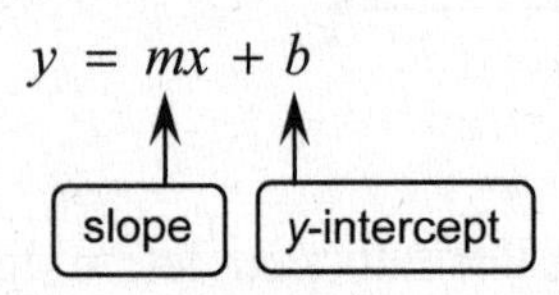

y
(0, b)
y = mx + b
x

Notes:

Name____ Date__________

Extra Practice

In Exercise 1–3, describe the slope of the line. Then find the slope.

1. 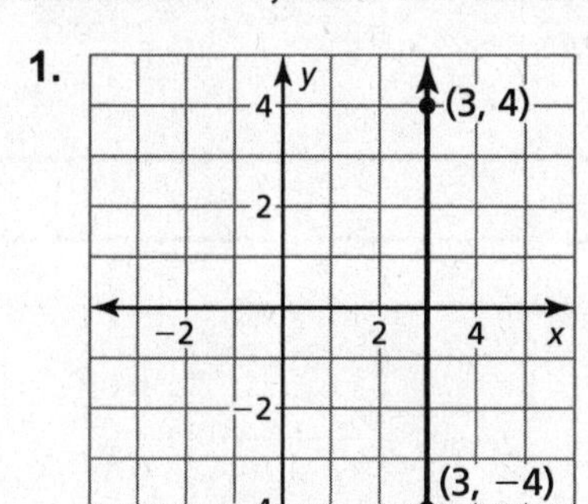

2. 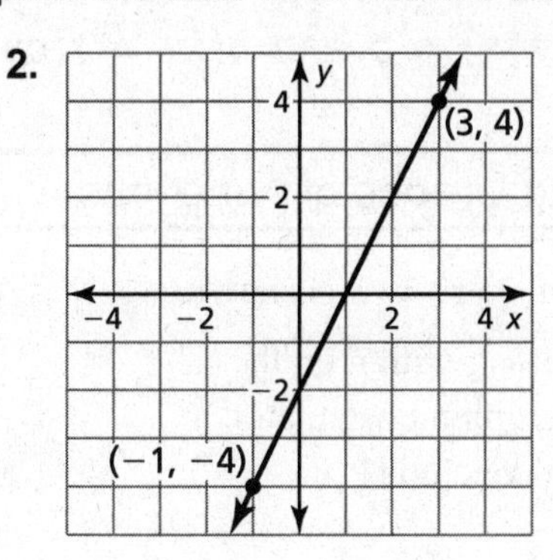

3. 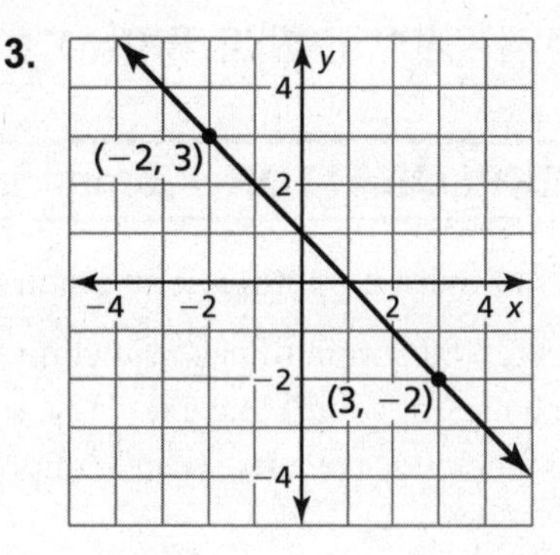

In Exercise 4 and 5, the points represented by the table lie on a line. Find the slope of the line.

4.

x	1	2	3	4
y	–2	–2	–2	–2

5.

x	–3	–1	1	3
y	11	3	–5	–13

In Exercise 6–8, find the slope and the *y*-intercept of the graph of the linear equation.

6. $6x + 4y = 24$

7. $y = -\frac{3}{4}x + 2$

8. $y = 5x$

9. A linear function f models a relationship in which the dependent variable decreases 6 units for every 3 units the independent variable decreases. The value of the function at 0 is 4. Graph the function. Identify the slope, y-intercept, and x-intercept of the graph.

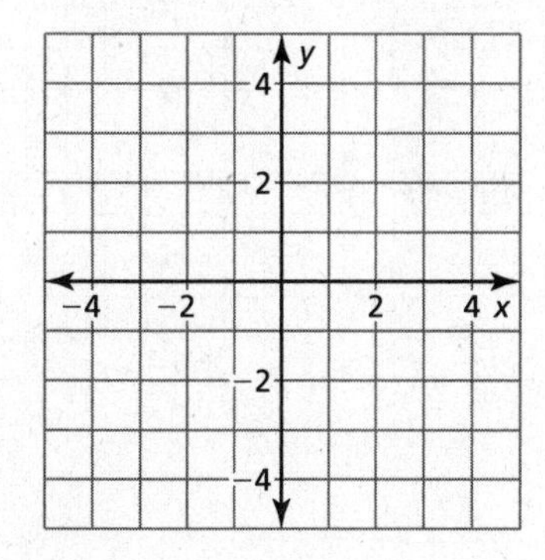

Name ______________________________ Date __________

3.6 Transformations of Graphs of Linear Functions

For use with Exploration 3.6

Essential Question How does the graph of the linear function $f(x) = x$ compare to the graphs of $g(x) = f(x) + c$ and $h(x) = f(cx)$?

1 EXPLORATION: Comparing Graphs of Functions

Work with a partner. The graph of $f(x) = x$ is shown.

Sketch the graph of each function, along with f, on the same set of coordinate axes. Use a graphing calculator to check your results. What can you conclude?

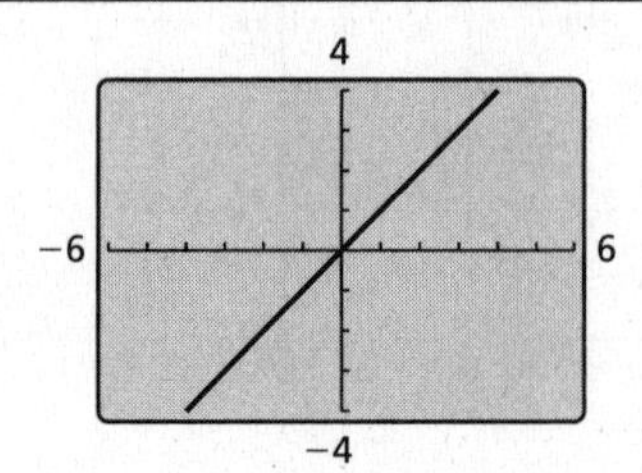

a. $g(x) = x + 4$ **b.** $g(x) = x + 2$ **c.** $g(x) = x - 2$ **d.** $g(x) = x - 4$

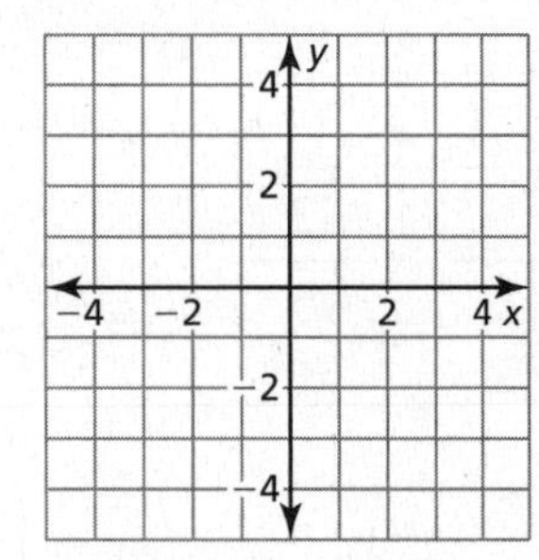
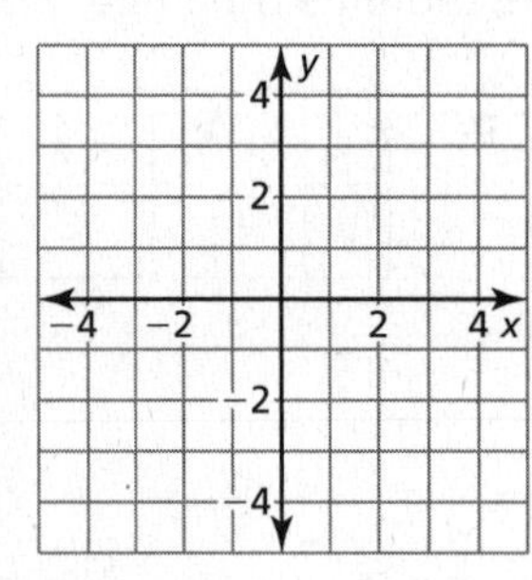
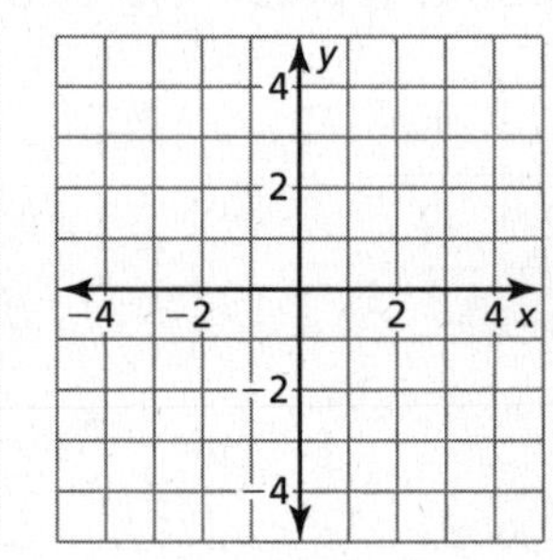
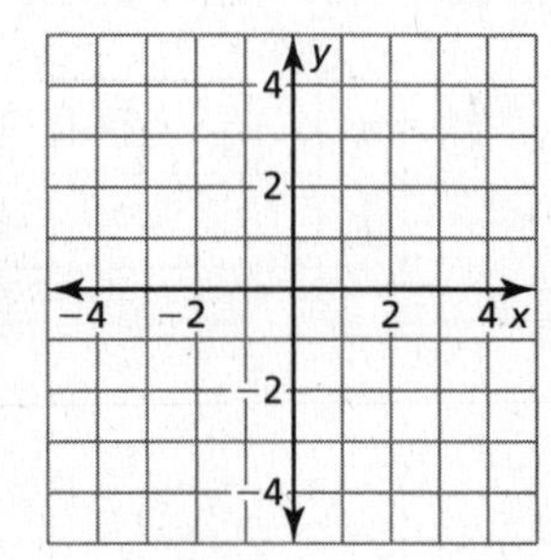

2 EXPLORATION: Comparing Graphs of Functions

Work with a partner. Sketch the graph of each function, along with $f(x) = x$, on the same set of coordinate axes. Use a graphing calculator to check your results. What can you conclude?

a. $h(x) = \frac{1}{2}x$ **b.** $h(x) = 2x$ **c.** $h(x) = -\frac{1}{2}x$ **d.** $h(x) = -2x$

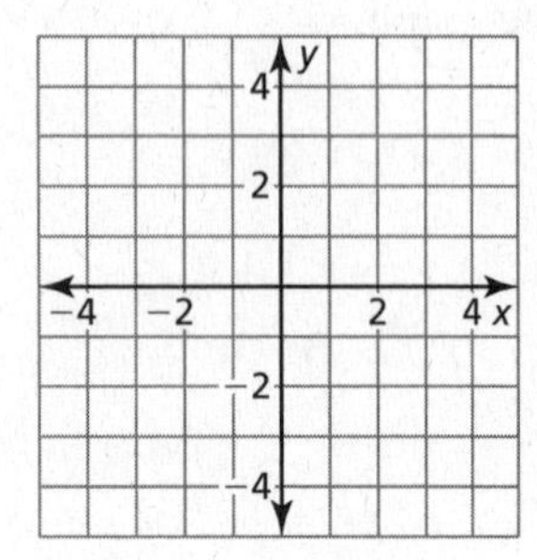
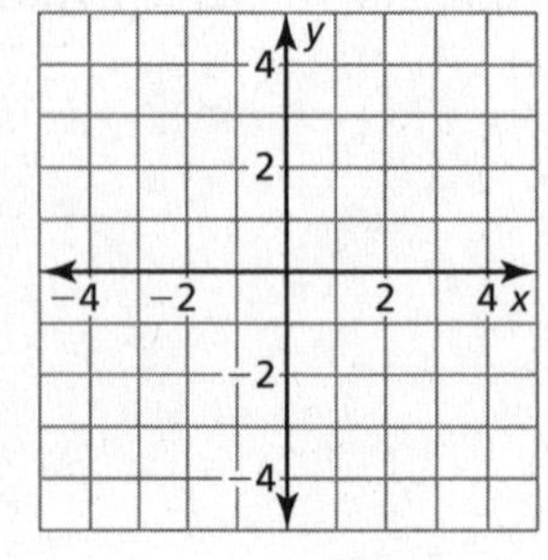
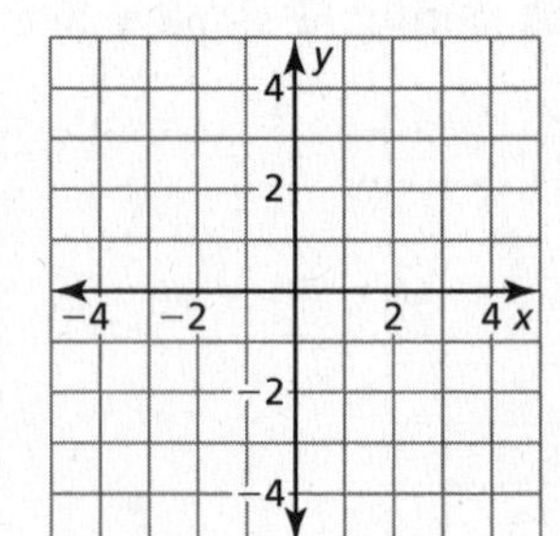
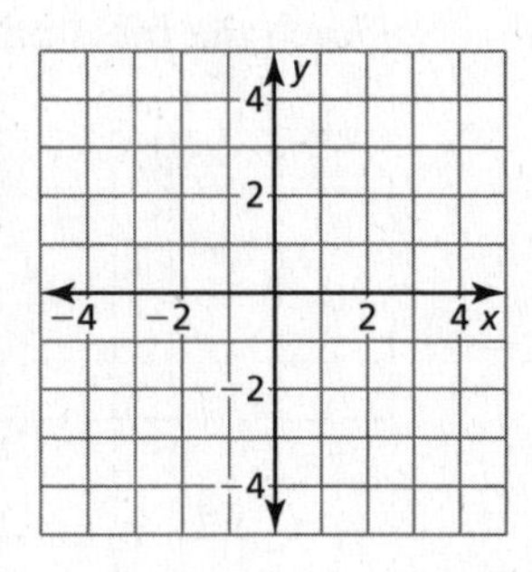

Name______________________________ Date__________

3 EXPLORATION: Matching Functions with Their Graphs

Work with a partner. Match each function with its graph. Use a graphing calculator to check your results. Then use the results of Explorations 1 and 2 to compare the graph of k to the graph of $f(x) = x$.

a. $k(x) = 2x - 4$

b. $k(x) = -2x + 2$

c. $k(x) = \frac{1}{2}x + 4$

d. $k(x) = -\frac{1}{2}x - 2$

A.

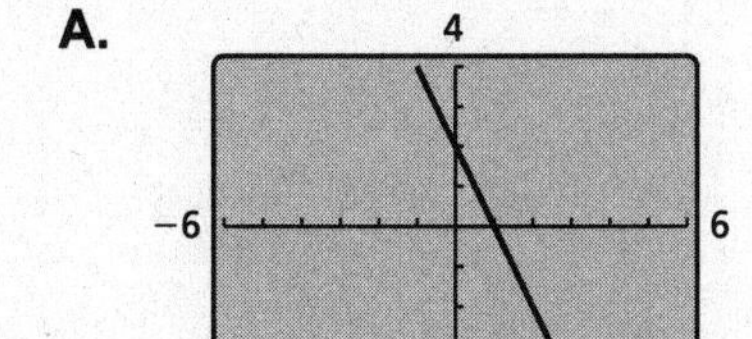

B.

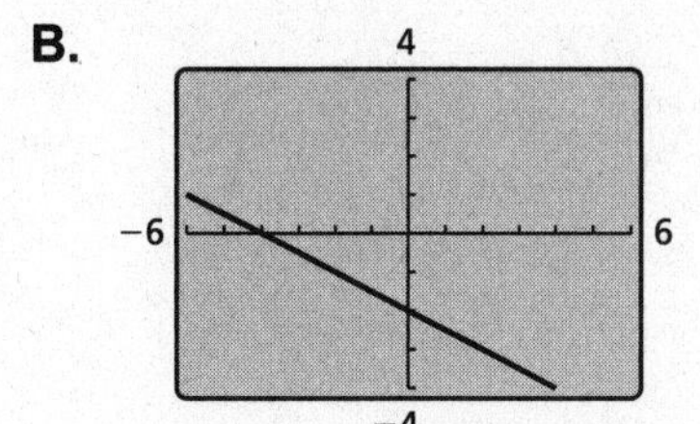

C.

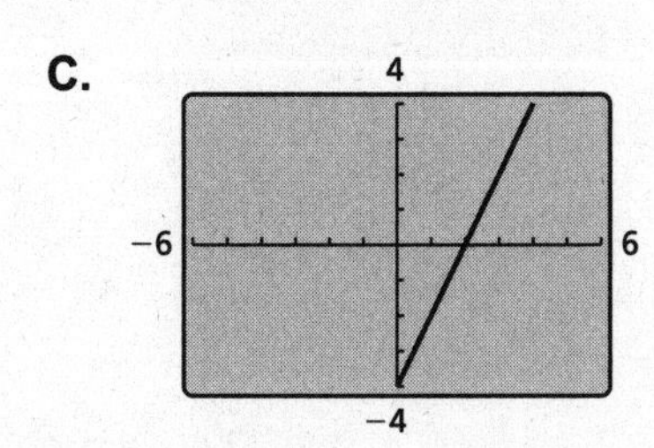

D.

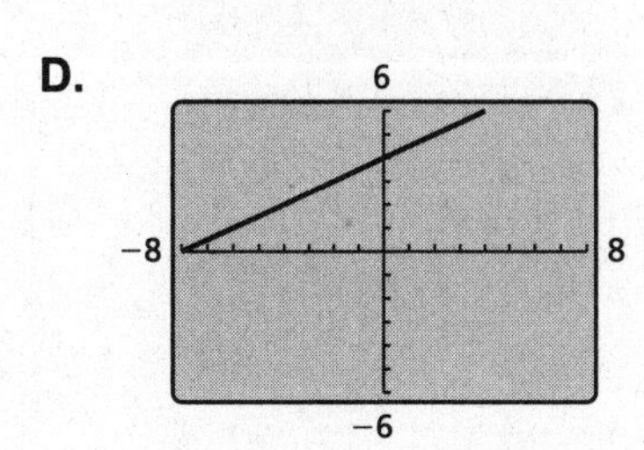

Communicate Your Answer

4. How does the graph of the linear function $f(x) = x$ compare to the graphs of $g(x) = f(x) + c$ and $h(x) = f(cx)$?

Name ______________________________ Date __________

3.6 Notetaking with Vocabulary
For use after Lesson 3.6

In your own words, write the meaning of each vocabulary term.

family of functions

parent function

transformation

translation

reflection

horizontal shrink

horizontal stretch

vertical stretch

vertical shrink

Notes:

Name____ Date

Core Concepts

A **translation** is a transformation that shifts a graph horizontally or vertically but does not change the size, shape, or orientation of the graph.

Horizontal Translations

The graph of $y = f(x - h)$ is a horizontal translation of the graph of $y = f(x)$, where $h \neq 0$.

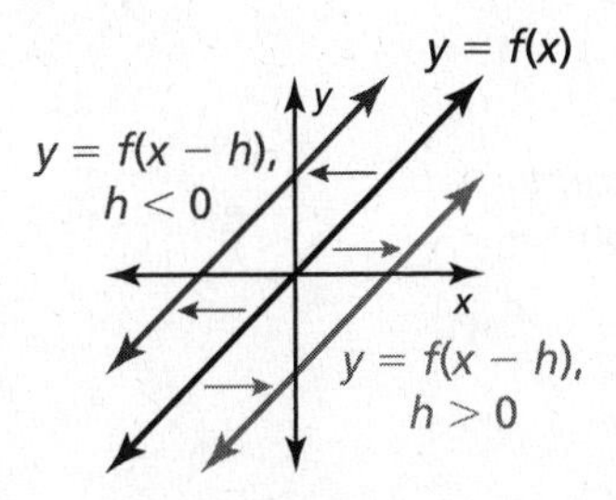

Subtracting h from the *inputs* before evaluating the function shifts the graph left when $h < 0$ and right when $h > 0$.

Vertical Translations

The graph of $y = f(x) + k$ is a vertical translation of the graph of $y = f(x)$, where $k \neq 0$.

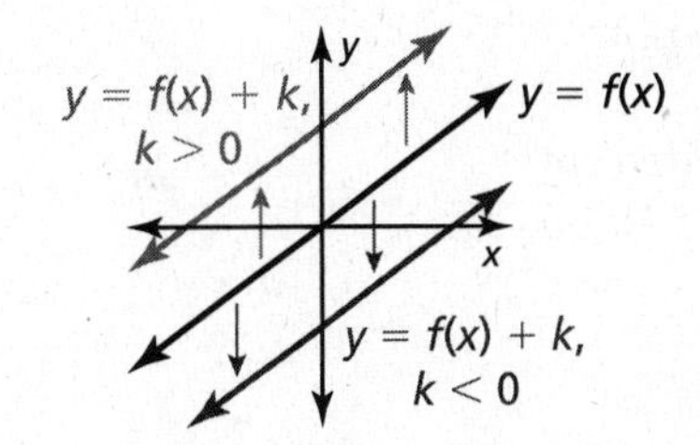

Adding k to the *outputs* shifts the graph down when $k < 0$ and up when $k > 0$.

Notes:

A **reflection** is a transformation that flips a graph over a line called the *line of reflection.*

Reflections in the *x*-axis

The graph of $y = -f(x)$ is a reflection in the x-axis of the graph of $y = f(x)$.

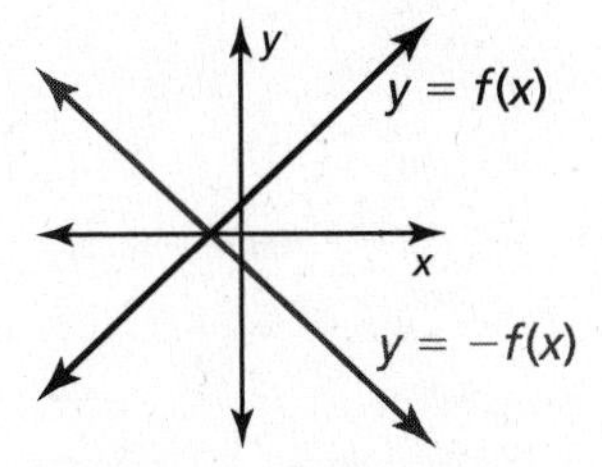

Multiplying the outputs by −1 changes their signs.

Reflections in the *y*-axis

The graph of $y = f(-x)$ is a reflection in the y-axis of the graph of $y = f(x)$.

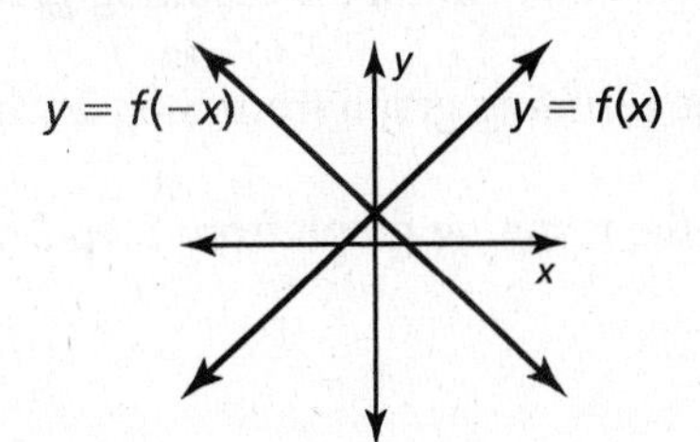

Multiplying the inputs by −1 changes their signs.

Notes:

3.6 Notetaking with Vocabulary (continued)

Horizontal Stretches and Shrinks

The graph of $y = f(ax)$ is a horizontal stretch or shrink by a factor of $\frac{1}{a}$ of the graph of $y = f(x)$, where $a > 0$ and $a \neq 1$.

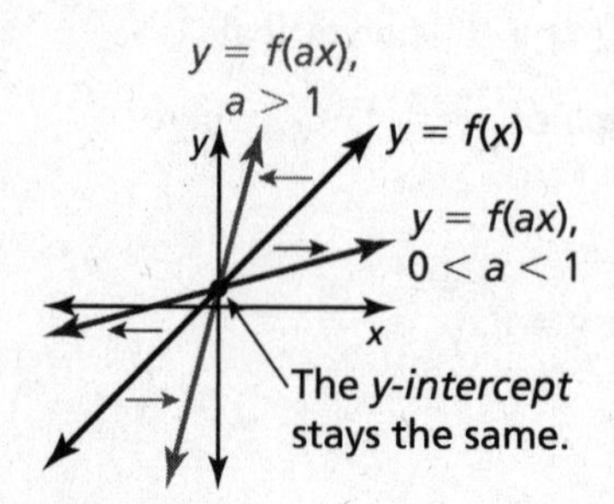

Vertical Stretches and Shrinks

The graph of $y = a \bullet f(x)$ is a vertical stretch or shrink by a factor of a of the graph of $y = f(x)$, where $a > 0$ and $a \neq 1$.

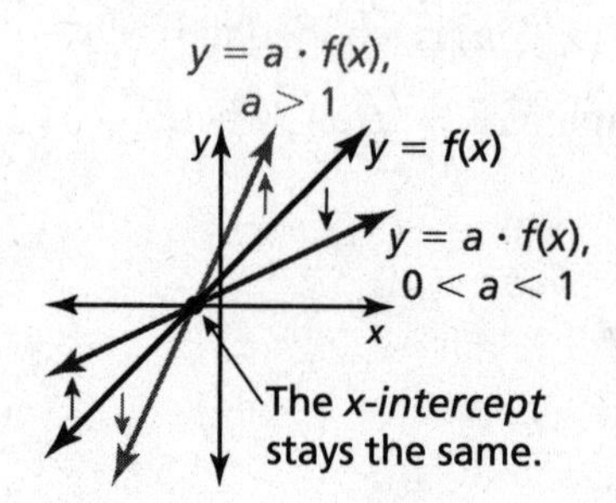

Notes:

Transformations of Graphs

The graph of $y = a \bullet f(x - h) + k$ or the graph of $y = f(ax - h) + k$ can be obtained from the graph of $y = f(x)$ by performing these steps.

Step 1 Translate the graph of $y = f(x)$ horizontally h units.

Step 2 Use a to stretch or shrink the resulting graph from Step 1.

Step 3 Reflect the resulting graph from Step 2 when $a < 0$.

Step 4 Translate the resulting graph from Step 3 vertically k units.

Notes:

Extra Practice

In Exercises 1–6, use the graphs of *f* and *g* to describe the transformation from the graph of *f* to the graph of *g*.

1.

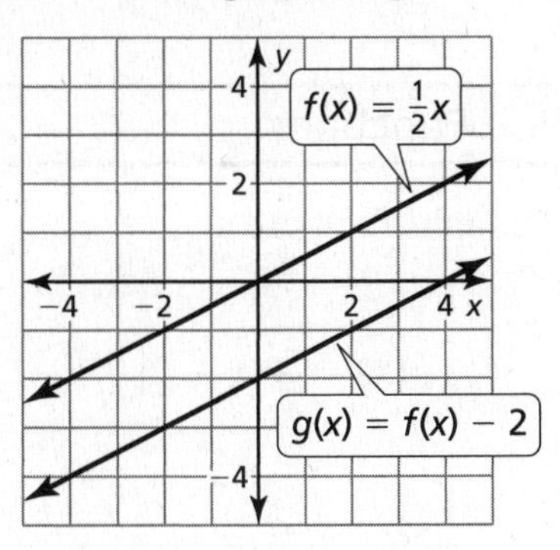

2.

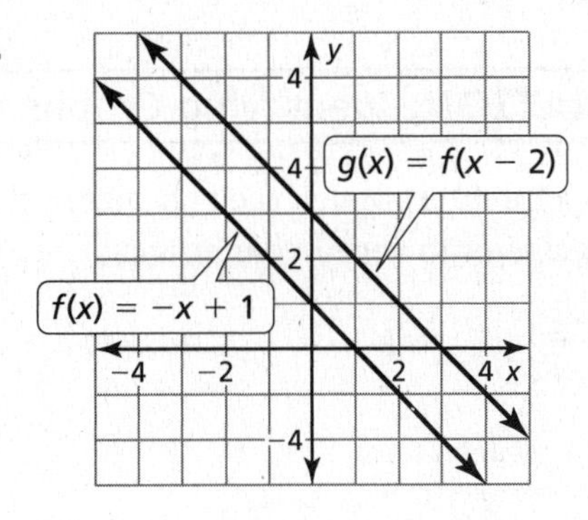

3.

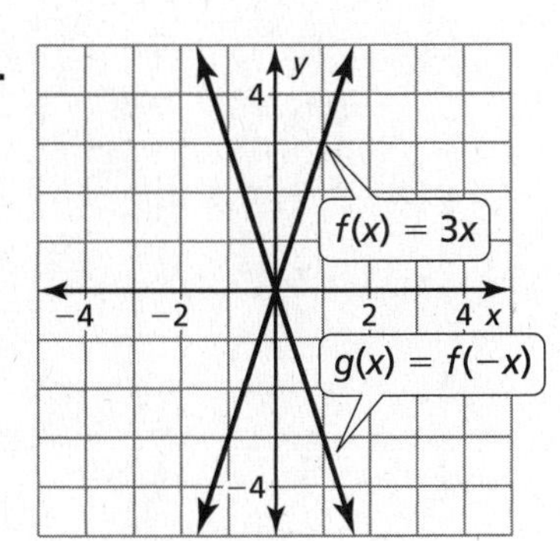

4.

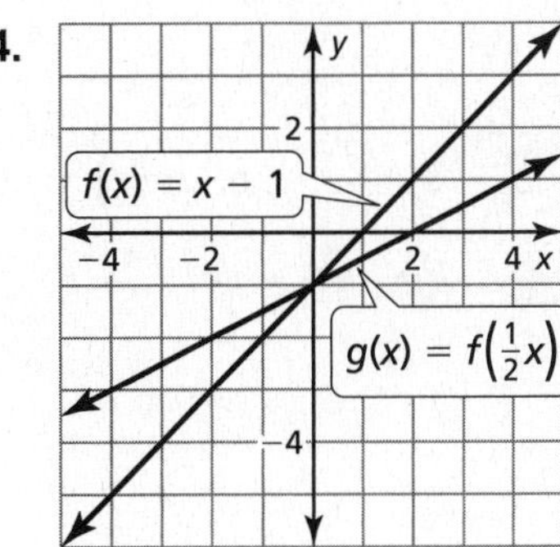

5.

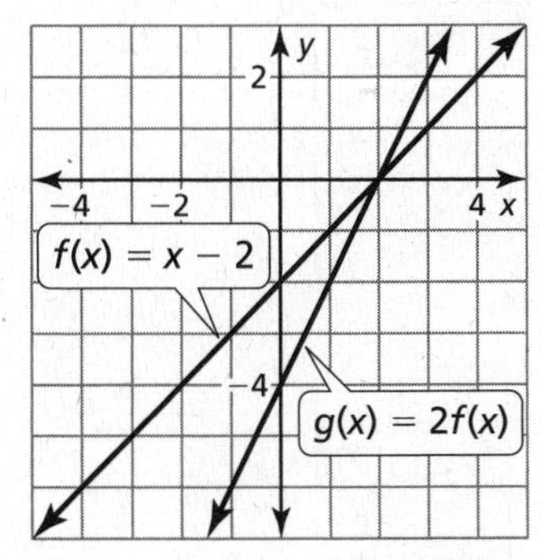

6.

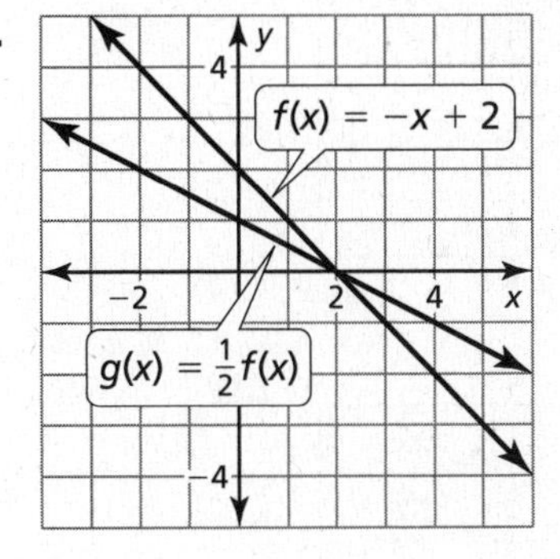

7. Graph $f(x) = x$ and $g(x) = 3x - 2$.

Describe the transformations from the graph of *f* to the graph of *g*.

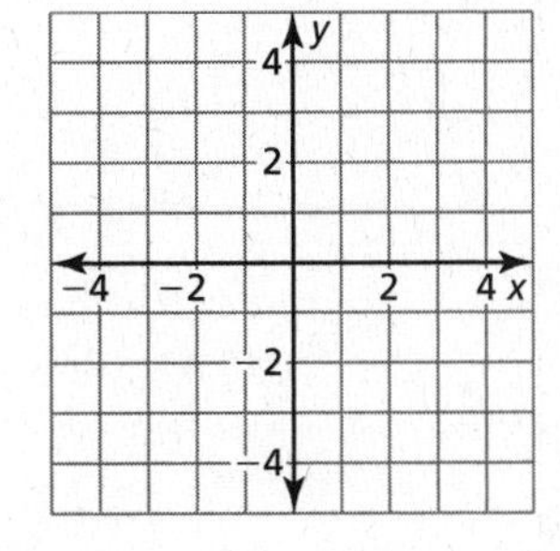

Name ______________________________ Date __________

3.7 Graphing Absolute Value Functions

For use with Exploration 3.7

Essential Question How do the values of a, h, and k affect the graph of the absolute value function $g(x) = a|x - h| + k$?

1 EXPLORATION: Identifying Graphs of Absolute Value Functions

Work with a partner. Match each absolute value function with its graph. Then use a graphing calculator to verify your answers.

a. $g(x) = -|x - 2|$

b. $g(x) = |x - 2| + 2$

c. $g(x) = -|x + 2| - 2$

d. $g(x) = |x - 2| - 2$

e. $g(x) = 2|x - 2|$

f. $g(x) = -|x + 2| + 2$

A.

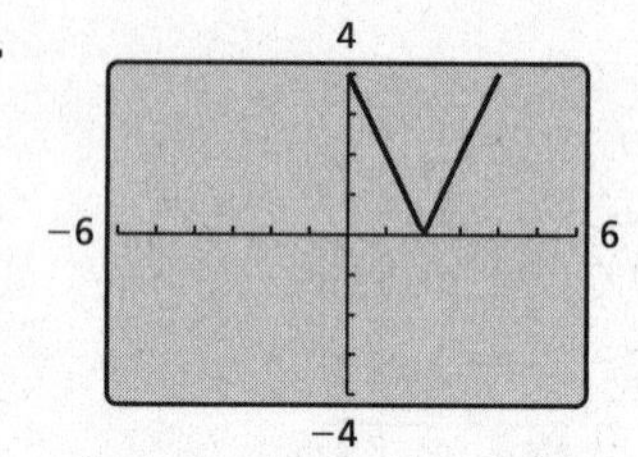

B.

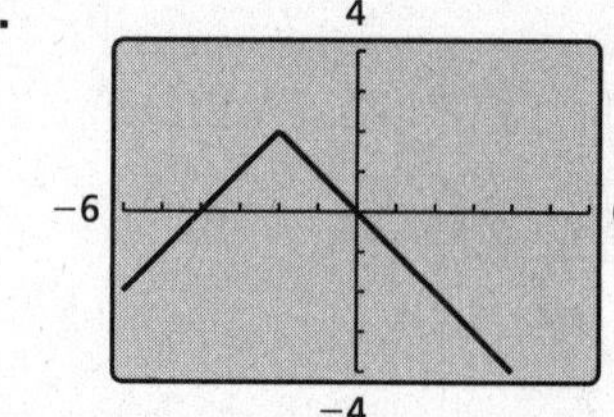

C.

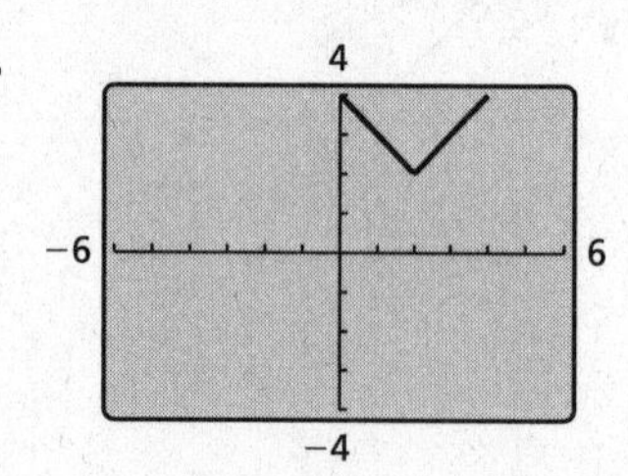

D.

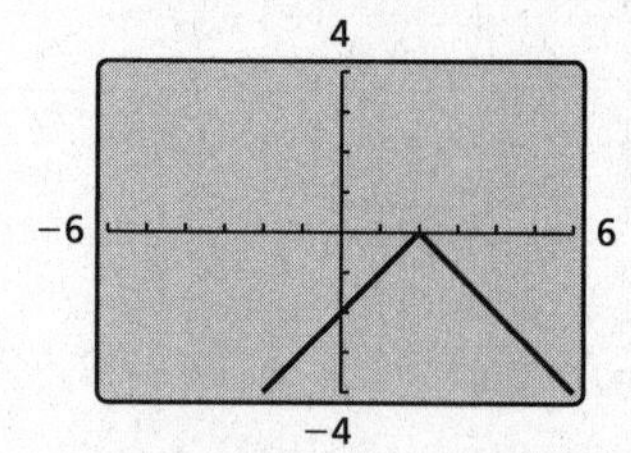

E.

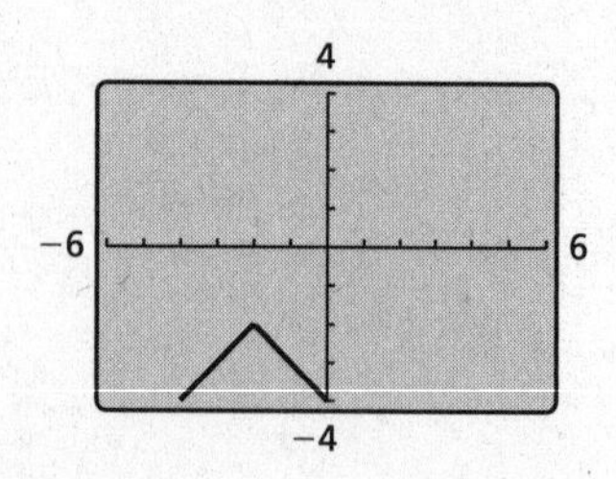

F.

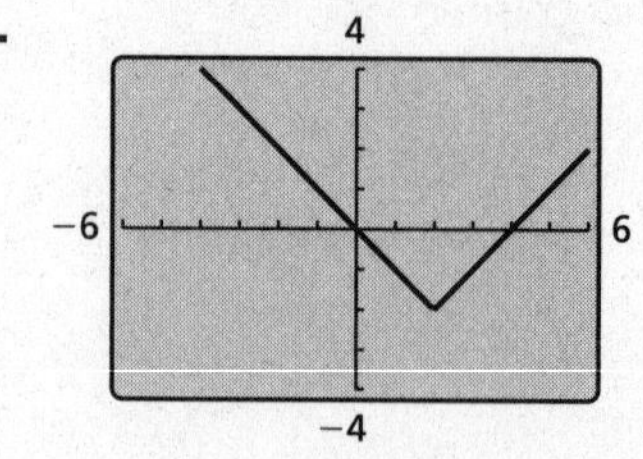

Communicate Your Answer

2. How do the values of a, h, and k affect the graph of the absolute value function $g(x) = a|x - h| + k$?

3. Write the equation of the absolute value function whose graph is shown. Use a graphing calculator to verify your equation.

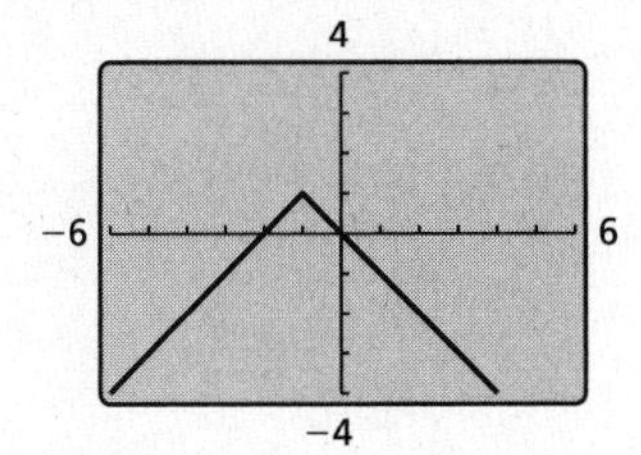

Name ______________________________ Date __________

3.7 Notetaking with Vocabulary
For use after Lesson 3.7

In your own words, write the meaning of each vocabulary term.

absolute value function

vertex

vertex form

Notes:

Core Concepts

Absolute Value Function

An **absolute value function** is a function that contains an absolute value expression. The parent absolute value function is $f(x) = |x|$. The graph of $f(x) = |x|$ is V-shaped and symmetric about the y-axis. The **vertex** is the point where the graph changes direction. The vertex of the graph of $f(x) = |x|$ is $(0, 0)$.

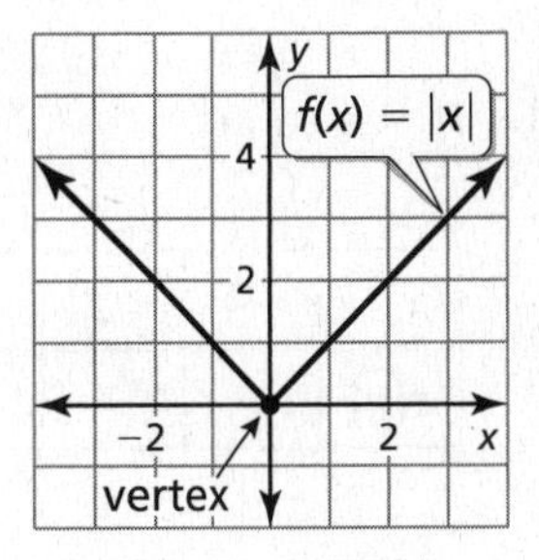

The domain of $f(x) = |x|$ is all real numbers.

The range is $y \geq 0$.

Notes:

Vertex Form of an Absolute Value Function

An absolute value function written in the form $g(x) = a|x - h| + k$, where $a \neq 0$, is in **vertex form**. The vertex of the graph of g is (h, k).

Any absolute value function can be written in vertex form, and its graph is symmetric about the line $x = h$.

Notes:

Extra Practice

In Exercises 1–4, graph the function. Compare the graph to the graph of $f(x) = |x|$. Describe the domain and range.

1. $t(x) = \frac{1}{2}|x|$

x	-4	-2	0	2	4
$t(x)$					

2. $u(x) = -|x|$

x	-2	-1	0	1	2
$u(x)$					

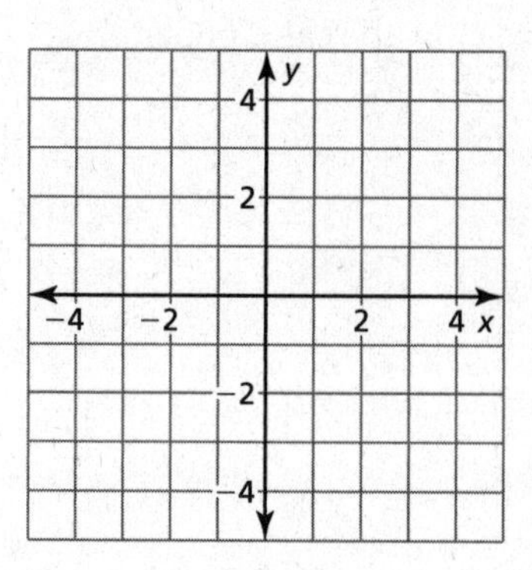

3. $p(x) = |x| - 3$

x	-2	-1	0	1	2
$p(x)$					

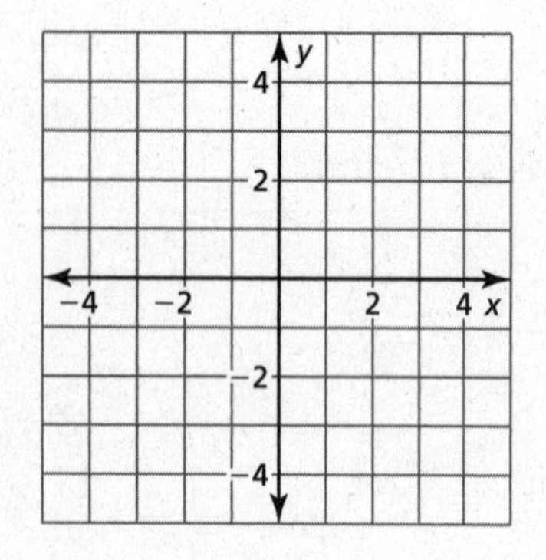

4. $r(x) = |x + 2|$

x	-4	-3	-2	-1	0
$r(x)$					

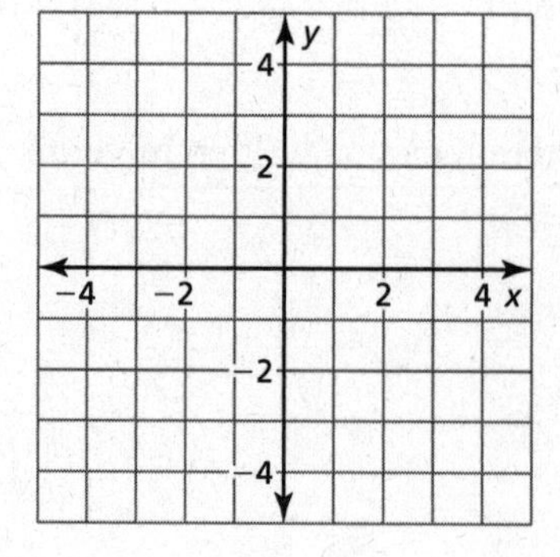

Name______________________________ Date__________

Chapter 4 Maintaining Mathematical Proficiency

Use the graph to answer the question.

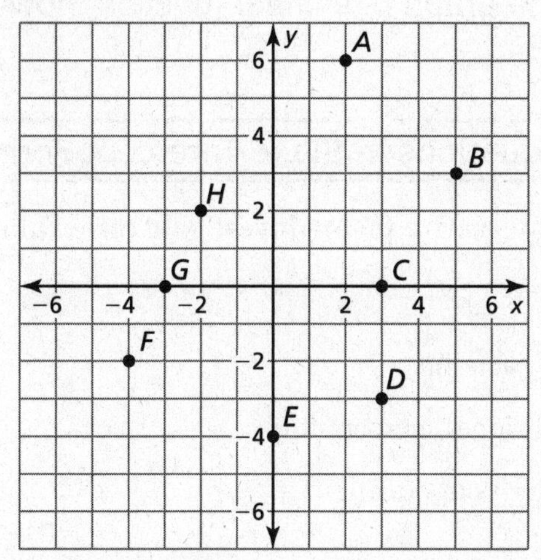

1. What ordered pair corresponds to point *A*?

2. What ordered pair corresponds to point *H*?

3. What ordered pair corresponds to point *E*?

4. Which point is located in Quadrant III?

5. Which point is located in Quadrant IV?

6. Which point is located on the negative *x*-axis?

Solve the equation for *y*.

7. $x - y = -12$

8. $8x + 4y = 16$

9. $3x - 5y + 15 = 0$

10. $0 = 3y - 6x + 12$

11. $y - 2 = 3x + 4y$

12. $6y + 3 - 2x = x$

13. Rectangle *ABCD* has vertices $A(4, -2)$, $B(4, 5)$, and $C(7, 5)$. What are the coordinates of vertex *D*?

Name ____ Date __________

4.1 Writing Equations in Slope-Intercept Form

For use with Exploration 4.1

Essential Question Given the graph of a linear function, how can you write an equation of the line?

1 EXPLORATION: Writing Equations in Slope-Intercept Form

Go to *BigIdeasMath.com* for an interactive tool to investigate this exploration.

Work with a partner.

- Find the slope and y-intercept of each line.
- Write an equation of each line in slope-intercept form.
- Use a graphing calculator to verify your equation.

a.

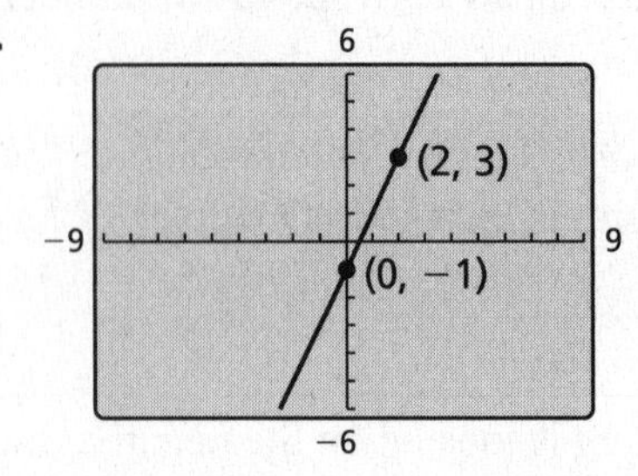

b.

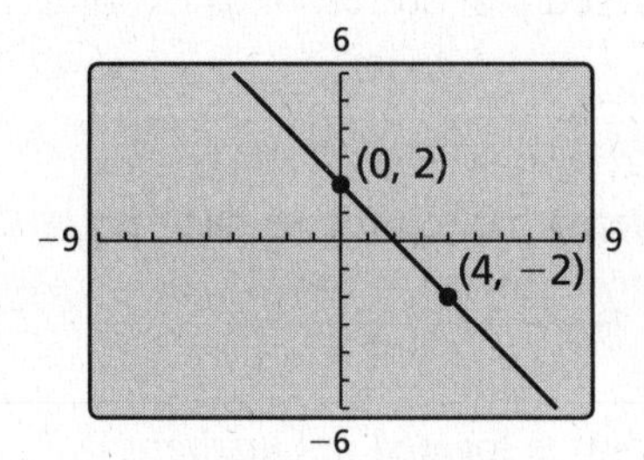

c.

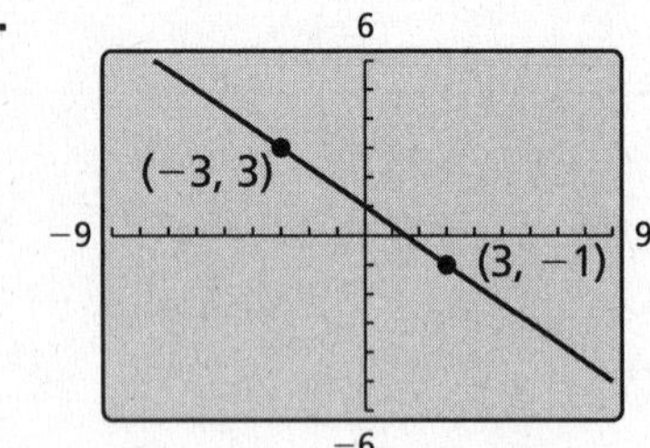

d.

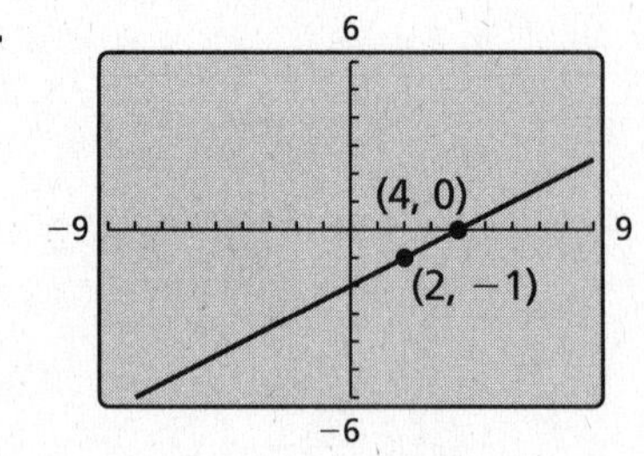

2 EXPLORATION: Mathematical Modeling

Work with a partner. The graph shows the cost of a smartphone plan.

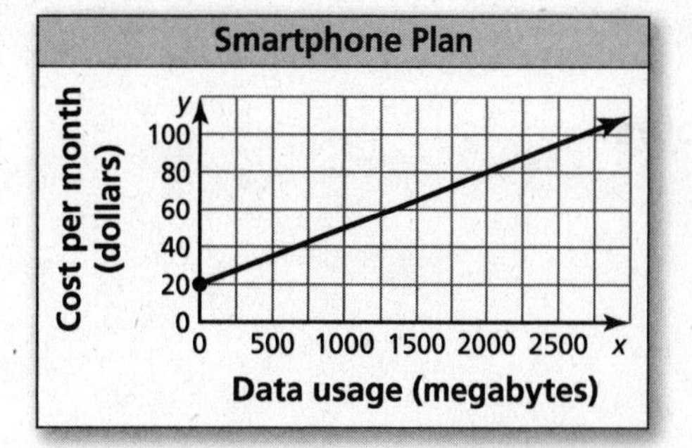

a. What is the y-intercept of the line? Interpret the y-intercept in the context of the problem.

b. Approximate the slope of the line. Interpret the slope in the context of the problem.

c. Write an equation that represents the cost as a function of data usage.

Communicate Your Answer

3. Given the graph of a linear function, how can you write an equation of the line?

4. Give an example of a graph of a linear function that is different from those above. Then use the graph to write an equation of the line.

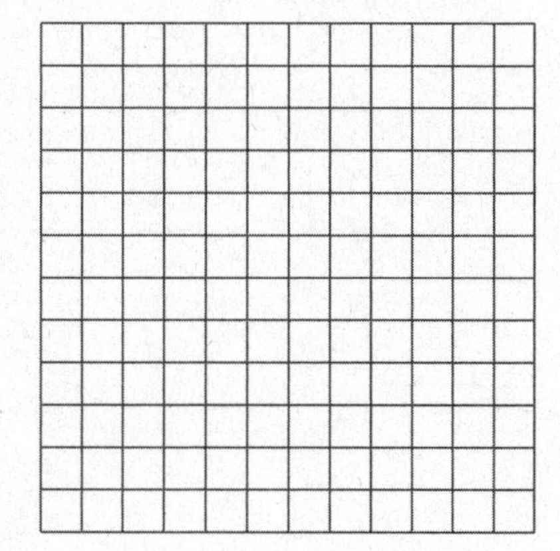

Name ______________________________ Date __________

4.1 Notetaking with Vocabulary
For use after Lesson 4.1

In your own words, write the meaning of each vocabulary term.

linear model

Notes:

Name____ Date______

Extra Practice

In Exercises 1–6, write an equation of the line with the given slope and *y*-intercept.

1. slope: 0
y-intercept: 9

2. slope: −1
y-intercept: 0

3. slope: 2
y-intercept: −3

4. slope: −3
y-intercept: 7

5. slope: 4
y-intercept: −2

6. slope: $\frac{1}{3}$
y-intercept: 2

In Exercises 7–12, write an equation of the line in slope-intercept form.

7.

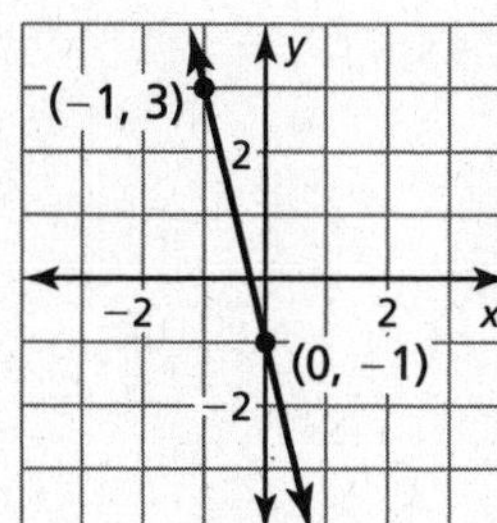

8.

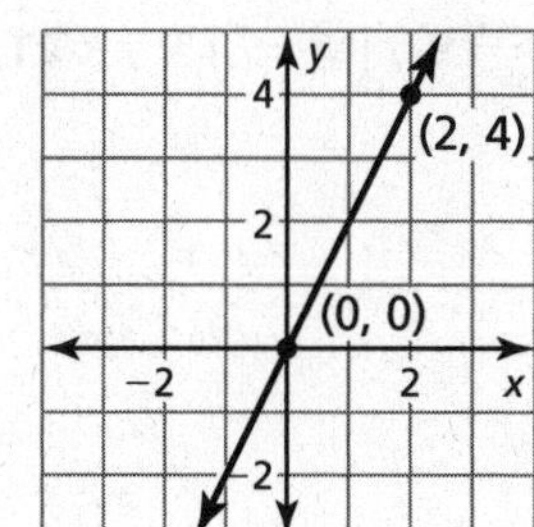

9.

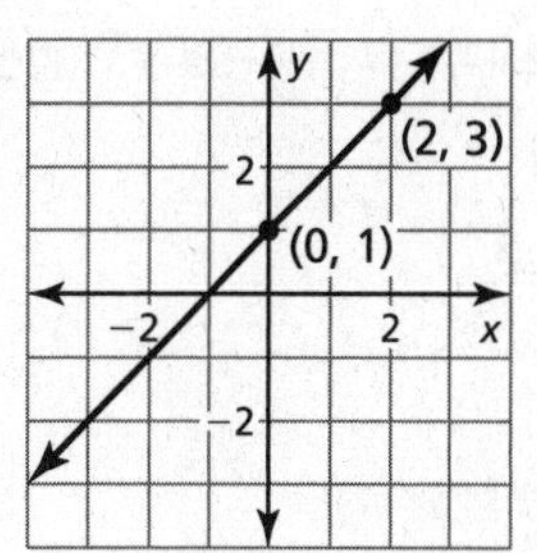

10.

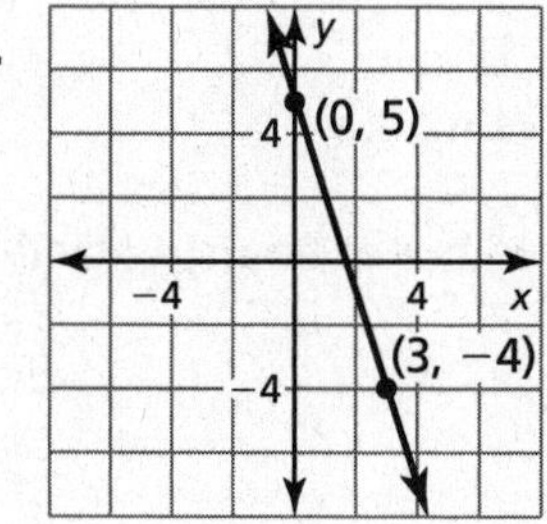

11.

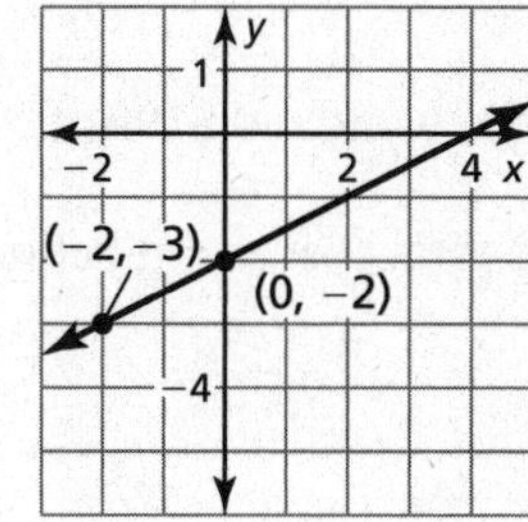

12.

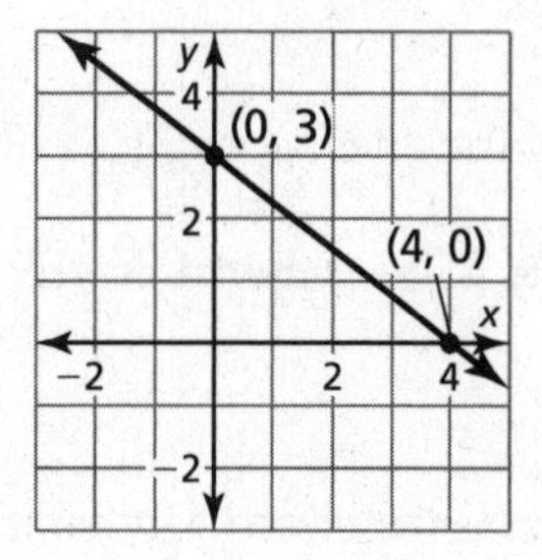

..etaking with Vocabulary (continued)

...ses 13–18, write an equation of the line that passes through the given po...

13. $(3, -1), (8, 4)$ **14.** $(2, 1), (3, 5)$ **15.** $(0, 2), (4, 3)$

16. $(-3, -2), (-4, -1)$ **17.** $(8, 0), (0, 8)$ **18.** $(-1, 7), (2, -5)$

In Exercises 19–24, write a linear function *f* with the given values.

19. $f(6) = -2, f(4) = -3$ **20.** $f(-5) = 5, f(5) = 15$ **21.** $f(8) = -3, f(9) = -4$

22. $f(2) = 6, f(7) = -4$ **23.** $f(-2) = -2, f(4) = 10$ **24.** $f(4) = 0, f(2) = 8$

25. An electrician charges \$120 after 2 hours of work and \$190 after 4 hours of work.

a. Write a linear model that represents the total cost as a function of the number of hours worked.

b. What is the electrician's initial fee?

c. How much does the electrician charge per hour?

Name____ Date________

4.2 Writing Equations in Point-Slope Form

For use with Exploration 4.2

Essential Question How can you write an equation of a line when you are given the slope and a point on the line?

1 EXPLORATION: Writing Equations of Lines

Go to *BigIdeasMath.com* for an interactive tool to investigate this exploration.

Work with a partner.

- Sketch the line that has the given slope and passes through the given point.
- Find the y-intercept of the line.
- Write an equation of the line.

a. $m = \frac{1}{2}$

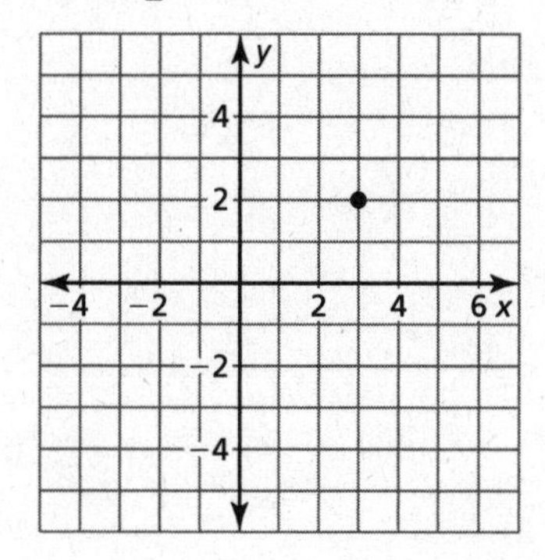

b. $m = -2$

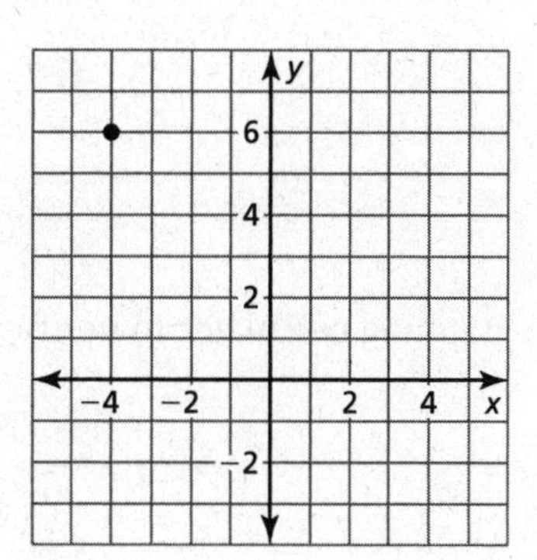

2 EXPLORATION: Writing a Formula

Work with a partner.

The point (x_1, y_1) is a given point on a nonvertical line. The point (x, y) is any other point on the line. Write an equation that represents the slope m of the line. Then rewrite this equation by multiplying each side by the difference of the x-coordinates to obtain the **point-slope form** of a linear equation.

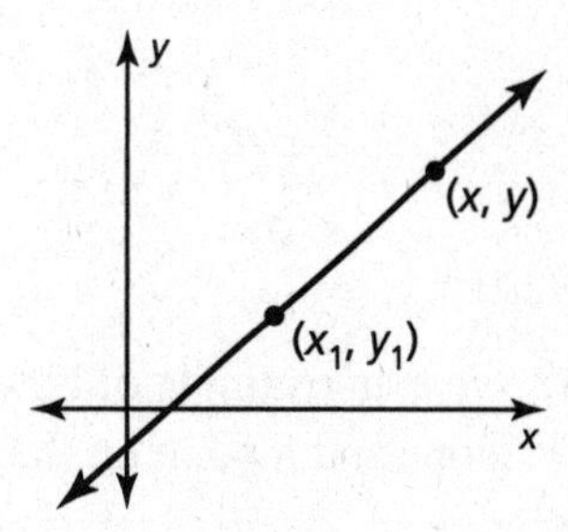

3 EXPLORATION: Writing an Equation

Go to *BigIdeasMath.com* for an interactive tool to investigate this exploration.

Work with a partner.

For four months, you have saved \$25 per month. You now have \$175 in your savings account.

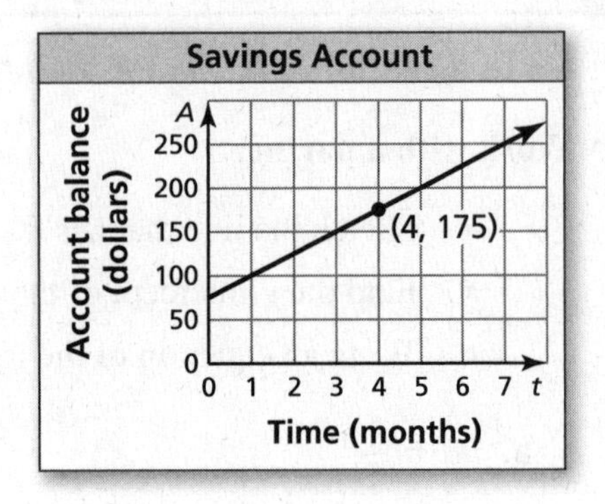

a. Use your result from Exploration 2 to write an equation that represents the balance A after t months.

b. Use a graphing calculator to verify your equation.

Communicate Your Answer

4. How can you write an equation of a line when you are given the slope and a point on the line?

5. Give an example of how to write an equation of a line when you are given the slope and a point on the line. Your example should be different from those above.

Name____ Date__________

4.2 Notetaking with Vocabulary

For use after Lesson 4.2

In your own words, write the meaning of each vocabulary term.

point-slope form

Core Concepts

Point-Slope Form

Words A linear equation written in the form $y - y_1 = m(x - x_1)$ is in **point-slope form.** The line passes through the point (x_1, y_1), and the slope of the line is m.

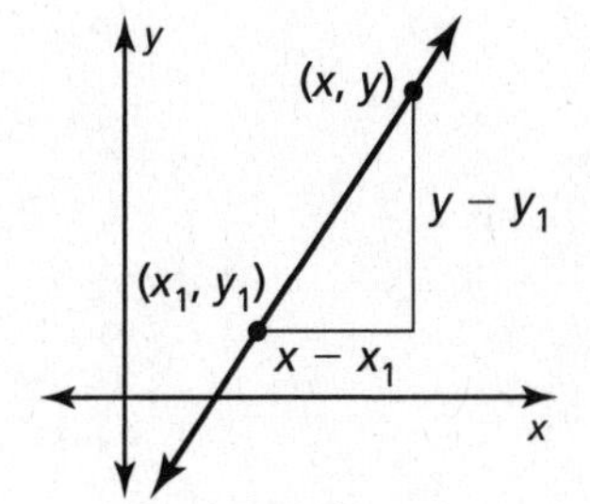

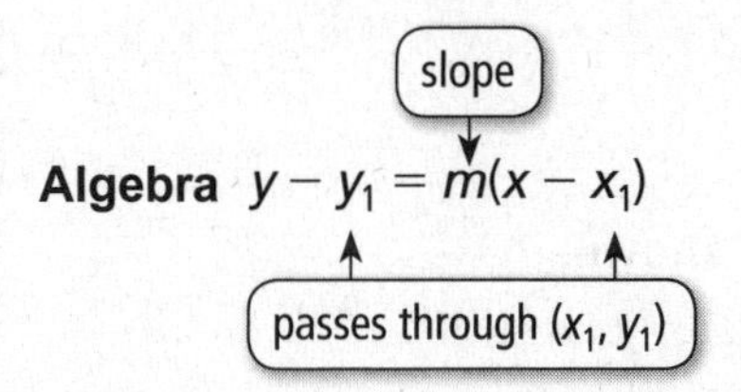

Algebra $y - y_1 = m(x - x_1)$

Notes:

Name ______________________________ Date __________

4.2 **Notetaking with Vocabulary (continued)**

Extra Practice

In Exercises 1–6, write an equation in point-slope form of the line that passes through the given point and has the given slope.

1. $(-2, 1); m = -3$

2. $(3, 5); m = 2$

3. $(-1, -2); m = -1$

4. $(5, 0); m = \frac{4}{3}$

5. $(0, 4); m = 7$

6. $(1, 2); m = -\frac{1}{2}$

In Exercises 7–12, write an equation in slope-intercept form of the line shown.

7.

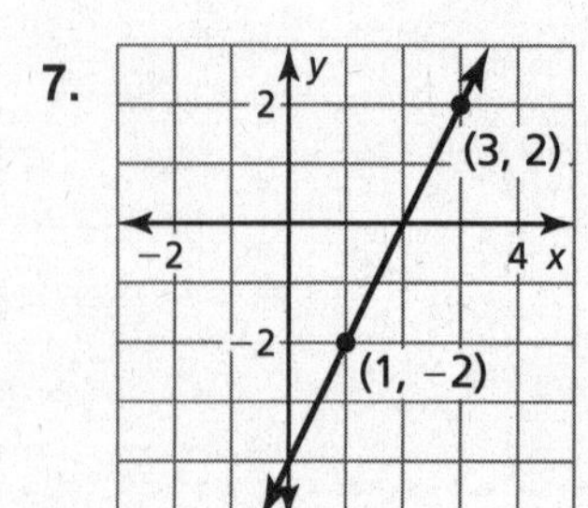

8.

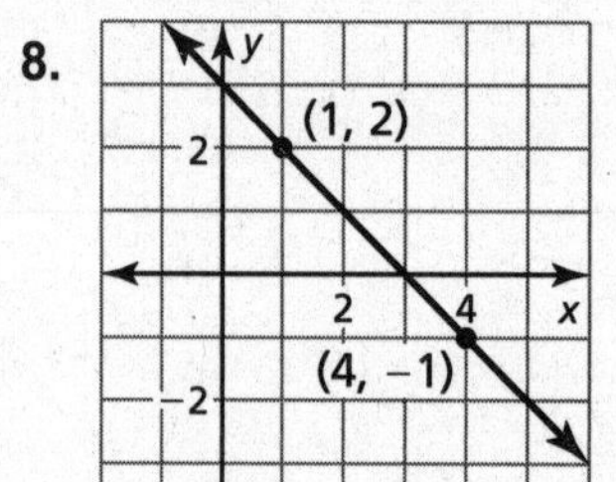

9.

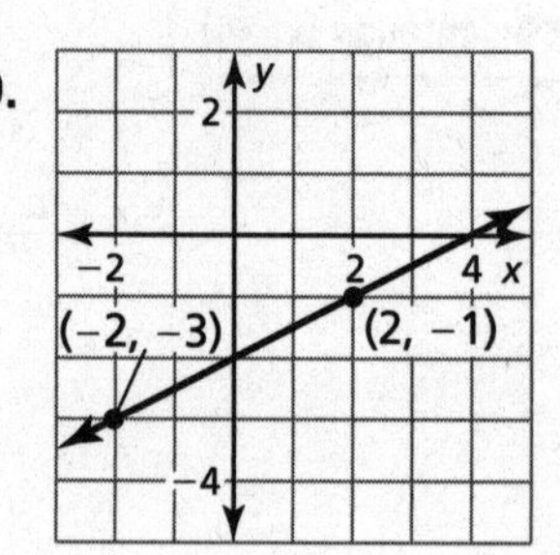

10.

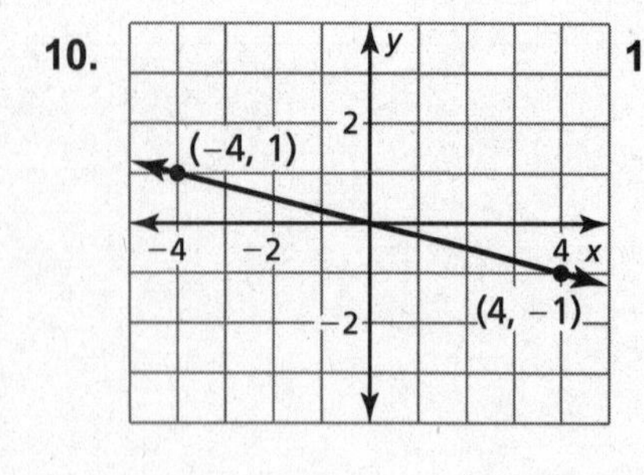

11.

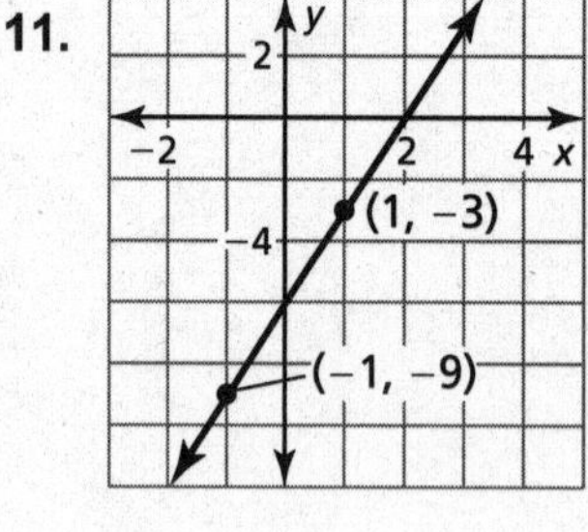

12.

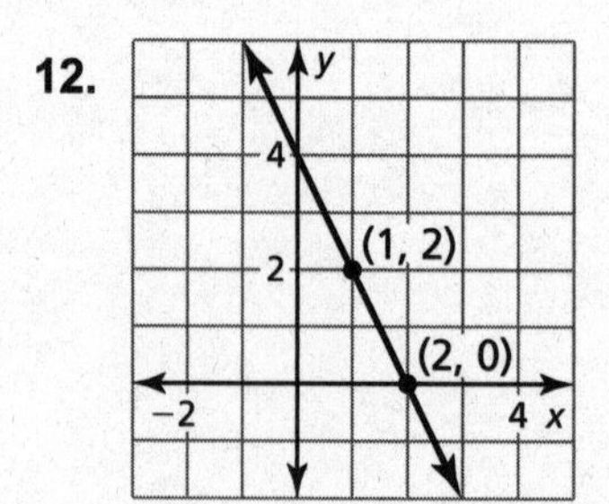

In Exercises 13–18, write a linear function *f* with the given values.

13. $f(-3) = -1, f(-2) = 4$ **14.** $f(-2) = 1, f(1) = 7$ **15.** $f(-1) = 2, f(3) = 3$

16. $f(0) = -2, f(4) = -1$ **17.** $f(1) = 0, f(0) = 8$ **18.** $f(3) = 5, f(2) = 6$

In Exercises 19 and 20, tell whether the data in the table can be modeled by a linear equation. Explain. If possible, write a linear equation that represents *y* as a function of *x*.

19.

x	–3	–1	0	1	3
y	–110	–60	–35	–10	40

20.

x	–3	–1	0	1	3
y	–98	18	8	62	142

21. Craig is driving at a constant speed of 60 miles per hour. After driving 3 hours, his odometer reads 265 miles. Write a linear function D that represents the miles driven after h hours. What does the odometer read after 7 hours of continuous driving?

Name ______________________________ Date __________

4.3 Writing Equations of Parallel and Perpendicular Lines

For use with Exploration 4.3

Essential Question How can you recognize lines that are parallel or perpendicular?

1 EXPLORATION: Recognizing Parallel Lines

Go to *BigIdeasMath.com* for an interactive tool to investigate this exploration.

Work with a partner. Write each linear equation in slope-intercept form. Then use a graphing calculator to graph the three equations in the same square viewing window. (The graph of the first equation is shown.) Which two lines appear parallel? How can you tell?

a. $3x + 4y = 6$

$3x + 4y = 12$

$4x + 3y = 12$

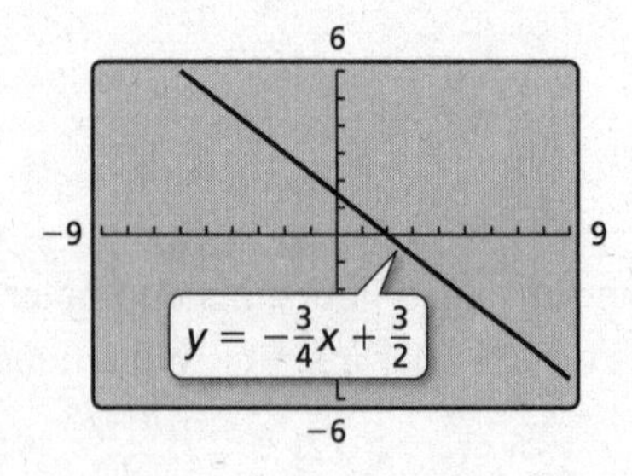

b. $5x + 2y = 6$

$2x + y = 3$

$2.5x + y = 5$

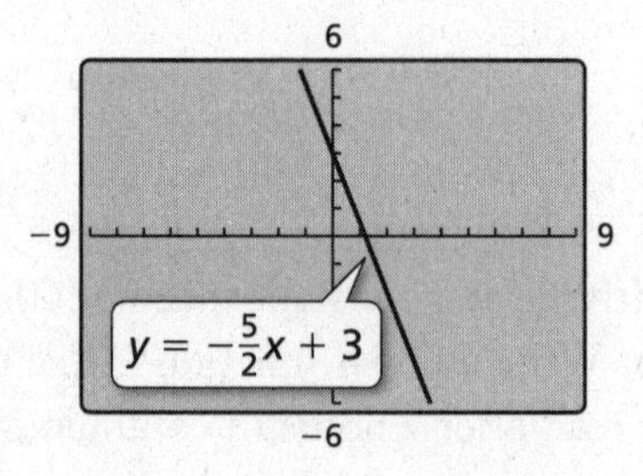

Name______________________________ Date__________

2 EXPLORATION: Recognizing Perpendicular Lines

Go to *BigIdeasMath.com* for an interactive tool to investigate this exploration.

Work with a partner. Write each linear equation in slope-intercept form. Then use a graphing calculator to graph the three equations in the same square viewing window. (The graph of the first equation is shown.) Which two lines appear perpendicular? How can you tell?

a. $3x + 4y = 6$

$3x - 4y = 12$

$4x - 3y = 12$

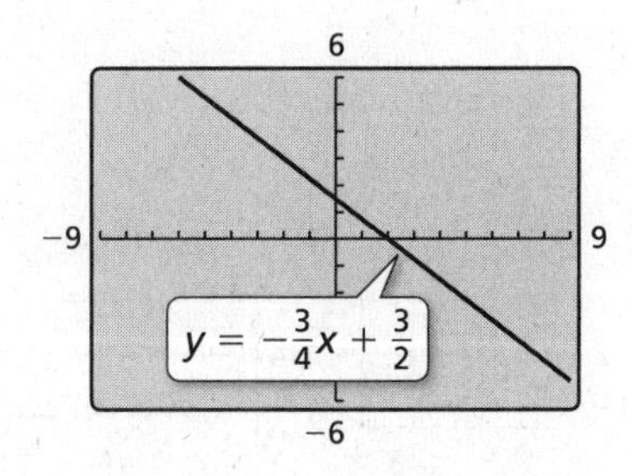

b. $2x + 5y = 10$

$-2x + y = 3$

$2.5x - y = 5$

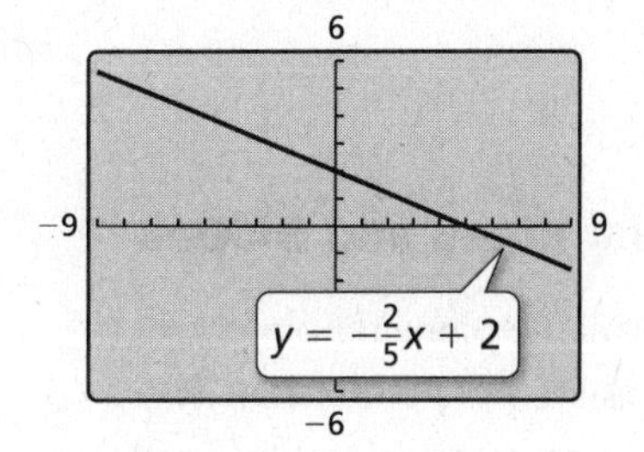

Communicate Your Answer

3. How can you recognize lines that are parallel or perpendicular?

4. Compare the slopes of the lines in Exploration 1. How can you use slope to determine whether two lines are parallel? Explain your reasoning.

5. Compare the slopes of the lines in Exploration 2. How can you use slope to determine whether two lines are perpendicular? Explain your reasoning.

Name ______________________________ Date __________

4.3 Notetaking with Vocabulary

For use after Lesson 4.3

In your own words, write the meaning of each vocabulary term.

parallel lines

perpendicular lines

Core Concepts

Parallel Lines and Slopes

Two lines in the same plane that never intersect are **parallel lines**. Two distinct nonvertical lines are parallel if and only if they have the same slope.

All vertical lines are parallel.

Notes:

Perpendicular Lines and Slopes

Two lines in the same plane that intersect to form right angles are **perpendicular lines**. Nonvertical lines are perpendicular if and only if their slopes are negative reciprocals.

Vertical lines are perpendicular to horizontal lines.

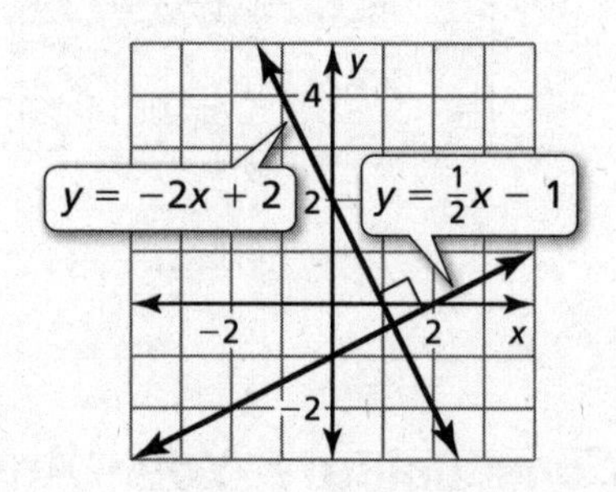

Notes:

Name____ Date__________

Extra Practice

In Exercises 1–6, determine which of the lines, if any, are parallel. Explain.

1.

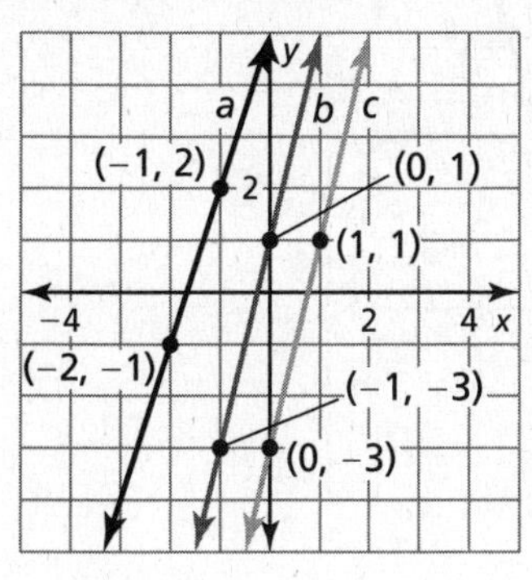

2.

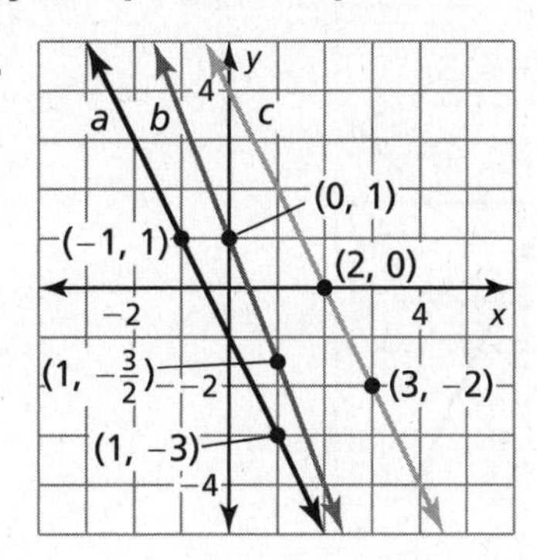

3. Line a passes through $(-4,-1)$ and $(2,2)$.
   Line b passes through $(-5,-3)$ and $(5,1)$.
   Line c passes through $(-2,-3)$ and $(2,-1)$.

4. Line a passes through $(-2,5)$ and $(2,1)$.
   Line b passes through $(-4,3)$ and $(3,4)$.
   Line c passes through $(-3,4)$ and $(2,-6)$.

5. Line a: $4x = -3y + 9$
   Line b: $8y = -6x + 16$
   Line c: $4y = -3x + 9$

6. Line a: $5y - x = 4$
   Line b: $5y = x + 7$
   Line c: $5y - 2x = 5$

In Exercises 7 and 8, write an equation of the line that passes through the given point and is parallel to the given line.

7. $(3,-1)$; $y = \frac{1}{3}x - 3$

8. $(1,-2)$; $y = -2x + 1$

Name ______________________________ Date __________

4.3 Notetaking with Vocabulary (continued)

In Exercises 9–14, determine which of the lines, if any, are parallel or perpendicular. Explain.

9.

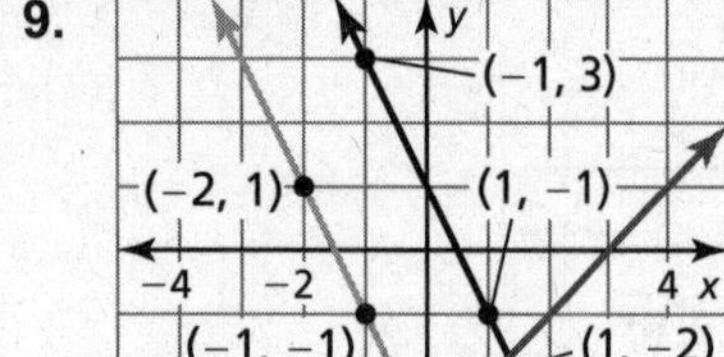

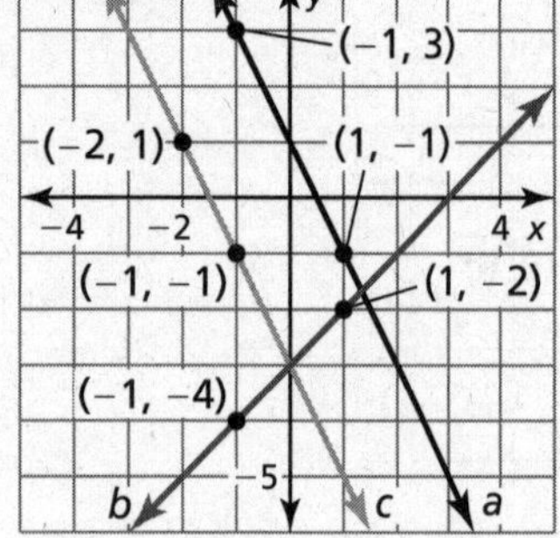

10.

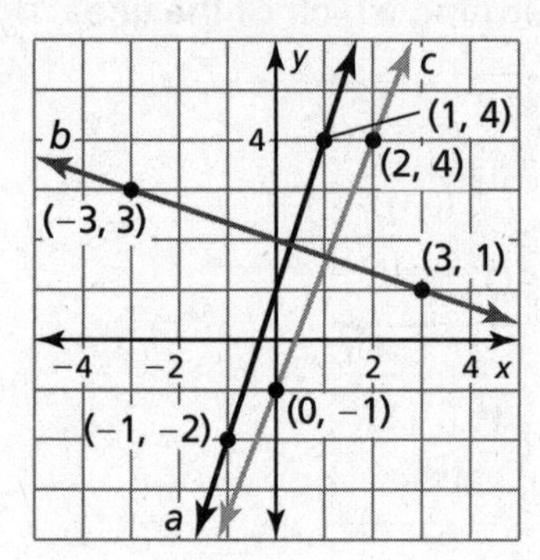

11. Line a passes through $(-2, 4)$ and $(1, 1)$.

Line b passes through $(2, 1)$ and $(4, 4)$.

Line c passes through $(1, -2)$ and $(-1, 4)$.

12. Line a passes through $(-2, -4)$ and $(-1, -1)$.

Line b passes through $(-1, -4)$ and $(1, 2)$.

Line c passes through $(2, 3)$ and $(4, 2)$.

13. Line a: $y = \frac{3}{4}x + 1$

Line b: $-3y = 4x - 3$

Line c: $4y = -3x + 9$

14. Line a: $5y - 2x = 1$

Line b: $y = \frac{5}{2}x - 1$

Line c: $y = \frac{2}{5}x + 3$

In Exercises 15 and 16, write an equation of the line that passes through the given point and is perpendicular to the given line.

15. $(-2, 2)$; $y = \frac{2}{3}x + 2$

16. $(3, 1)$; $2y = 4x - 3$

Name____ Date__________

4.4 Scatter Plots and Lines of Fit

For use with Exploration 4.4

Essential Question How can you use a scatter plot and a line of fit to make conclusions about data?

A **scatter plot** is a graph that shows the relationship between two data sets. The two data sets are graphed as ordered pairs in a coordinate plane.

1 EXPLORATION: Finding a Line of Fit

Go to *BigIdeasMath.com* for an interactive tool to investigate this exploration.

Work with a partner. A survey was taken of 179 married couples. Each person was asked his or her age. The scatter plot shows the results.

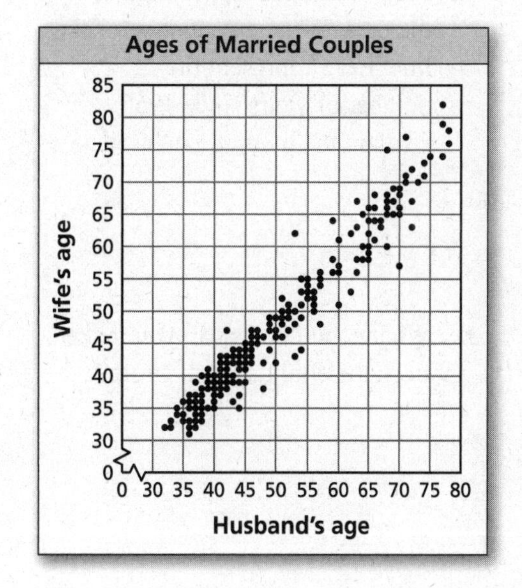

a. Draw a line that approximates the data. Write an equation of the line. Explain the method you used.

b. What conclusions can you make from the equation you wrote? Explain your reasoning.

2 EXPLORATION: Finding a Line of Fit

Go to *BigIdeasMath.com* for an interactive tool to investigate this exploration.

Work with a partner. The scatter plot shows the median ages of American women at their first marriage for selected years from 1960 through 2010.

a. Draw a line that approximates the data. Write an equation of the line. Let x represent the number of years since 1960. Explain the method you used.

b. What conclusions can you make from the equation you wrote?

c. Use your equation to predict the median age of American women at their first marriage in the year 2020.

Communicate Your Answer

3. How can you use a scatter plot and a line of fit to make conclusions about data?

4. Use the Internet or some other reference to find a scatter plot of real-life data that is different from those given above. Then draw a line that approximates the data and write an equation of the line. Explain the method you used.

Name____ Date__________

4.4 Notetaking with Vocabulary
For use after Lesson 4.4

In your own words, write the meaning of each vocabulary term.

scatter plot

correlation

line of fit

Core Concepts

Scatter Plot

A **scatter plot** is a graph that shows the relationship between two data sets. The two data sets are graphed as ordered pairs in a coordinate plane. Scatter plots can show trends in the data.

Notes:

Name ______________________________ Date __________

Using a Line of Fit to Model Data

Step 1 Make a scatter plot of the data.

Step 2 Decide whether the data can be modeled by a line.

Step 3 Draw a line that appears to fit the data closely. There should be approximately as many points above the line as below it.

Step 4 Write an equation using two points on the line. The points do not have to represent actual data pairs, but they must lie on the line of fit.

Notes:

Extra Practice

1. The scatter plot shows the weights (in pounds) of a baby over time.

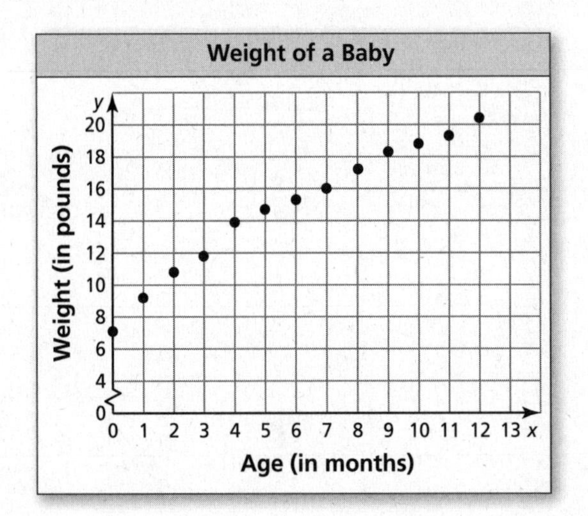

a. What is the weight of the baby when the baby is four months old?

b. What is the age of the baby when the baby weighs 17.2 pounds?

c. What tends to happen to weight of the baby as the age increases?

Name______________________________ Date__________

4.4 Notetaking with Vocabulary (continued)

In Exercises 2–5, tell whether x and y show a *positive*, a *negative*, or *no* correlation.

2.

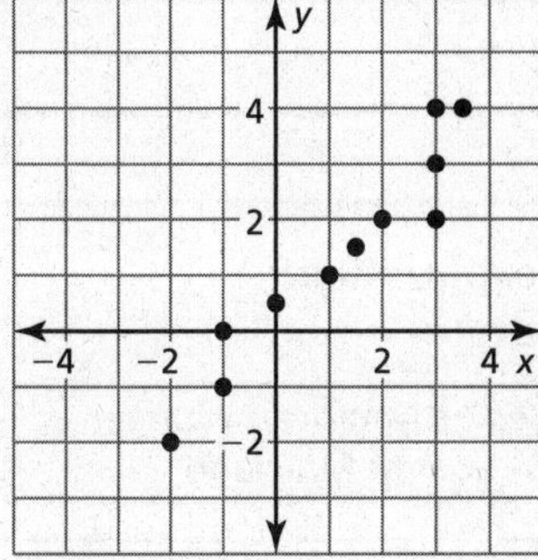

3.

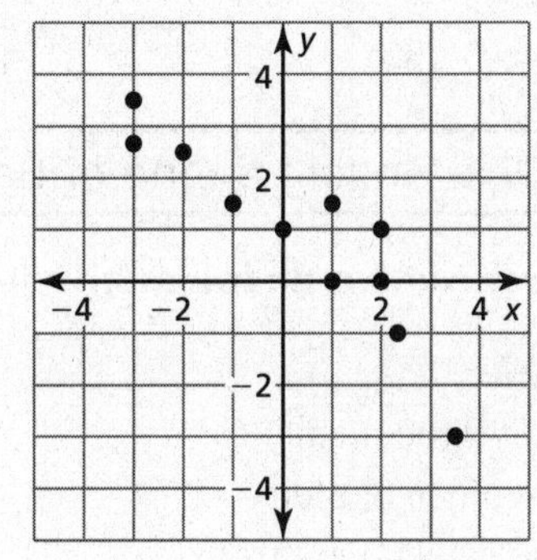

4.

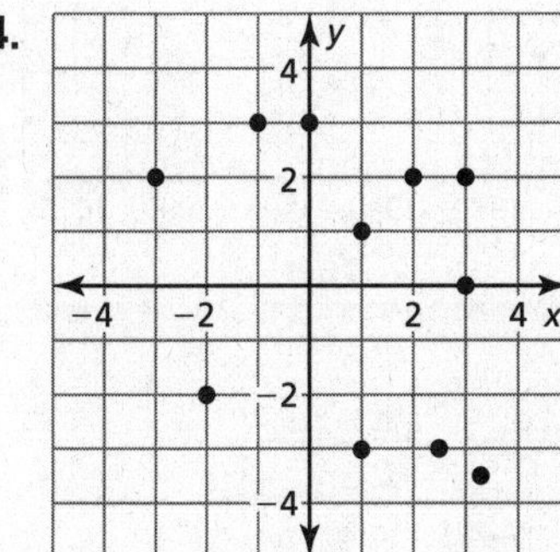

5.

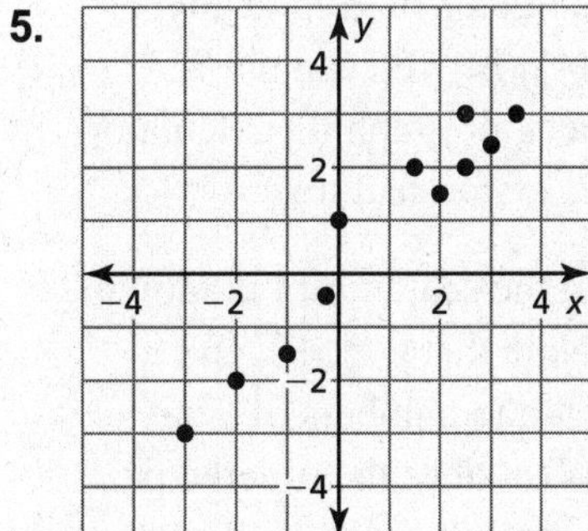

6. The table shows the depth y (in centimeters) of water filling a bathtub after x minutes.

Time (minutes), x	0	2	4	6	8	10	12
Depth (centimeters), y	6	8	11	14	17	20	24

a. Write an equation that models the depth of the water as a function of time.

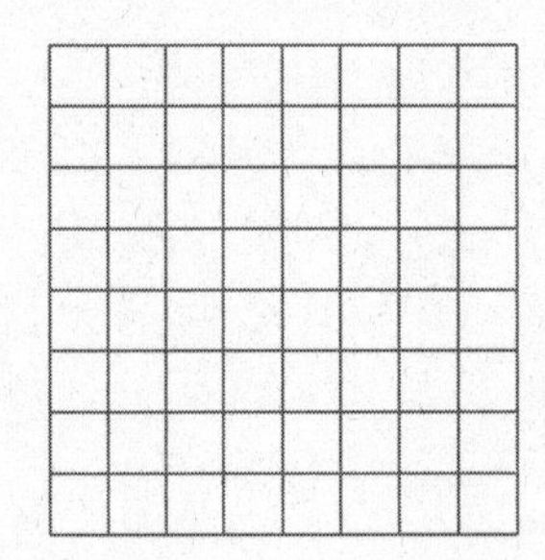

b. Interpret the slope and y-intercept of the line of fit.

Name ______________________________ Date __________

4.5 Analyzing Lines of Fit
For use with Exploration 4.5

Essential Question How can you *analytically* find a line of best fit for a scatter plot?

1 EXPLORATION: Finding a Line of Best Fit

Go to *BigIdeasMath.com* for an interactive tool to investigate this exploration.

Work with a partner.
The scatter plot shows the median ages of American women at their first marriage for selected years from 1960 through 2010. In Exploration 2 in Section 4.4, you approximated a line of fit graphically. To find the line of *best* fit, you can use a computer, spreadsheet, or graphing calculator that has a *linear regression* feature.

a. The data from the scatter plot is shown in the table. Note that 0, 5, 10, and so on represent the numbers of years since 1960. What does the ordered pair (25, 23.3) represent?

L1	L2	L3
0	20.3	
5	20.6	
10	20.8	
15	21.1	
20	22	
25	23.3	
30	23.9	
35	24.5	
40	25.1	
45	25.3	
50	26.1	

L1(55)=

b. Use the *linear regression* feature to find an equation of the line of best fit. You should obtain results such as those shown below.

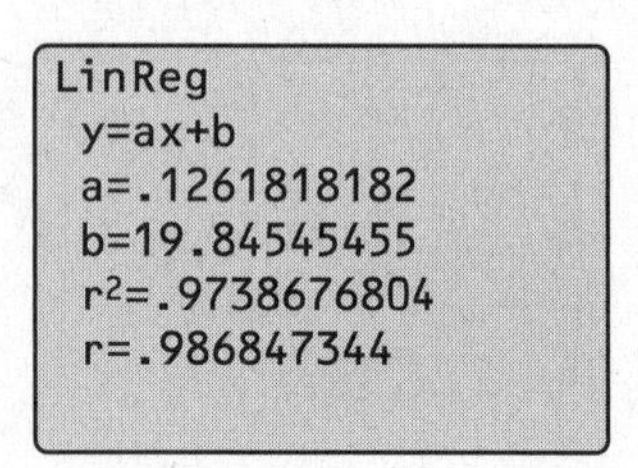

c. Write an equation of the line of best fit. Compare your result with the equation you obtained in Exploration 2 in Section 4.4.

Communicate Your Answer

2. How can you *analytically* find a line of best fit for a scatter plot?

3. The data set relates the number of chirps per second for striped ground crickets and the outside temperature in degrees Fahrenheit. Make a scatter plot of the data. Then find an equation of the line of best fit. Use your result to estimate the outside temperature when there are 19 chirps per second.

Chirps per second	20.0	16.0	19.8	18.4	17.1
Temperature (°F)	88.6	71.6	93.3	84.3	80.6

Chirps per second	14.7	15.4	16.2	15.0	14.4
Temperature (°F)	69.7	69.4	83.3	79.6	76.3

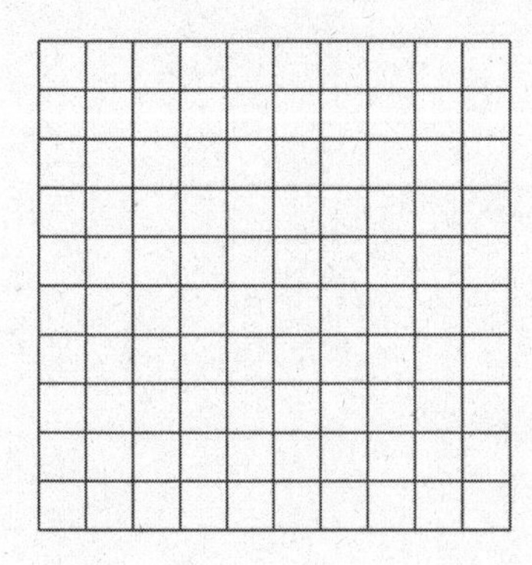

Name ______________________ Date ________

4.5 Notetaking with Vocabulary
For use after Lesson 4.5

In your own words, write the meaning of each vocabulary term.

residual

linear regression

line of best fit

correlation coefficient

interpolation

extrapolation

causation

Notes:

Name______________________________ Date__________

4.5 Notetaking with Vocabulary (continued)

Core Concepts

Residuals

A **residual** is the difference of the y-value of a data point and the corresponding y-value found using the line of fit. A residual can be positive, negative, or zero.

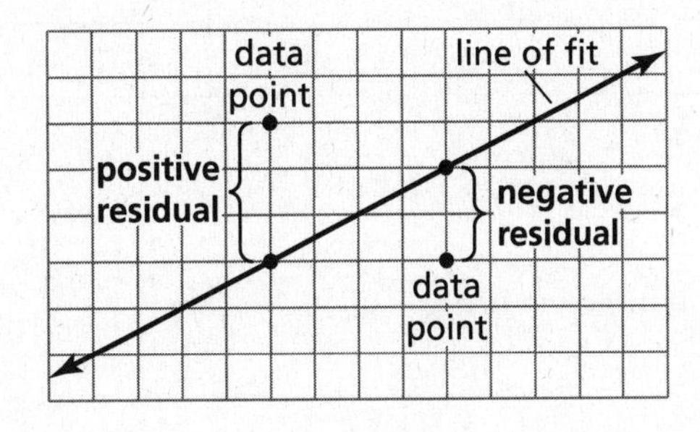

A scatter plot of the residuals shows how well a model fits a data set. If the model is a good fit, then the absolute values of the residuals are relatively small, and the residual points will be more or less evenly dispersed about the horizontal axis. If the model is not a good fit, then the residual points will form some type of pattern that suggests the data are not linear. Wildly scattered residual points suggest that the data might have no correlation.

Notes:

Extra Practice

In Exercises 1 and 2, use residuals to determine whether the model is a good fit for the data in the table. Explain.

1. $y = -3x + 2$

x	−4	−3	−2	−1	0	1	2	3	4
y	13	11	8	6	3	0	−4	−8	−10

2. $y = -0.5x + 1$

x	0	1	2	3	4	5	6	7	8
y	2	0	–3	–5	–7	–6	–4	–3	–1

3. The table shows the number of visitors y to a particular beach for average daily temperatures x.

Average Daily Temperature (°F)	Number of Beach Visitors
80	100
82	150
83	145
85	190
86	215
88	263
89	300
90	350

a. Use a graphing calculator to find an equation of the line of best fit. Then plot the data and graph the equation in the same viewing window.

b. Identify and interpret the correlation coefficient.

c. Interpret the slope and y-intercept of the line of best fit.

Name____ Date__________

4.6 Arithmetic Sequences
For use with Exploration 4.6

Essential Question How can you use an arithmetic sequence to describe a pattern?

An **arithmetic sequence** is an ordered list of numbers in which the difference between each pair of consecutive **terms**, or numbers in the list, is the same.

1 EXPLORATION: Describing a Pattern

Go to *BigIdeasMath.com* for an interactive tool to investigate this exploration.

Work with a partner. Use the figures to complete the table. Plot the points given by your completed table. Describe the pattern of the y-values.

a. $n = 1$ $n = 2$ $n = 3$ $n = 4$ $n = 5$

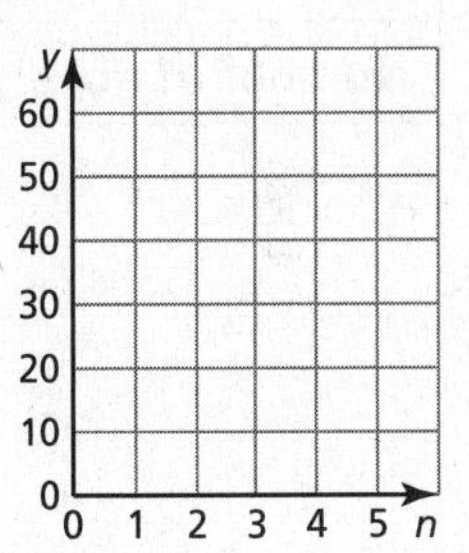

Number of stars, *n*	1	2	3	4	5
Number of sides, *y*					

b. $n = 1$ $n = 2$ $n = 3$ $n = 4$ $n = 5$

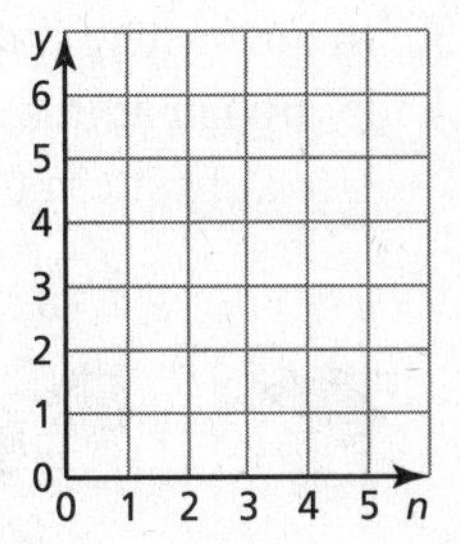

n	1	2	3	4	5
Number of circles, *y*					

1 EXPLORATION: Describing a Pattern (continued)

c. $n = 1$ $n = 2$ $n = 3$ $n = 4$ $n = 5$

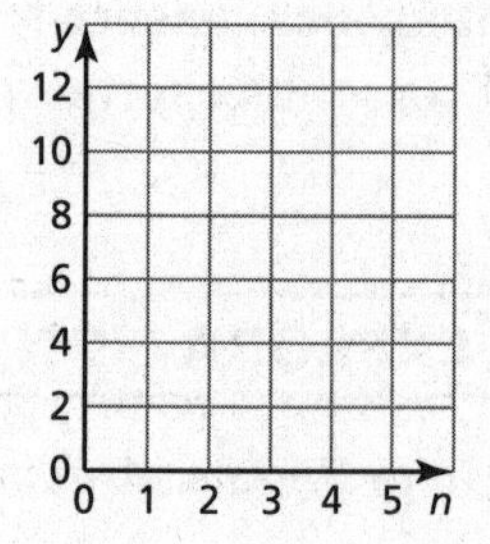

Number of rows, n	1	2	3	4	5
Number of dots, y					

Communicate Your Answer

2. How can you use an arithmetic sequence to describe a pattern? Give an example from real life.

3. In chemistry, water is called H_2O because each molecule of water has two hydrogen atoms and one oxygen atom. Describe the pattern shown below. Use the pattern to determine the number of atoms in 23 molecules.

$n = 1$ $n = 2$ $n = 3$ $n = 4$ $n = 5$

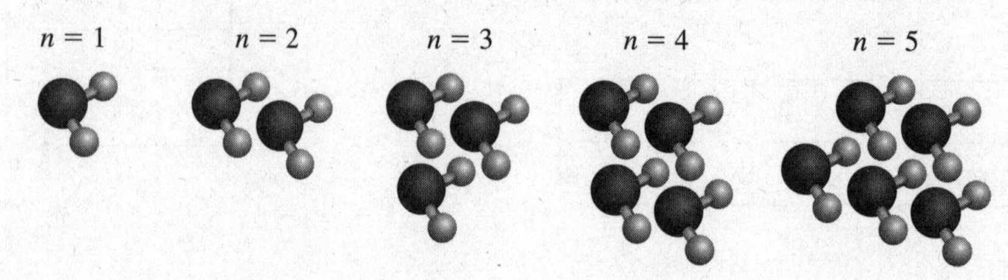

Name______________________ Date__________

4.6 Notetaking with Vocabulary

For use after Lesson 4.6

In your own words, write the meaning of each vocabulary term.

sequence

term

arithmetic sequence

common difference

Core Concepts

Arithmetic Sequence

In an **arithmetic sequence**, the difference between each pair of consecutive terms is the same. This difference is called the **common difference**. Each term is found by adding the common difference to the previous term.

5, 10, 15, 20, . . . Terms of an arithmetic sequence

+5 +5 +5 ← common difference

Notes:

Equation for an Arithmetic Sequence

Let a_n be the nth term of an arithmetic sequence with first term a_1 and common difference d. The nth term is given by

$$a_n = a_1 + (n - 1)d.$$

Notes:

Name ______________________________ Date __________

Extra Practice

In Exercises 1–6, write the next three terms of the arithmetic sequence.

1. 1, 8, 15, 22, …

2. 20, 14, 8, 2, …

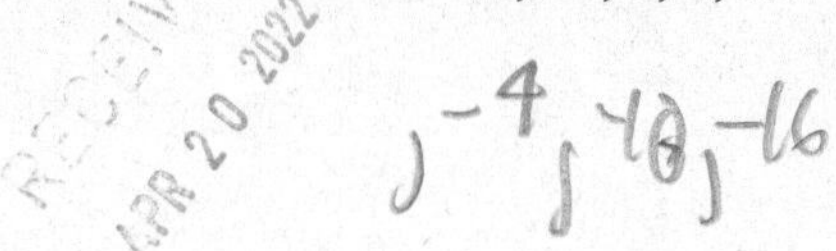

3. 12, 21, 30, 39, …

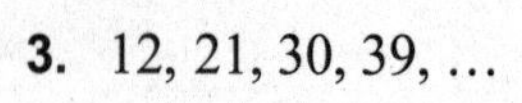

4. 5, 12, 19, 26, …

5. 3, 7, 11, 15, …

6. 2, 14, 26, 38, …

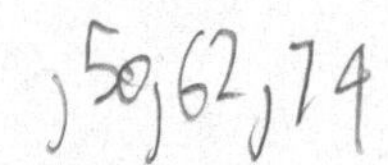

In Exercises 7–12, graph the arithmetic sequence.

7. 1, 3, 5, 7, …

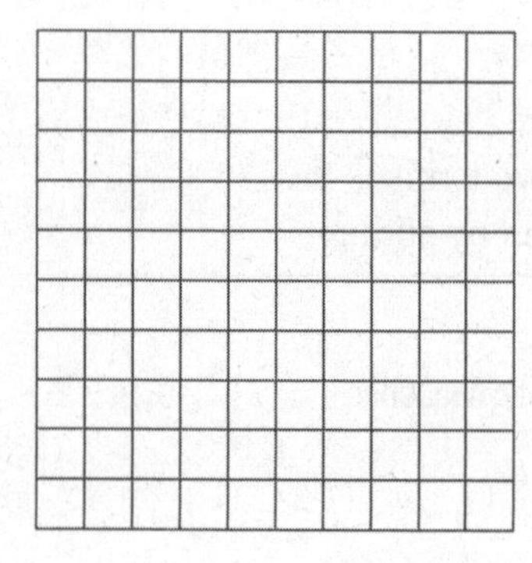

8. 9, 6, 3, 0, …

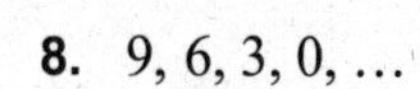

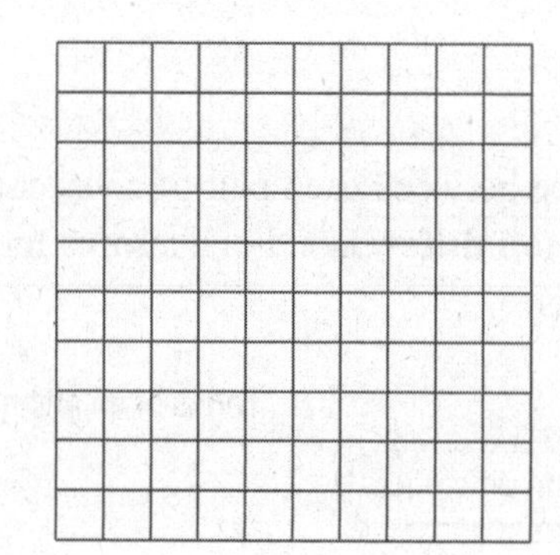

9. $\frac{15}{2}, \frac{13}{2}, \frac{11}{2}, \frac{9}{2}, \ldots$

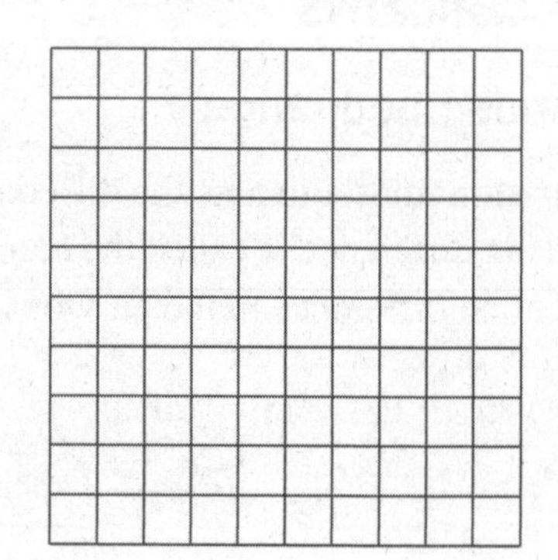

10. 1, 2.5, 4, 5.5, …

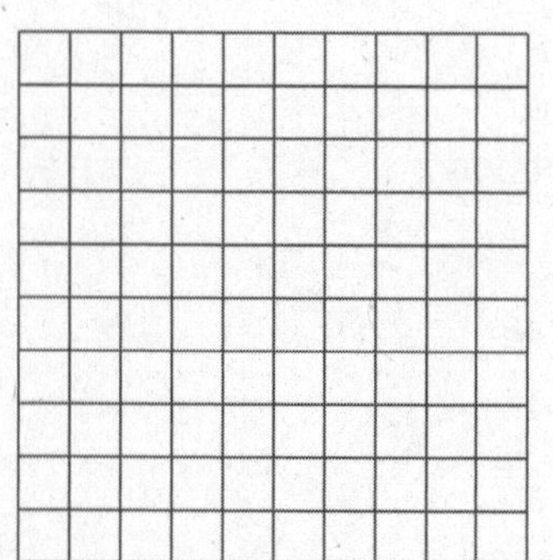

11. 1, 4, 7, 10, …

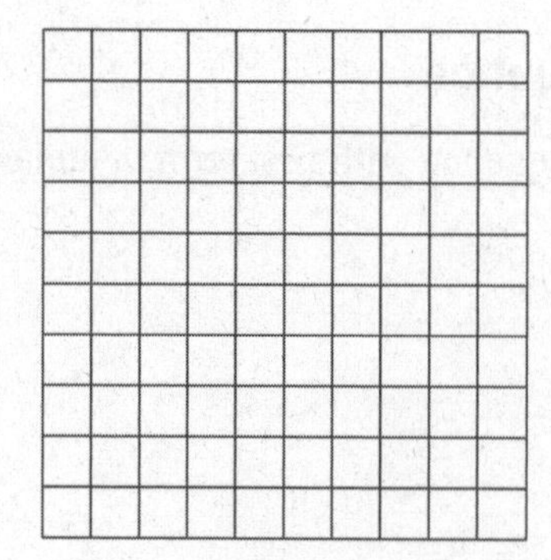

12. $\frac{1}{4}, \frac{5}{4}, \frac{9}{4}, \frac{13}{4}, \ldots$

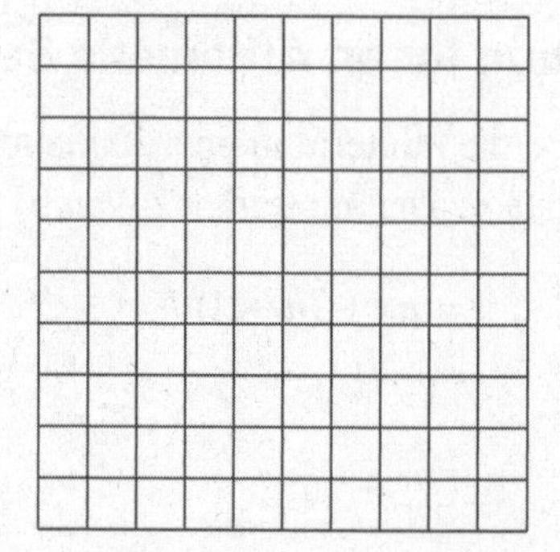

Name____ Date__________

In Exercises 13–15, determine whether the graph represents an arithmetic sequence. Explain.

13.

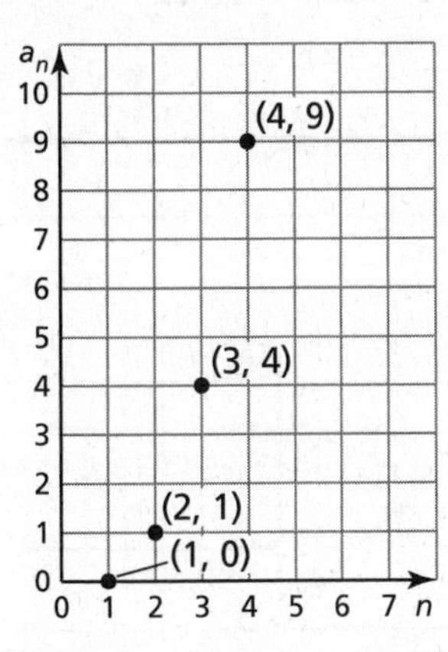

14.

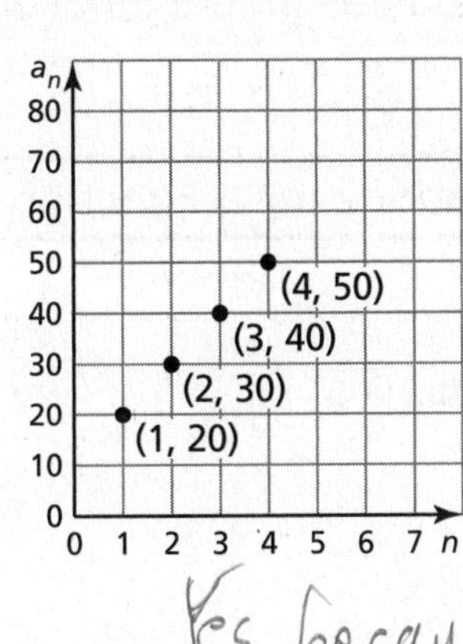

Yes because the common difference = 10

15.

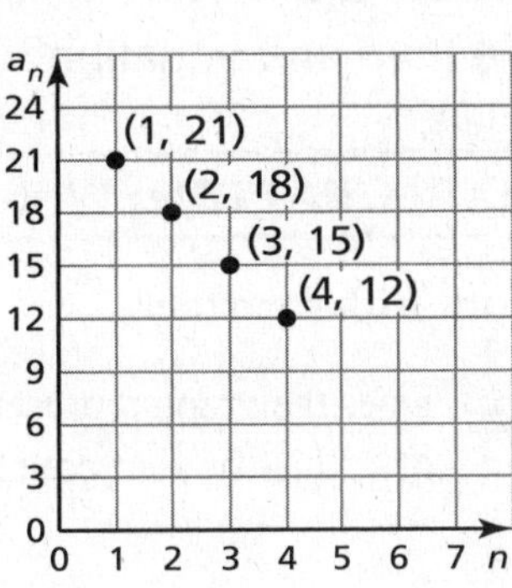

In Exercises 16–21, write an equation for the *n*th term of the arithmetic sequence. Then find a_{10}.

16. −5.4, −6.6, −7.8, −9.0, …

$a_n = a_1 + (n-1)d$

$a_n = -5.4 - 1.2n - 1.2$

$a_n = -1.2n - 4.2$ | -16.2

17. 43, 38, 33, 28, …

18. 6, 10, 14, 18, …

$a_n = 6 + 4n - 4$

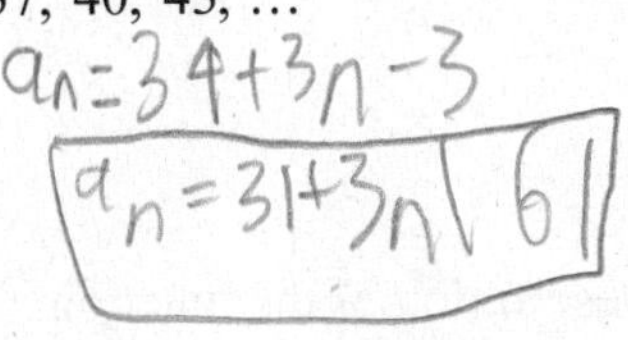

19. −11, −9, −7, −5, …

20. 34, 37, 40, 43, …

$a_n = 34 + 3n - 3$

$a_n = 31 + 3n$ | 61

21. $\frac{9}{4}, \frac{7}{4}, \frac{5}{4}, \frac{3}{4}, \ldots$

22. In an auditorium, the first row of seats has 30 seats. Each row behind the first row has 4 more seats than the row in front of it. How many seats are in the 25th row?

$a_n = 30 + (n-1)4$

$a_n = 26 + 4n$ $n = 25$

$a_n = 126$ seats

Name ____________________ Date ________

4.7 Piecewise Functions
For use with Exploration 4.7

Essential Question How can you describe a function that is represented by more than one equation?

1 EXPLORATION: Writing Equations for a Function

Work with a partner.

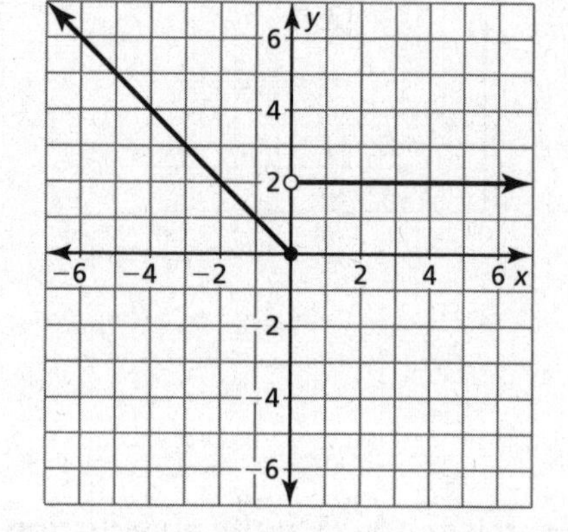

a. Does the graph represent y as a function of x? Justify your conclusion.

b. What is the value of the function when $x = 0$? How can you tell?

c. Write an equation that represents the values of the function when $x \leq 0$.

$f(x) =$ ________, if $x \leq 0$

d. Write an equation that represents the values of the function when $x > 0$.

$f(x) =$ ________, if $x > 0$

e. Combine the results of parts (c) and (d) to write a single description of the function.

$$f(x) = \begin{cases} \_\_\_\_\_\_\_\_, & \text{if } x \leq 0 \\ \_\_\_\_\_\_\_\_, & \text{if } x > 0 \end{cases}$$

Name____ Date__________

2 EXPLORATION: Writing Equations for a Function

Work with a partner.

a. Does the graph represent y as a function of x? Justify your conclusion.

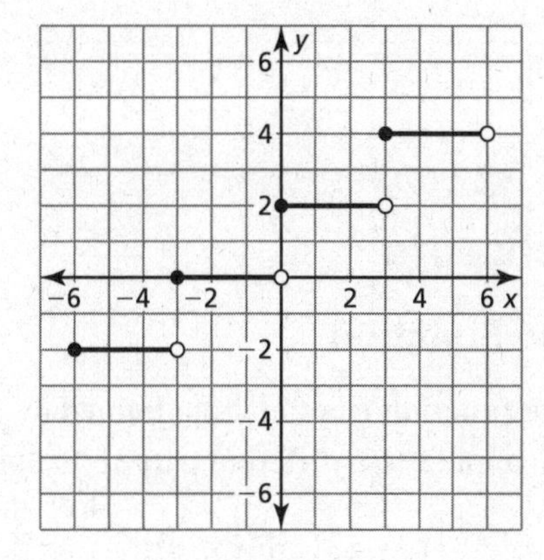

b. Describe the values of the function for the following intervals.

$$f(x) = \begin{cases} \_\_\_\_\_\_, & \text{if } -6 \le x < -3 \\ \_\_\_\_\_\_, & \text{if } -3 \le x < 0 \\ \_\_\_\_\_\_, & \text{if } 0 \le x < 3 \\ \_\_\_\_\_\_, & \text{if } 3 \le x < 6 \end{cases}$$

Communicate Your Answer

3. How can you describe a function that is represented by more than one equation?

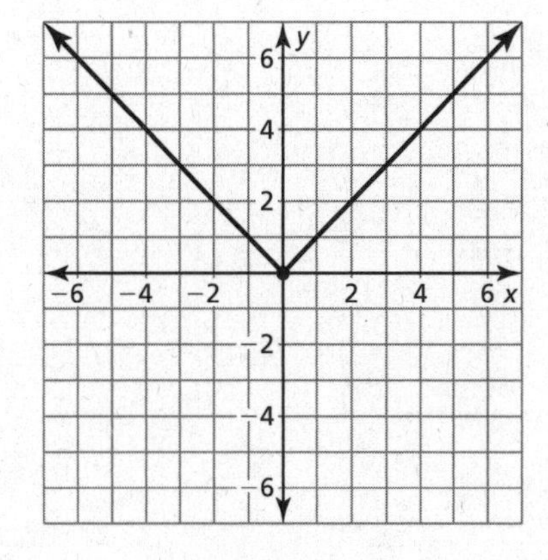

4. Use two equations to describe the function represented by the graph?

Name ______________________________ Date __________

4.7 Notetaking with Vocabulary

For use after Lesson 4.7

In your own words, write the meaning of each vocabulary term.

piecewise function

step function

Core Concepts

Piecewise Function

A **piecewise function** is a function defined by two or more equations. Each "piece" of the function applies to a different part of its domain. An example is shown below.

$$f(x) = \begin{cases} x - 2, & \text{if } x \leq 0 \\ 2x + 1, & \text{if } x > 0 \end{cases}$$

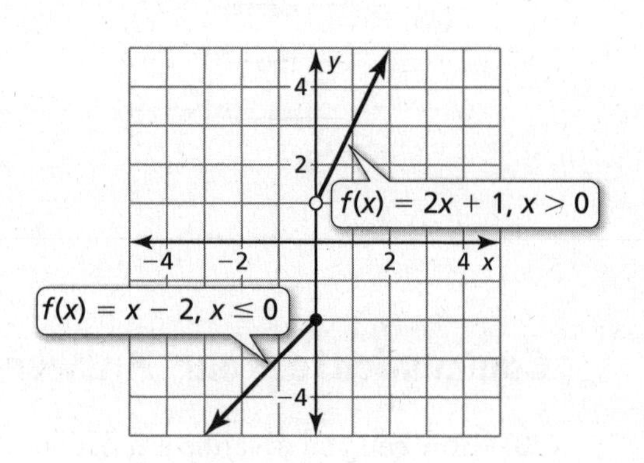

- The expression $x - 2$ represents the value of f when x is less than or equal to 0.
- The expression $2x + 1$ represents the value of f when x is greater than 0.

Name______________________________ Date__________

Extra Practice

In Exercise 1–9, evaluate the function.

$$f(x) = \begin{cases} 3x - 1, & \text{if } x \le 1 \\ 1 - 2x, & \text{if } x > 1 \end{cases}$$

$$g(x) = \begin{cases} 3x - 1, & \text{if } x \le -3 \\ 2, & \text{if } -3 < x < 1 \\ -3x, & \text{if } x \ge 1 \end{cases}$$

1. $f(0)$

2. $f(1)$

3. $f(5)$

4. $f(-4)$

5. $g(0)$

6. $g(-3)$

7. $g(1)$

8. $g(3)$

9. $g(-5)$

In Exercise 10–13, graph the function. Describe the domain and range.

10. $y = \begin{cases} -4x, & \text{if } x \le 0 \\ 4, & \text{if } x > 0 \end{cases}$

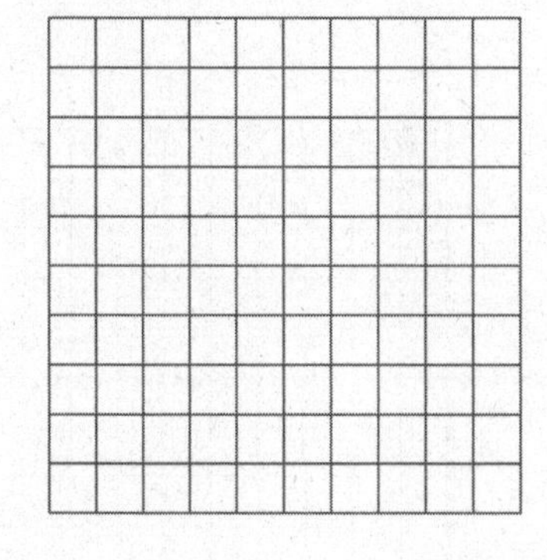

11. $y = \begin{cases} 4 - x, & \text{if } x < 2 \\ x + 3, & \text{if } x \ge 2 \end{cases}$

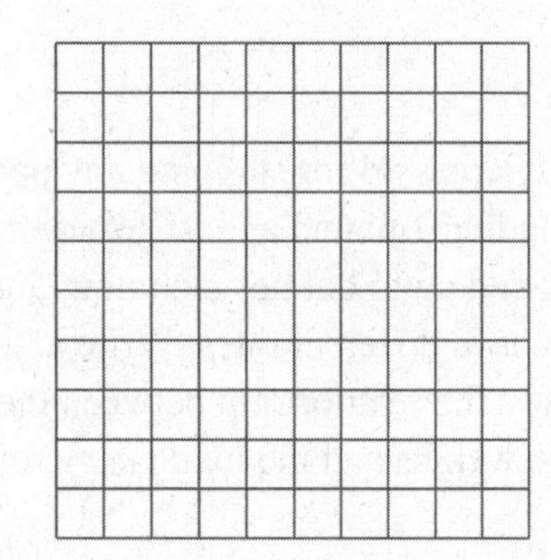

4.7 Notetaking with Vocabulary (continued)

12. $y = \begin{cases} 2x, & \text{if } x < -2 \\ 2, & \text{if } -2 \le x < 2 \\ -2x, & \text{if } x \ge 2 \end{cases}$

13. $y = \begin{cases} -1, & \text{if } x \le -1 \\ 0, & \text{if } -1 < x < 2 \\ 1, & \text{if } x \ge 2 \end{cases}$

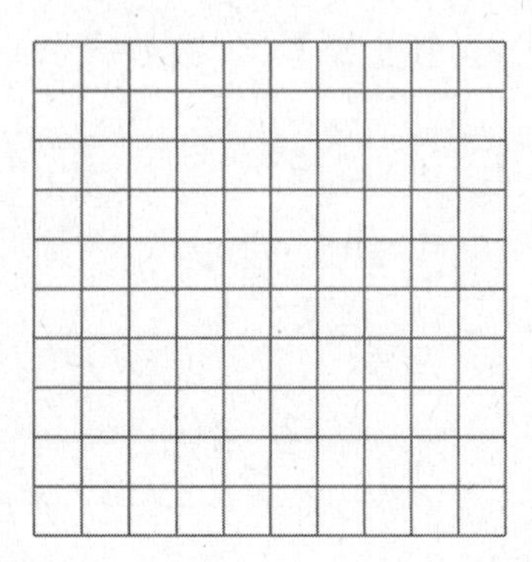

In Exercise 14 and 15, write a piecewise function for the graph.

14.

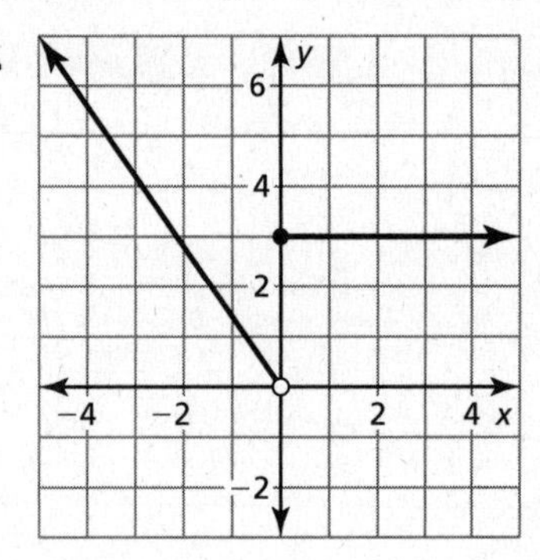

15.

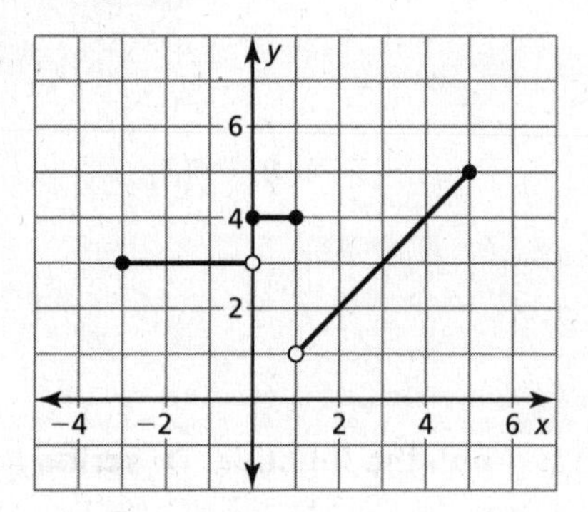

16. A postal service charges \$4 for shipping any package weighing up to but not including 1 pound and \$1 for each additional pound or portion of a pound up to but not including 5 pounds. Packages 5 pounds or over have different rates. Write and graph a step function that shows the relationship between the number x of pounds a package weighs and the total cost y for postage.

Name____ Date__________

Chapter 5 Maintaining Mathematical Proficiency

Graph the equation.

1. $y + 2 = x$

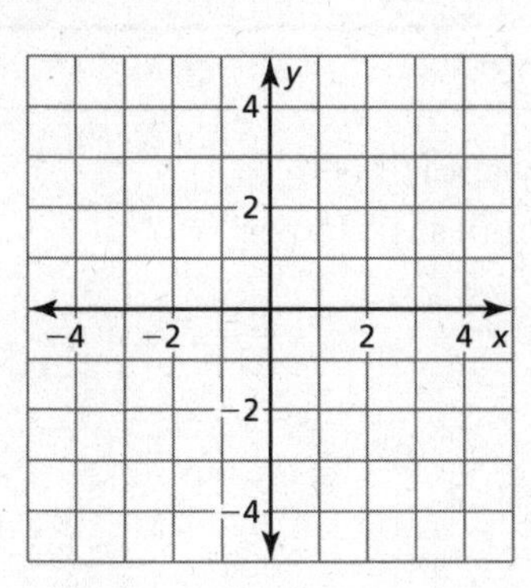

2. $2x - y = 3$

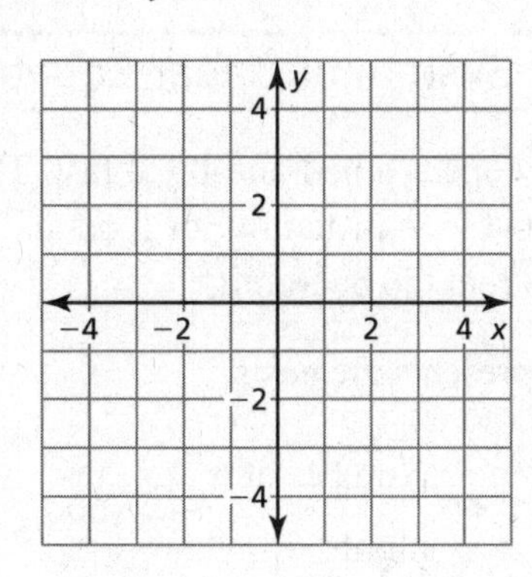

3. $5x + 2y = 10$

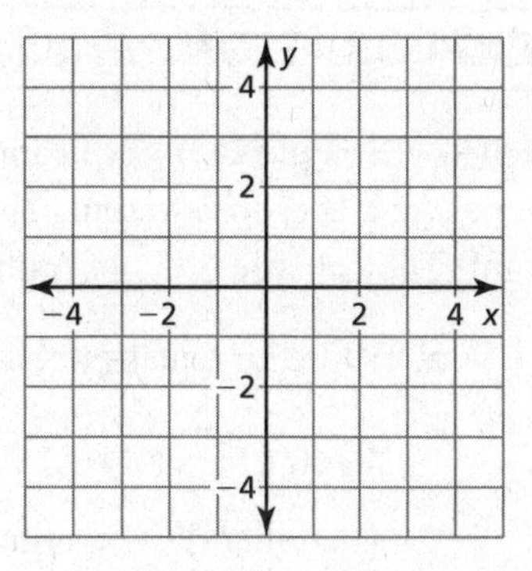

4. $y - 3 = x$

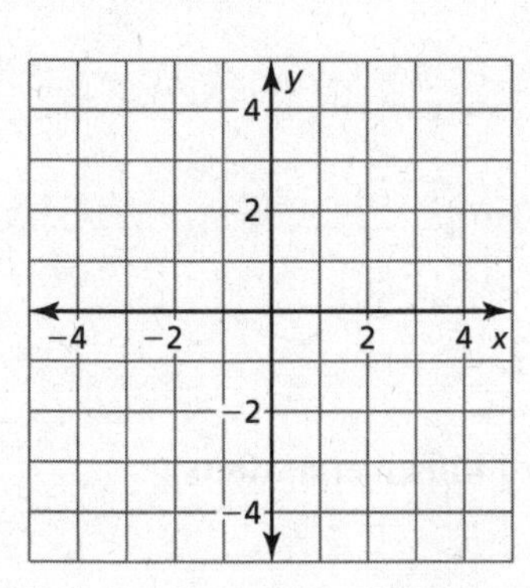

5. $3x - y = -2$

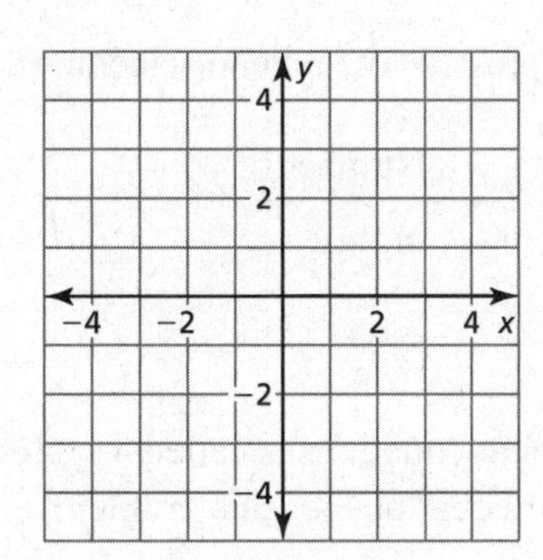

6. $3x + 4y = 12$

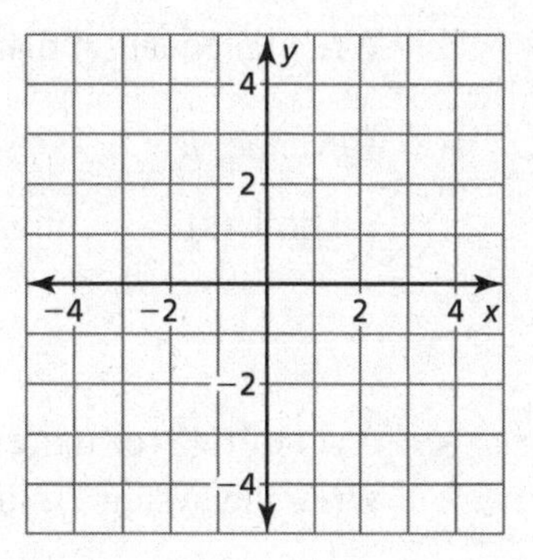

Solve the inequality. Graph the solution.

7. $a - 3 > -2$

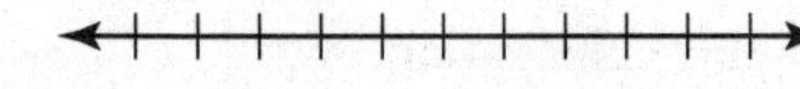

8. $-4 \geq -2c$

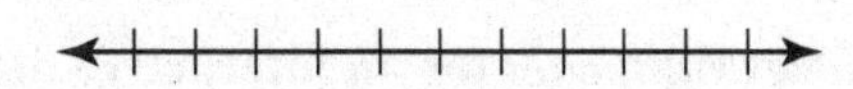

9. $2d - 5 < -3$

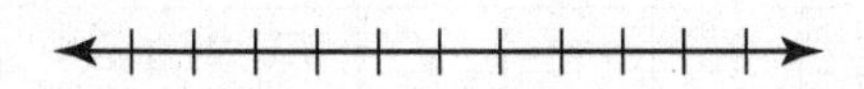

10. $8 - 3r \leq 5 - 2r$

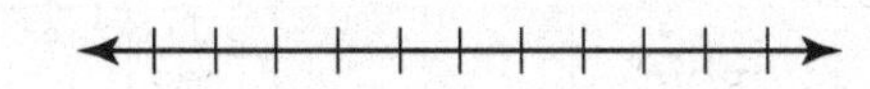

Name ____ Date __________

5.1 Solving Systems of Linear Equations by Graphing

For use with Exploration 5.1

Essential Question How can you solve a system of linear equations?

1 EXPLORATION: Writing a System of Linear Equations

Work with a partner. Your family opens a bed-and-breakfast. They spend \$600 preparing a bedroom to rent. The cost to your family for food and utilities is \$15 per night. They charge \$75 per night to rent the bedroom.

a. Write an equation that represents the costs.

Cost, C (in dollars) = \$15 per night • Number of nights, x + \$600

b. Write an equation that represents the revenue (income).

Revenue, R (in dollars) = \$75 per night • Number of nights, x

c. A set of two (or more) linear equations is called a **system of linear equations.** Write the system of linear equations for this problem.

2 EXPLORATION: Using a Table or Graph to Solve a System

Go to *BigIdeasMath.com* for an interactive tool to investigate this exploration.

Work with a partner. Use the cost and revenue equations from Exploration 1 to determine how many nights your family needs to rent the bedroom before recovering the cost of preparing the bedroom. This is the *break-even point*.

a. Complete the table.

***x* (nights)**	0	1	2	3	4	5	6	7	8	9	10	11
***C* (dollars)**												
***R* (dollars)**												

2 EXPLORATION: Using a Table or Graph to Solve a System (continued)

b. How many nights does your family need to rent the bedroom before breaking even?

c. In the same coordinate plane, graph the cost equation and the revenue equation from Exploration 1.

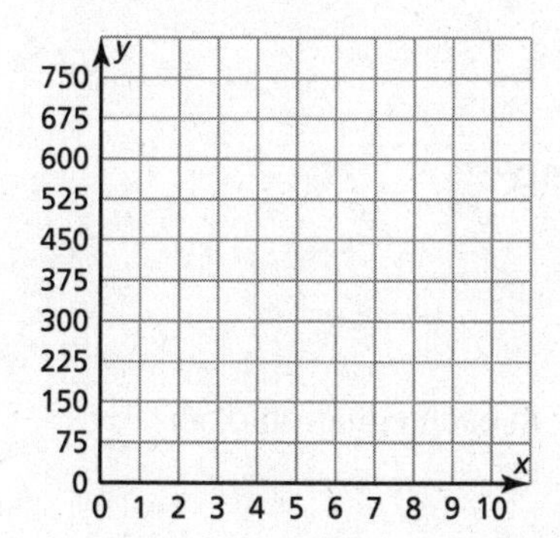

d. Find the point of intersection of the two graphs. What does this point represent? How does this compare to the break-even point in part (b)? Explain.

Communicate Your Answer

3. How can you solve a system of linear equations? How can you check your solution?

4. Solve each system by using a table or sketching a graph. Explain why you chose each method. Use a graphing calculator to check each solution.

a. $y = -4.3x - 1.3$
$y = 1.7x + 4.7$

b. $y = x$
$y = -3x + 8$

c. $y = -x - 1$
$y = 3x + 5$

Name ______________________________ Date __________

5.1 Notetaking with Vocabulary
For use after Lesson 5.1

In your own words, write the meaning of each vocabulary term.

system of linear equations

solution of a system of linear equations

Core Concepts

Solving a System of Linear Equations by Graphing

Step 1 Graph each equation in the same coordinate plane.

Step 2 Estimate the point of intersection.

Step 3 Check the point from Step 2 by substituting for x and y in each equation of the original system.

Notes:

Name____ Date__________

Extra Practice

In Exercises 1–6, tell whether the ordered pair is a solution of the system of linear equations.

1. $(3, 1);\ x + y = 4$
   $2x - y = 3$

2. $(1, 3);\ x - y = -2$
   $2x + y = 5$

3. $(2, 0);\ y = x - 2$
   $y = -3x + 6$

4. $(-1, -2);\ x - 2y = 3$
   $2x - y = 0$

5. $(-2, 3);\ 3x - 2y = -12$
   $2x + 4y = 9$

6. $(4, -3);\ 2x + 2y = 2$
   $3x - 3y = 21$

In Exercises 7–9, use the graph to solve the system of linear equations. Check your solution.

7. $3x - 2y = 10$
   $x + y = 0$

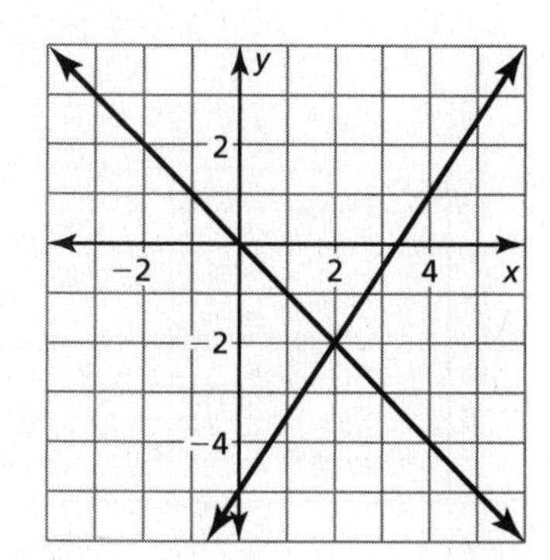

8. $x - 2y = 5$
   $2x + y = -5$

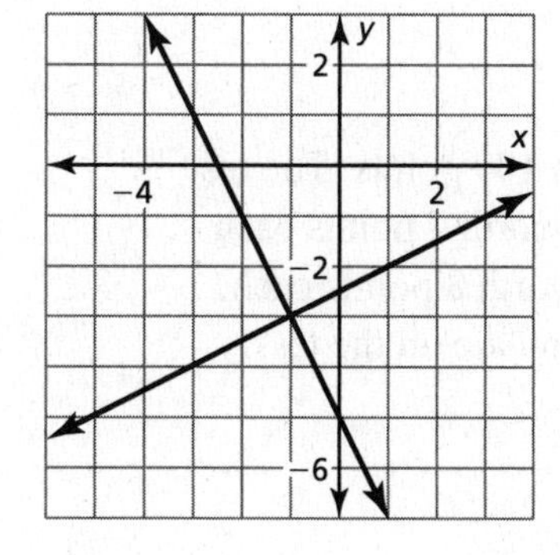

9. $x + 2y = 8$
   $3x - 2y = 8$

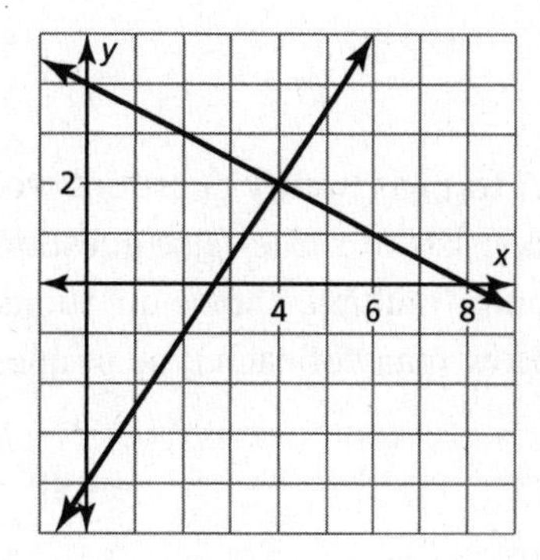

Name ____________________ Date __________

5.1 Notetaking with Vocabulary (continued)

In Exercises 10–15, solve the system of linear equations by graphing.

10. $y = -x + 3$

$y = x + 5$

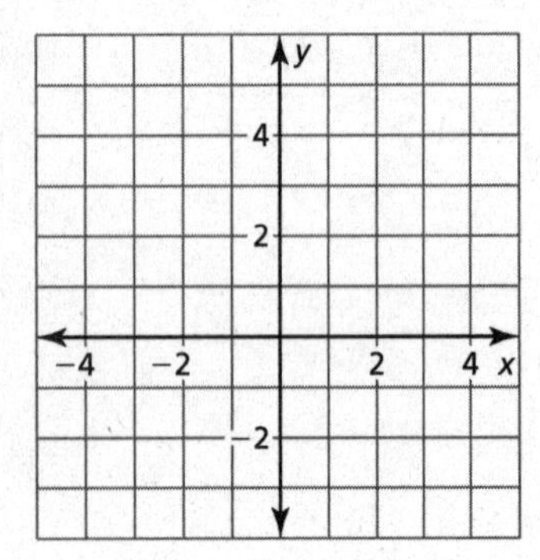

11. $y = \frac{1}{2}x + 2$

$y = -\frac{1}{2}x + 4$

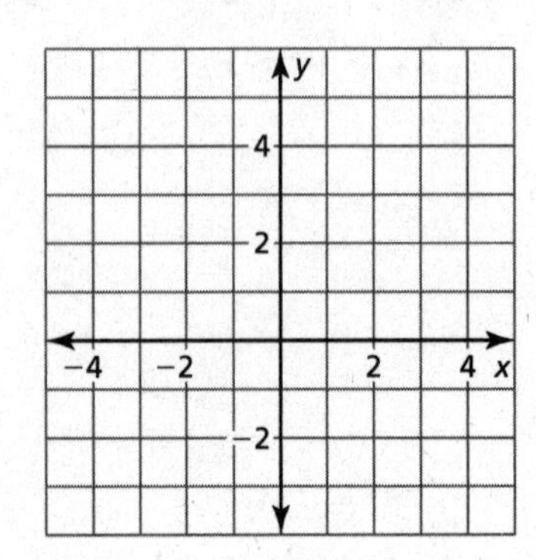

12. $3x - 2y = 6$

$y = -3$

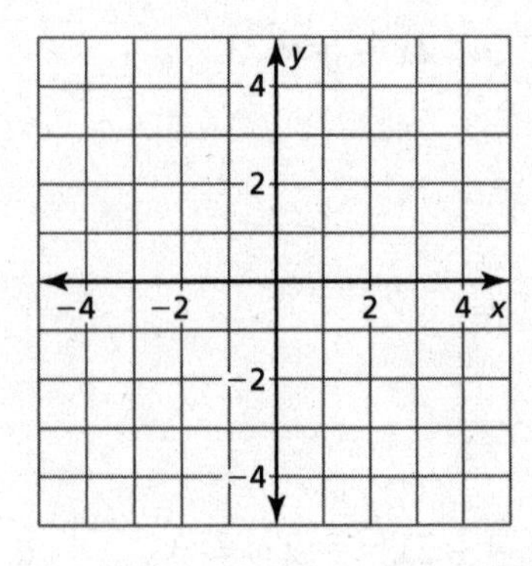

13. $y = 4x$

$y = -4x + 8$

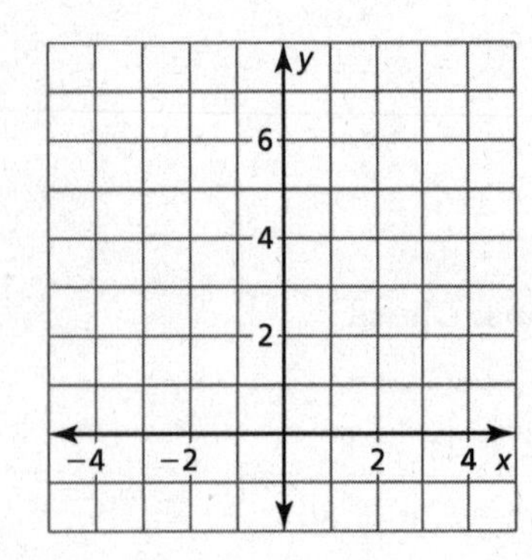

14. $y = \frac{1}{4}x + 3$

$y = \frac{3}{4}x + 5$

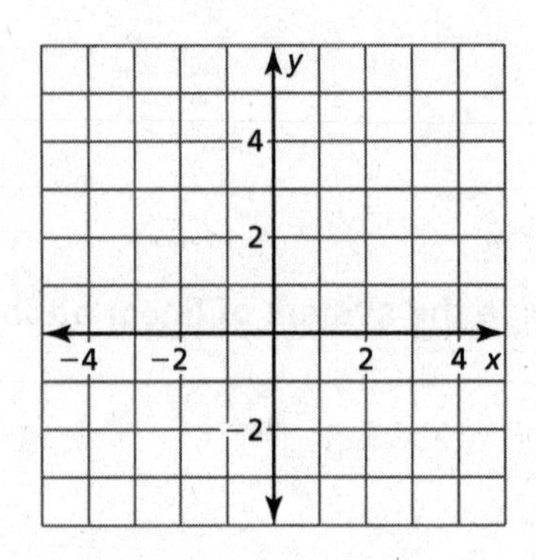

15. $3x - 4y = 7$

$5x + 2y = 3$

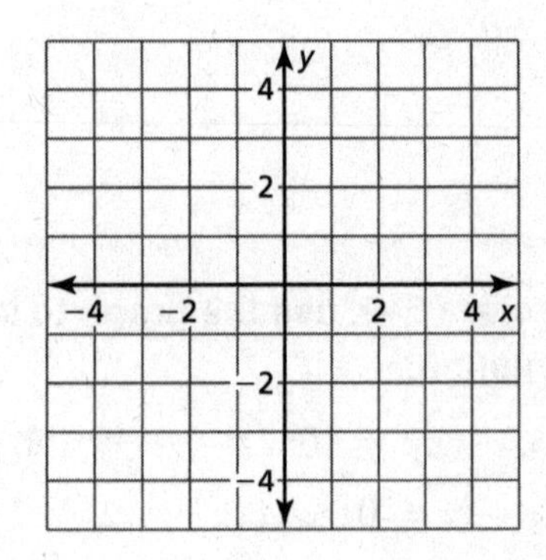

16. A test has twenty questions worth 100 points. The test consists of x true-false questions worth 4 points each and y multiple choice questions worth 8 points each. How many of each type of question are on the test?

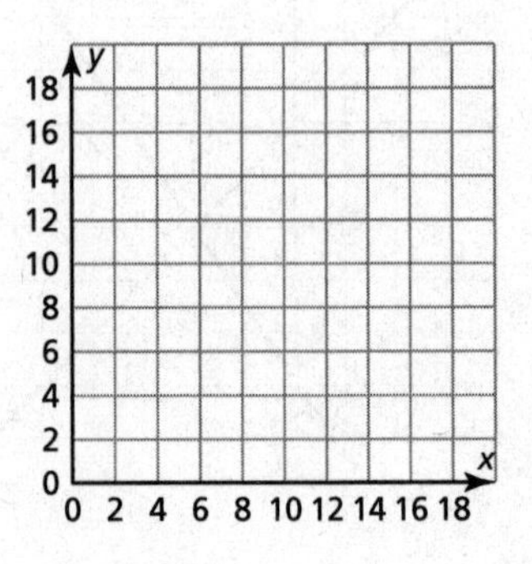

Name____ Date__________

5.2 Solving Systems of Linear Equations by Substitution

For use with Exploration 5.2

Essential Question How can you use substitution to solve a system of linear equations?

1 EXPLORATION: Using Substitution to Solve Systems

Work with a partner. Solve each system of linear equations using two methods.

Method 1 **Solve for x first.**

Solve for x in one of the equations. Substitute the expression for x into the other equation to find y. Then substitute the value of y into one of the original equations to find x.

Method 2 **Solve for y first.**

Solve for y in one of the equations. Substitute the expression for y into the other equation to find x. Then substitute the value of x into one of the original equations to find y.

Is the solution the same using both methods? Explain which method you would prefer to use for each system.

a. $x + y = -7$
$-5x + y = 5$

b. $x - 6y = -11$
$3x + 2y = 7$

c. $4x + y = -1$
$3x - 5y = -18$

Name ______________________________ Date __________

2 EXPLORATION: Writing and Solving a System of Equations

Go to *BigIdeasMath.com* for an interactive tool to investigate this exploration.

Work with a partner.

a. Write a random ordered pair with integer coordinates. One way to do this is to use a graphing calculator. The ordered pair generated at the right is $(-2, -3)$.

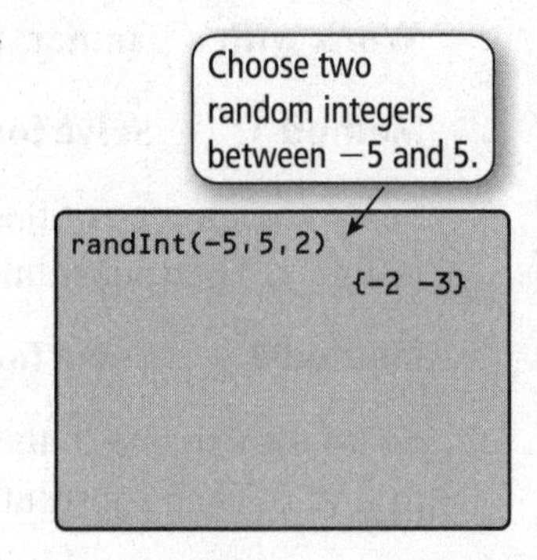

b. Write a system of linear equations that has your ordered pair as its solution.

c. Exchange systems with your partner and use one of the methods from Exploration 1 to solve the system. Explain your choice of method.

Communicate Your Answer

3. How can you use substitution to solve a system of linear equations?

4. Use one of the methods from Exploration 1 to solve each system of linear equations. Explain your choice of method. Check your solutions.

a. $x + 2y = -7$
$2x - y = -9$

b. $x - 2y = -6$
$2x + y = -2$

c. $-3x + 2y = -10$
$-2x + y = -6$

d. $3x + 2y = 13$
$x - 3y = -3$

e. $3x - 2y = 9$
$-x - 3y = 8$

f. $3x - y = -6$
$4x + 5y = 11$

Name____ Date__________

5.2 Notetaking with Vocabulary
For use after Lesson 5.2

In your own words, write the meaning of each vocabulary term.

system of linear equations

solution of a system of linear equations

Core Concepts

Solving a System of Linear Equations by Substitution

Step 1 Solve one of the equations for one of the variables.

Step 2 Substitute the expression from Step 1 into the other equation and solve for the other variable.

Step 3 Substitute the value from Step 2 into one of the original equations and solve.

Notes:

Extra Practice

In Exercises 1–18, solve the system of linear equations by substitution. Check your solution.

1. $2x + 2y = 10$
$y = 5 + x$

2. $2x - y = 3$
$x = -2y - 1$

3. $x - 3y = -1$
$x = y$

4. $x - 2y = -3$
$y = x + 1$

5. $2x + y = 3$
$x = 3y + 5$

6. $3x + y = -5$
$y = 2x + 5$

7. $y = 2x + 8$
$y = -2x$

8. $y = \frac{3}{4}x + 1$
$y = \frac{1}{4}x + 3$

9. $2x - 3y = 0$
$y = 4$

5.2 Notetaking with Vocabulary (continued)

10. $x + y = 3$
$2x + 4y = 8$

11. $y = \frac{1}{2}x + 1$
$y = -\frac{1}{2}x + 9$

12. $3x - 2y = 3$
$4x - y = 4$

13. $7x - 4y = 8$
$5x - y = 2$

14. $y = \frac{3}{5}x - 12$
$y = \frac{1}{3}x - 8$

15. $3x - 4y = -1$
$5x + 2y = 7$

16. $y = -x + 3$
$x + 2y = 0$

17. $y - 5x = -2$
$-4x + y = 2$

18. $4x - 8y = 3$
$8x + 4y = 1$

19. An adult ticket to a museum costs $3 more than a children's ticket. When 200 adult tickets and 100 children's tickets are sold, the total revenue is $2100. What is the cost of a children's ticket?

Name ______________________________ Date __________

5.3 Solving Systems of Linear Equations by Elimination

For use with Exploration 5.3

Essential Question How can you use elimination to solve a system of linear equations?

1 EXPLORATION: Writing and Solving a System of Equations

Work with a partner. You purchase a drink and a sandwich for \$4.50. Your friend purchases a drink and five sandwiches for \$16.50. You want to determine the price of a drink and the price of a sandwich.

a. Let x represent the price (in dollars) of one drink. Let y represent the price (in dollars) of one sandwich. Write a system of equations for the situation. Use the following verbal model.

Number of drinks	•	Price per drink	+	Number of sandwiches	•	Price per sandwich	=	Total price

Label one of the equations Equation 1 and the other equation Equation 2.

b. Subtract Equation 1 from Equation 2. Explain how you can use the result to solve the system of equations. Then find and interpret the solution.

2 EXPLORATION: Using Elimination to Solve Systems

Work with a partner. Solve each system of linear equations using two methods.

Method 1 **Subtract.** Subtract Equation 2 from Equation 1.Then use the result to solve the system.

Method 2 **Add.** Add the two equations. Then use the result to solve the system.

Is the solution the same using both methods? Which method do you prefer?

a. $3x - y = 6$
$3x + y = 0$

b. $2x + y = 6$
$2x - y = 2$

c. $x - 2y = -7$
$x + 2y = 5$

3 EXPLORATION: Using Elimination to Solve a System

Work with a partner.

$2x + y = 7$ Equation 1

$x + 5y = 17$ Equation 2

a. Can you eliminate a variable by adding or subtracting the equations as they are? If not, what do you need to do to one or both equations so that you can?

b. Solve the system individually. Then exchange solutions with your partner and compare and check the solutions.

Communicate Your Answer

4. How can you use elimination to solve a system of linear equations?

5. When can you add or subtract the equations in a system to solve the system? When do you have to multiply first? Justify your answers with examples.

6. In Exploration 3, why can you multiply an equation in the system by a constant and not change the solution of the system? Explain your reasoning.

Name ______________________ Date __________

5.3 Notetaking with Vocabulary

For use after Lesson 5.3

In your own words, write the meaning of each vocabulary term.

coefficient

Core Concepts

Solving a System of Linear Equations by Elimination

Step 1 Multiply, if necessary, one or both equations by a constant so at least one pair of like terms has the same or opposite coefficients.

Step 2 Add or subtract the equations to eliminate one of the variables.

Step 3 Solve the resulting equation.

Step 4 Substitute the value from Step 3 into one of the original equations and solve for the other variable.

Notes:

Extra Practice

In Exercises 1–18, solve the system of linear equations by elimination. Check your solution.

1. $x + 3y = 17$
$-x + 2y = 8$

2. $2x - y = 5$
$5x + y = 16$

3. $2x + 3y = 10$
$-2x - y = -2$

4. $4x + 3y = 6$
$-x - 3y = 3$

5. $5x + 2y = -28$
$-5x + 3y = 8$

6. $2x - 5y = 8$
$3x + 5y = -13$

7. $2x + y = 12$
$3x - 18 = y$

8. $4x + 3y = 14$
$2y = 6 + 4x$

9. $-4x = -2 + 4y$
$-4y = 1 - 4x$

10. $x + 2y = 20$
$2x + y = 19$

11. $3x - 2y = -2$
$4x - 3y = -4$

12. $9x + 4y = 11$
$3x - 10y = -2$

13. $4x + 3y = 21$
$5x + 2y = 21$

14. $-3x - 5y = -7$
$-4x - 3y = -2$

15. $8x + 4y = 12$
$7x + 3y = 10$

16. $4x + 3y = -7$
$-2x - 5y = 7$

17. $8x - 3y = -9$
$5x + 4y = 12$

18. $-3x + 5y = -2$
$2x - 2y = 1$

19. The sum of two numbers is 22. The difference is 6. What are the two numbers?

Name ______________________________ Date __________

5.4 Solving Special Systems of Linear Equations

For use with Exploration 5.4

Essential Question Can a system of linear equations have no solution or infinitely many solutions?

1 EXPLORATION: Using a Table to Solve a System

Go to *BigIdeasMath.com* for an interactive tool to investigate this exploration.

Work with a partner. You invest $450 for equipment to make skateboards. The materials for each skateboard cost $20. You sell each skateboard for $20.

a. Write the cost and revenue equations. Then complete the table for your cost C and your revenue R.

x (skateboards)	0	1	2	3	4	5	6	7	8	9	10
C (dollars)											
R (dollars)											

b. When will your company break even? What is wrong?

2 EXPLORATION: Writing and Analyzing a System

Go to *BigIdeasMath.com* for an interactive tool to investigate this exploration.

Work with a partner. A necklace and matching bracelet have two types of beads. The necklace has 40 small beads and 6 large beads and weighs 10 grams. The bracelet has 20 small beads and 3 large beads and weighs 5 grams. The threads holding the beads have no significant weight.

a. Write a system of linear equations that represents the situation. Let x be the weight (in grams) of a small bead and let y be the weight (in grams) of a large bead.

b. Graph the system in the coordinate plane shown. What do you notice about the two lines?

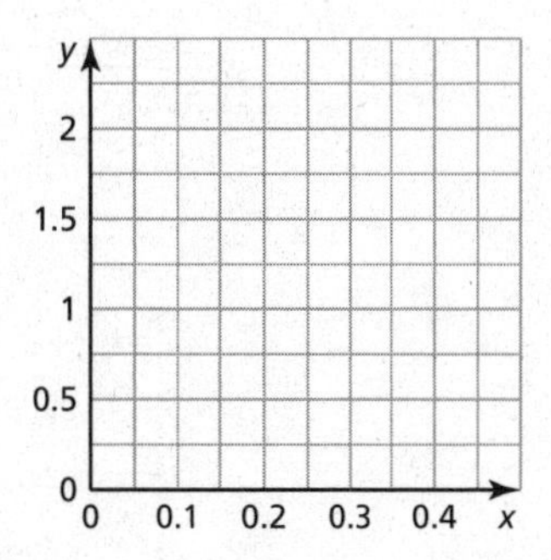

c. Can you find the weight of each type of bead? Explain your reasoning.

Communicate Your Answer

3. Can a system of linear equations have no solution or infinitely many solutions? Give examples to support your answers.

4. Does the system of linear equations represented by each graph have *no solution*, *one solution*, or *infinitely many solutions*? Explain.

a.

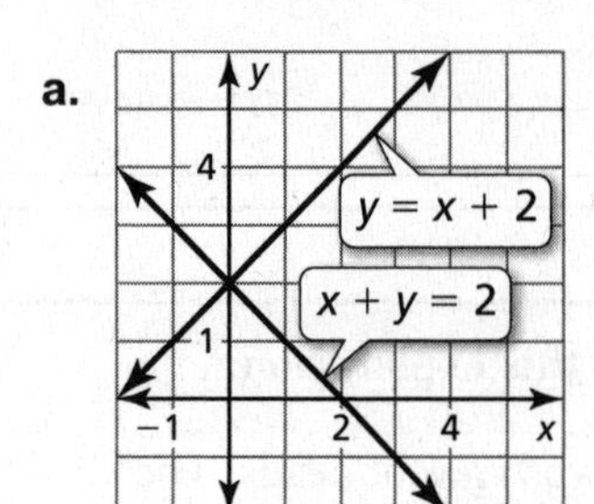

b.

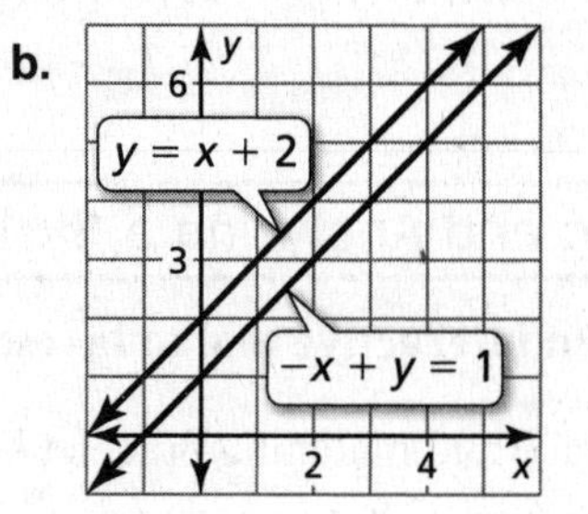

c.

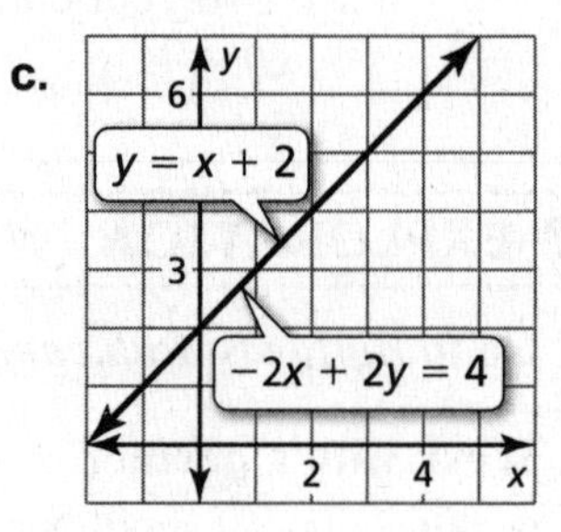

Name____ Date__________

5.4 Notetaking with Vocabulary

For use after Lesson 5.4

In your own words, write the meaning of each vocabulary term.

parallel

Core Concepts

Solutions of Systems of Linear Equations

A system of linear equations can have *one solution*, *no solution*, or *infinitely many solutions*.

One solution

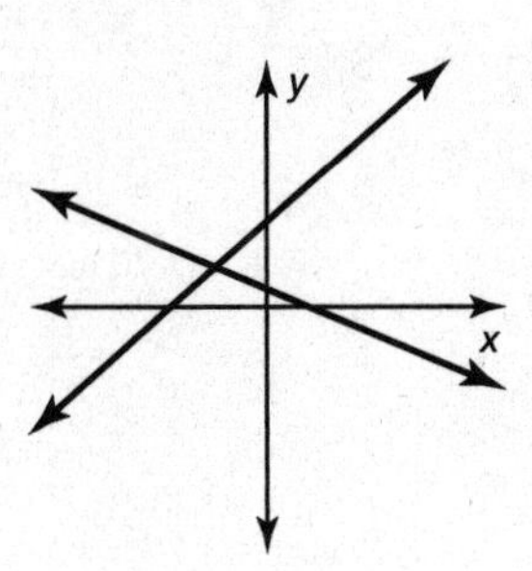

The lines intersect.

No solution

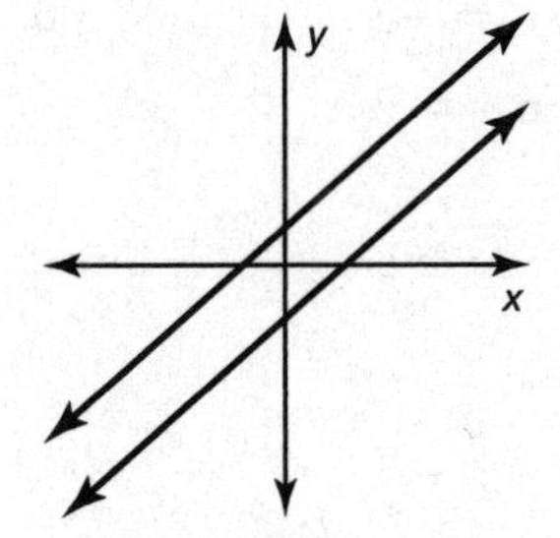

The lines are parallel.

Infinitely many solutions

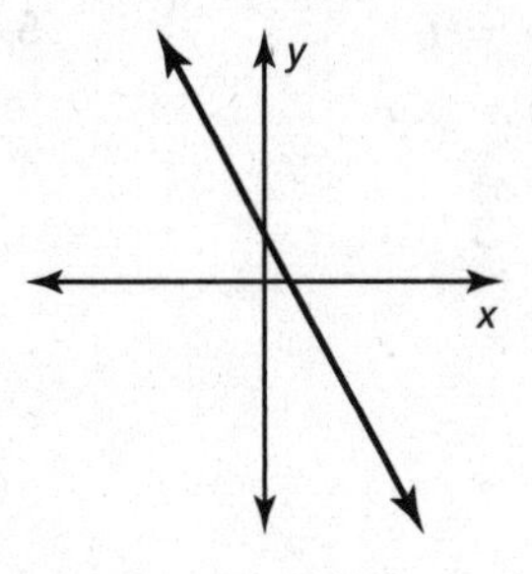

The lines are the same.

Notes:

Name ______________________________ Date __________

Extra Practice

In Exercises 1–18, solve the system of linear equations.

1. $y = 3x - 7$
$y = 3x + 4$

2. $y = 5x - 1$
$y = -5x + 5$

3. $2x - 3y = 10$
$-2x + 3y = -10$

4. $x + 3y = 6$
$-x - 3y = 3$

5. $6x + 6y = -3$
$-6x - 6y = 3$

6. $2x - 5y = -3$
$3x + 5y = 8$

7. $2x + 3y = 1$
$-2x + 3y = -7$

8. $4x + 3y = 17$
$-8x - 6y = 34$

9. $3x - 2y = 6$
$-9x + 6y = -18$

10. $-2x + 5y = -21$
$2x - 5y = 21$

11. $3x - 8y = 3$
$8x - 3y = 8$

12. $18x + 12y = 24$
$3x + 2y = 6$

13. $15x - 6y = 9$
$5x - 2y = 27$

14. $-3x - 5y = 8$
$6x + 10y = -16$

15. $2x - 4y = 2$
$-2x - 4y = 6$

16. $5x + 7y = 7$
$7x + 5y = 5$

17. $y = \frac{2}{3}x + 7$
$y = \frac{2}{3}x - 5$

18. $-3x + 5y = 15$
$9x - 15y = -45$

19. You have \$15 in savings. Your friend has \$25 in savings. You both start saving \$5 per week. Write a system of linear equations that represents this situation. Will you ever have the same amount of savings as your friend? Explain.

Name ____ Date __________

5.5 Solving Equations by Graphing

For use with Exploration 5.5

Essential Question How can you use a system of linear equations to solve an equation with variables on both sides?

1 EXPLORATION: Solving an Equation by Graphing

Go to *BigIdeasMath.com* for an interactive tool to investigate this exploration.

Work with a partner. Solve $2x - 1 = -\frac{1}{2}x + 4$ by graphing.

a. Use the left side to write a linear equation. Then use the right side to write another linear equation.

b. Graph the two linear equations from part (a). Find the x-value of the point of intersection. Check that the x-value is the solution of

$$2x - 1 = -\frac{1}{2}x + 4.$$

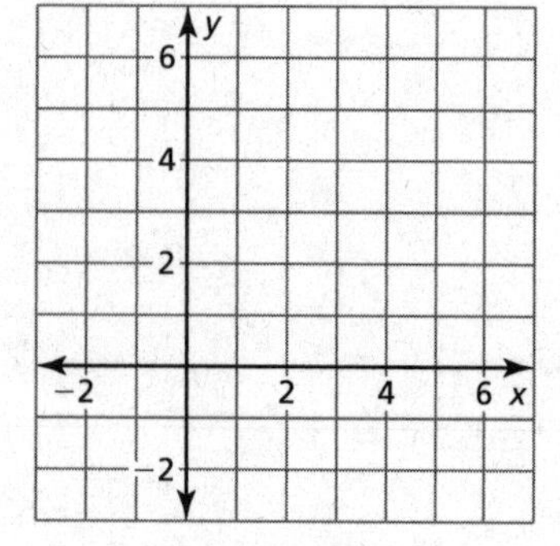

c. Explain why this "graphical method" works.

2 EXPLORATION: Solving Equations Algebraically and Graphically

Go to *BigIdeasMath.com* for an interactive tool to investigate this exploration.

Work with a partner. Solve each equation using two methods.

Method 1 Use an algebraic method.

Method 2 Use a graphical method.

Is the solution the same using both methods?

a. $\frac{1}{2}x + 4 = -\frac{1}{4}x + 1$

b. $\frac{2}{3}x + 4 = \frac{1}{3}x + 3$

2 EXPLORATION: Solving Equations Algebraically and Graphically (continued)

c. $-\frac{2}{3}x - 1 = \frac{1}{3}x - 4$

d. $\frac{4}{5}x + \frac{7}{5} = 3x - 3$

e. $-x + 2.5 = 2x - 0.5$

f. $-3x + 1.5 = x + 1.5$

Communicate Your Answer

3. How can you use a system of linear equations to solve an equation with variables on both sides?

4. Compare the algebraic method and the graphical method for solving a linear equation with variables on both sides. Describe the advantages and disadvantages of each method.

Name ______________________ Date __________

5.5 Notetaking with Vocabulary
For use after Lesson 5.5

In your own words, write the meaning of each vocabulary term.

absolute value equation

Core Concepts

Solving Linear Equations by Graphing

Step 1 To solve the equation $ax + b = cx + d$, write two linear equations.

$$ax + b = cx + d$$

$y = ax + b$ and $y = cx + d$

Step 2 Graph the system of linear equations. The x-value of the solution of the system of linear equations is the solution of the equation $ax + b = cx + d$.

Notes:

Name________________________________ Date__________

5.5 Notetaking with Vocabulary (continued)

Extra Practice

In Exercises 1–9, solve the equation by graphing. Check your solution(s).

1. $2x - 7 = -2x + 9$

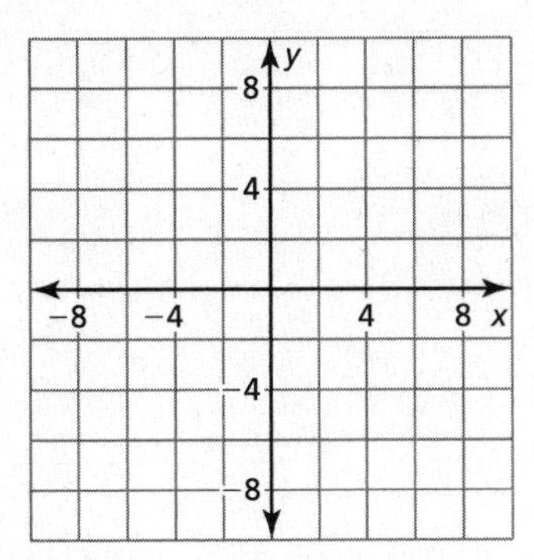

2. $3x = x - 4$

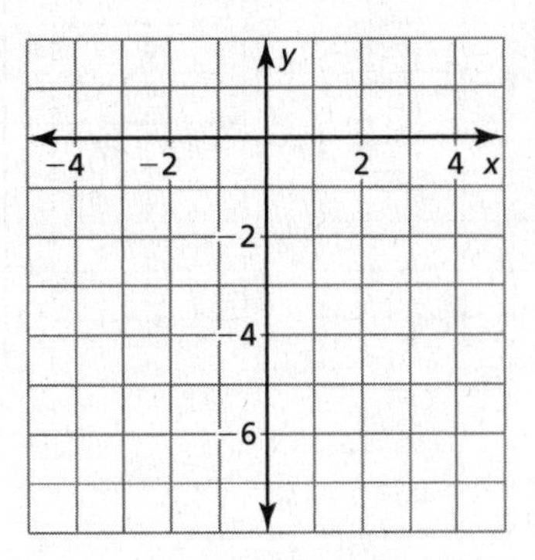

3. $4x + 1 = -2x - 5$

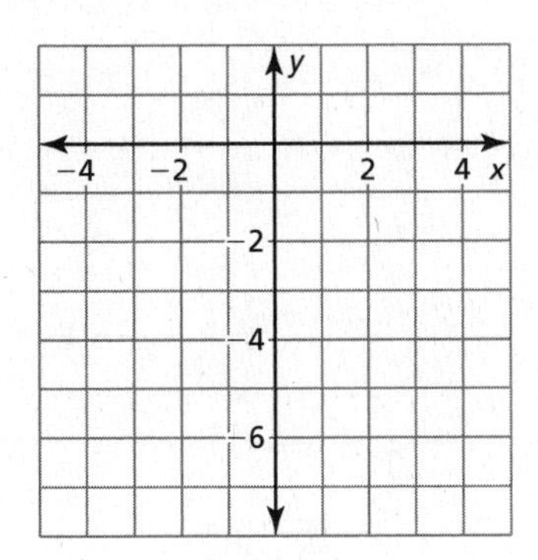

4. $-x - 4 = 3(x - 4)$

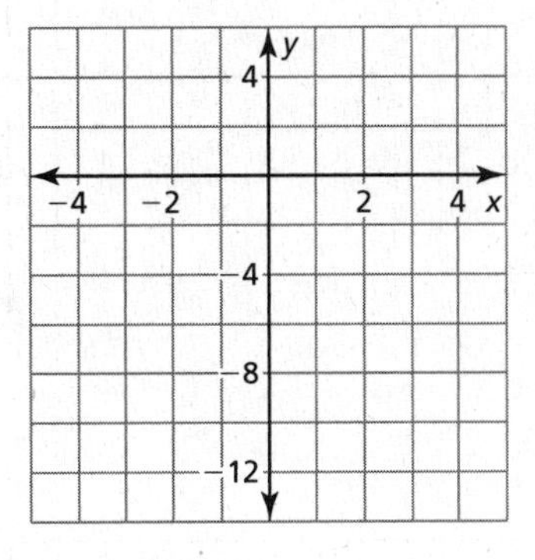

5. $-3x - 5 = 6 - 3x$

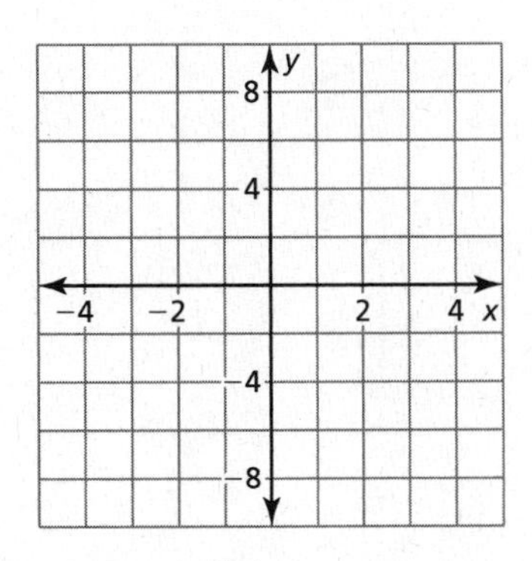

6. $7x - 14 = -7(2 - x)$

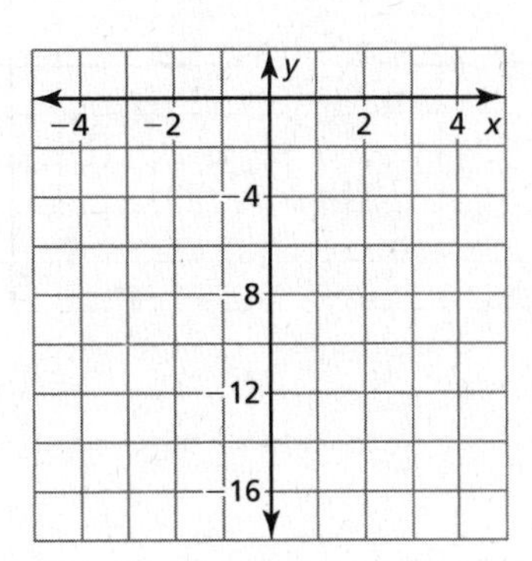

Name ____________________ Date __________

7. $|3x| = |2x + 10|$

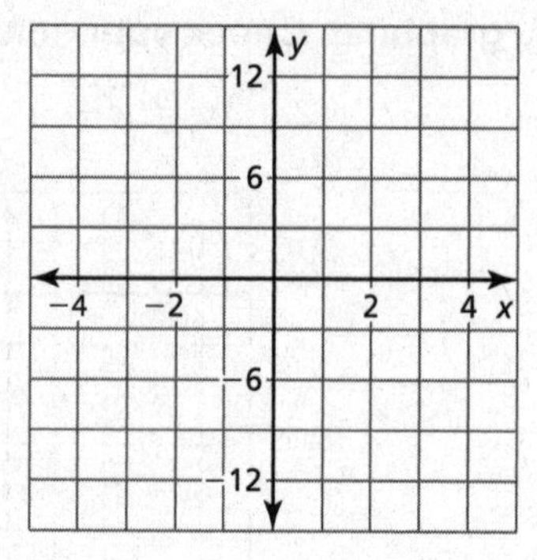

8. $|x - 1| = |x + 3|$

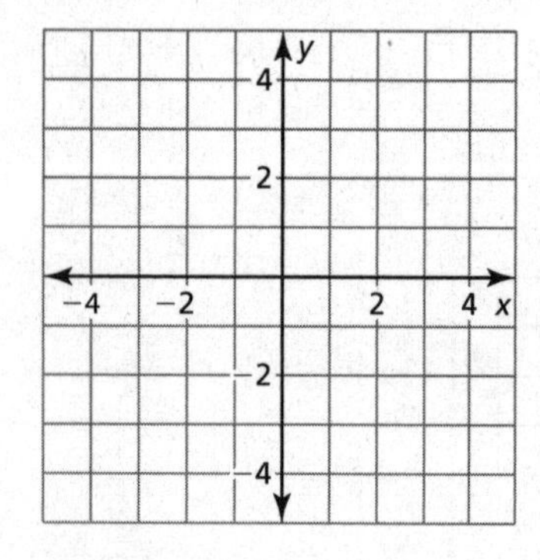

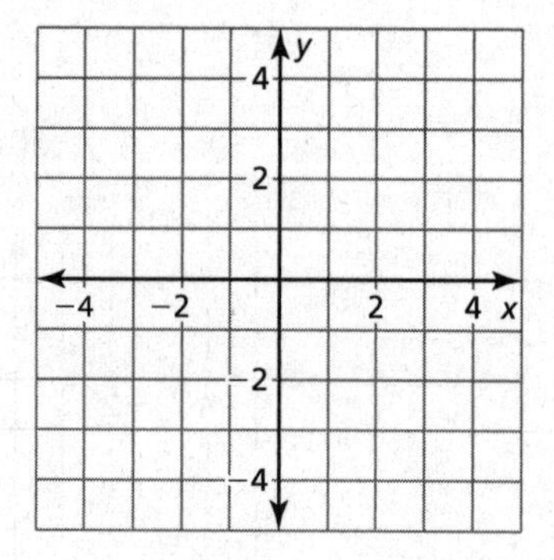

9. $|x + 4| = |2 - x|$

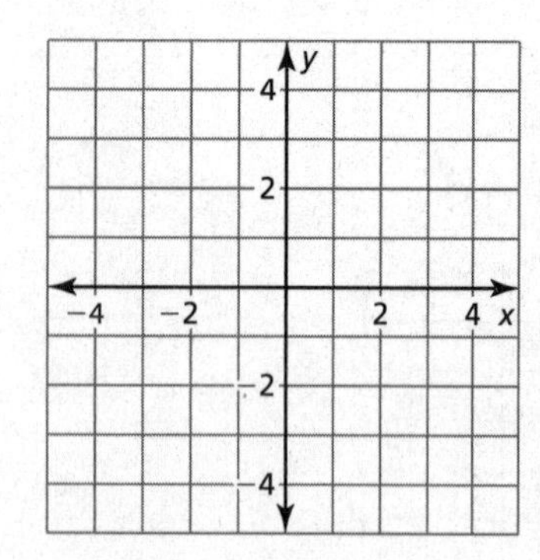

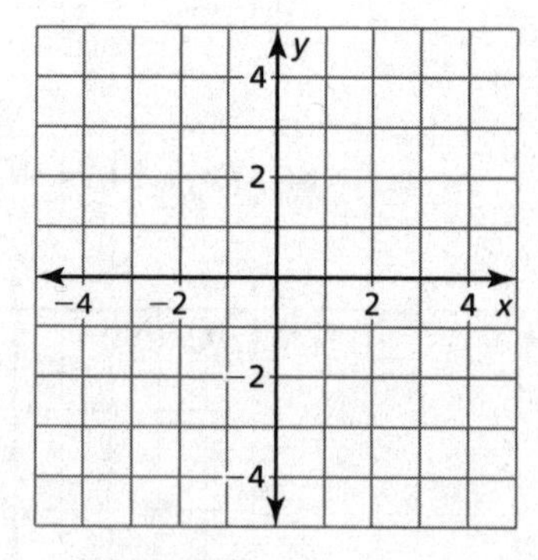

Name______________________________ Date__________

5.6 Graphing Linear Inequalities in Two Variables

For use with Exploration 5.6

Essential Question How can you graph a linear inequality in two variables?

A **solution of a linear inequality in two variables** is an ordered pair (x, y) that makes the inequality true. The **graph of a linear inequality** in two variables shows all the solutions of the inequality in a coordinate plane.

1 EXPLORATION: Writing a Linear Inequality in Two Variables

Work with a partner.

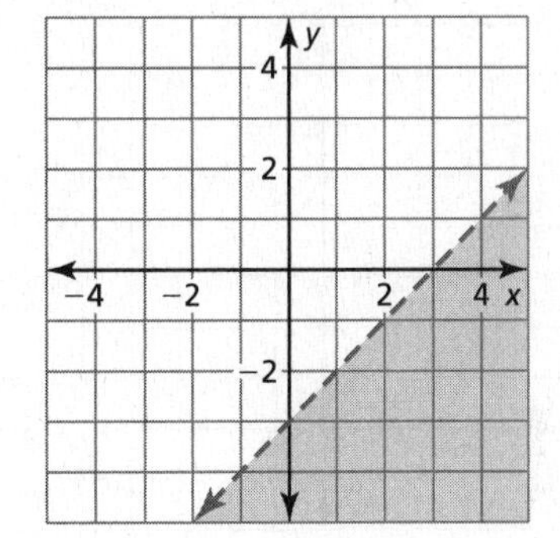

a. Write an equation represented by the dashed line.

b. The solutions of an inequality are represented by the shaded region. In words, describe the solutions of the inequality.

c. Write an inequality represented by the graph. Which inequality symbol did you use? Explain your reasoning.

2 EXPLORATION: Using a Graphing Calculator

Go to *BigIdeasMath.com* for an interactive tool to investigate this exploration.

Work with a partner. Use a graphing calculator to graph $y \geq \frac{1}{4}x - 3$.

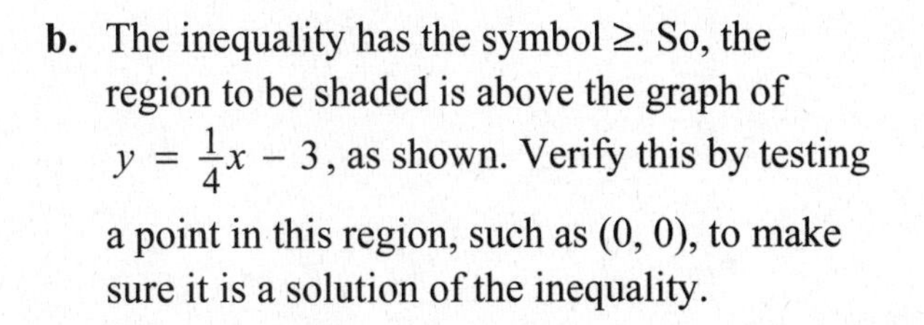

a. Enter the equation $y = \frac{1}{4}x - 3$ into your calculator.

b. The inequality has the symbol ≥. So, the region to be shaded is above the graph of $y = \frac{1}{4}x - 3$, as shown. Verify this by testing a point in this region, such as (0, 0), to make sure it is a solution of the inequality.

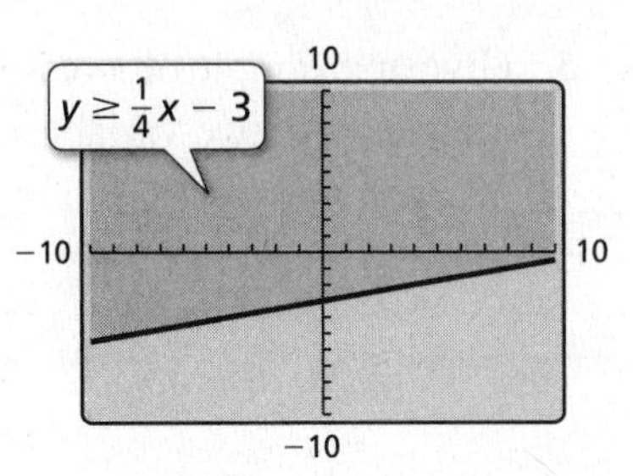

Because the inequality symbol is *greater than or equal to*, the line is solid and not dashed. Some graphing calculators always use a solid line when graphing inequalities. In this case, you have to determine whether the line should be solid or dashed, based on the inequality symbol used in the original inequality.

3 EXPLORATION: Graphing Linear Inequalities in Two Variables

Go to *BigIdeasMath.com* for an interactive tool to investigate this exploration.

Work with a partner. Graph each linear inequality in two variables. Explain your steps. Use a graphing calculator to check your graphs.

a. $y > x + 5$ **b.** $y \leq -\frac{1}{2}x + 1$ **c.** $y \geq -x - 5$

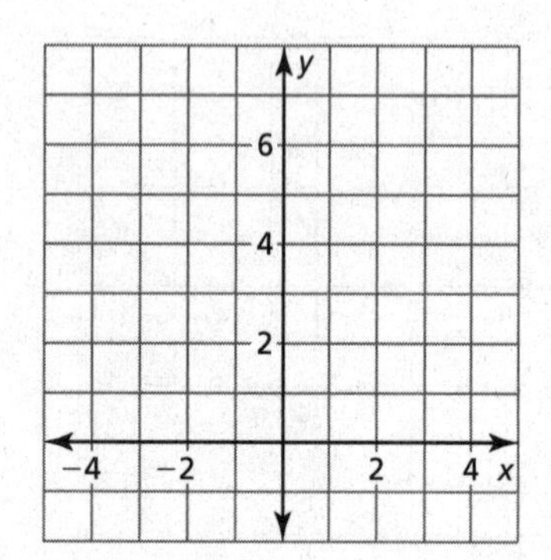

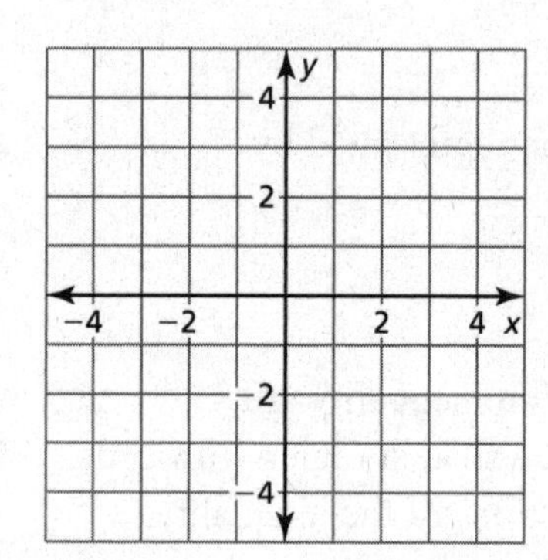

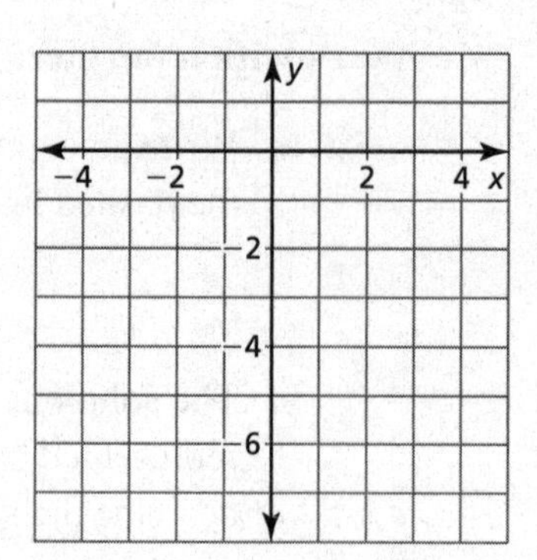

Communicate Your Answer

4. How can you graph a linear inequality in two variables?

5. Give an example of a real-life situation that can be modeled using a linear inequality in two variables.

Name____ Date__________

5.6 Notetaking with Vocabulary
For use after Lesson 5.6

In your own words, write the meaning of each vocabulary term.

linear inequality in two variables

solution of a linear inequality in two variables

graph of a linear inequality

half-planes

Core Concepts

Graphing a Linear Inequality in Two Variables

Step 1 Graph the boundary line for the inequality. Use a dashed line for $<$ or $>$. Use a solid line for $\leq$ or $\geq$.

Step 2 Test a point that is not on the boundary line to determine whether it is a solution of the inequality.

Step 3 When a test point is a solution, shade the half-plane that contains the point. When the test point is *not* a solution, shade the half-plane that does *not* contain the point.

Notes:

Name ____ Date __________

Extra Practice

In Exercises 1–6, tell whether the ordered pair is a solution of the inequality.

1. $x + y > 5; (3, 2)$

2. $x - y \geq 2; (5, 3)$

3. $x + 2y \leq 4; (-1, 2)$

4. $5x + y < 7; (2, -2)$

5. $3x - 4y > 6; (-1, -1)$

6. $-x - 2y \geq 5; (-2, -3)$

In Exercises 7–18, graph the inequality in a coordinate plane.

7. $y < 4$

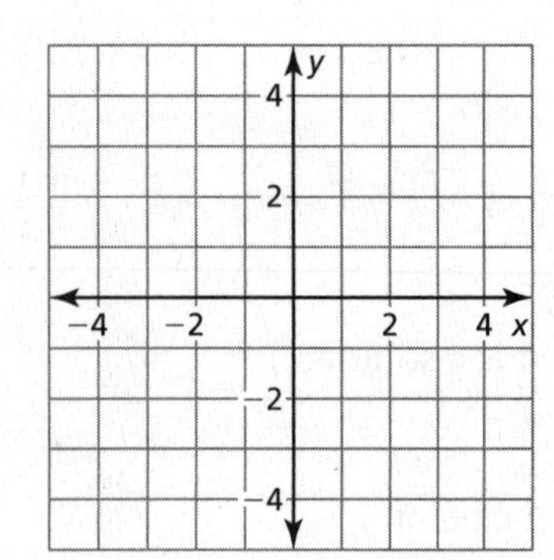

8. $y > -1$

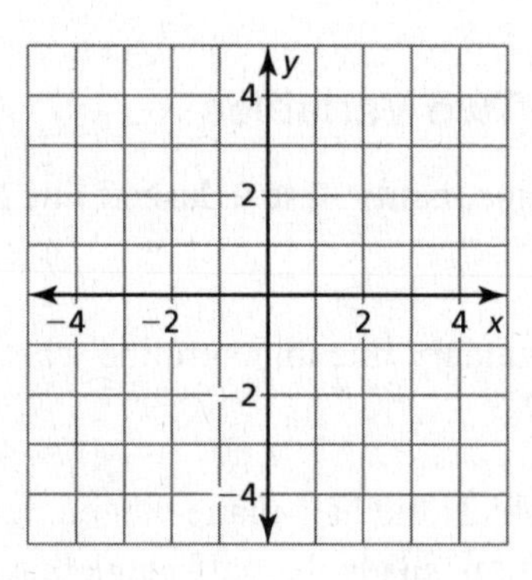

9. $x > 3$

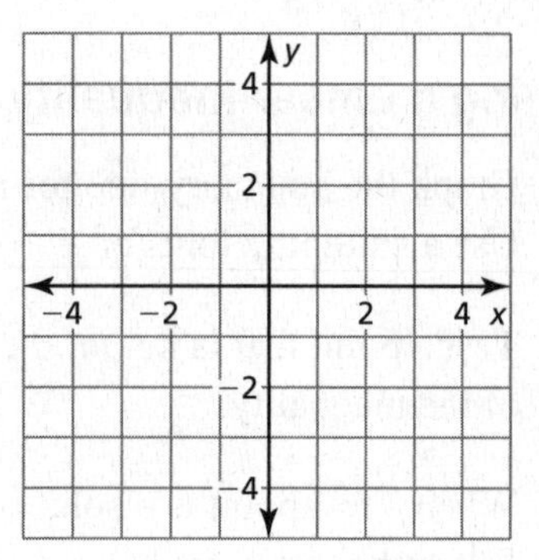

10. $x \leq -1$

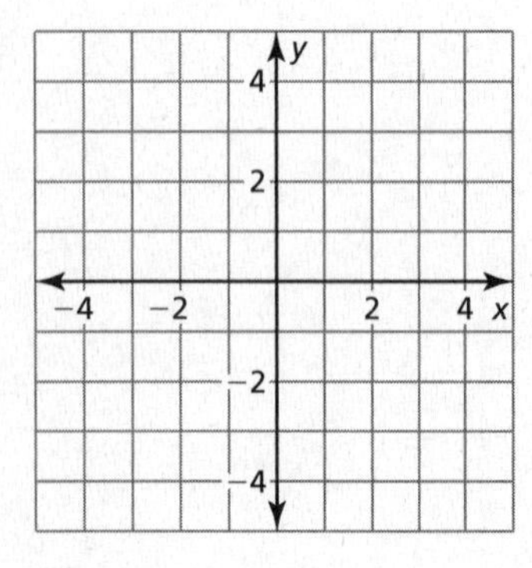

11. $y < -2$

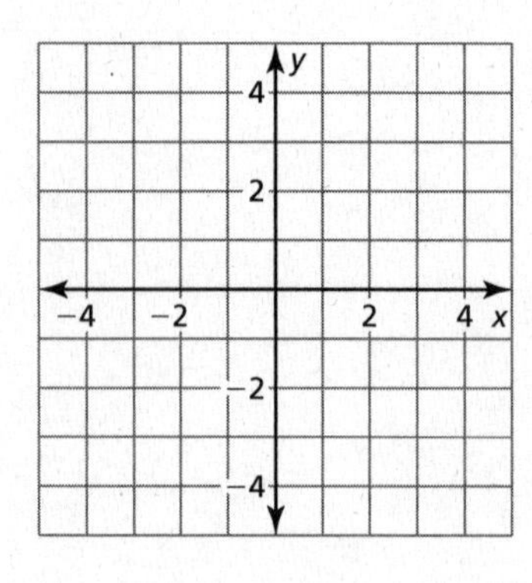

12. $x > -2$

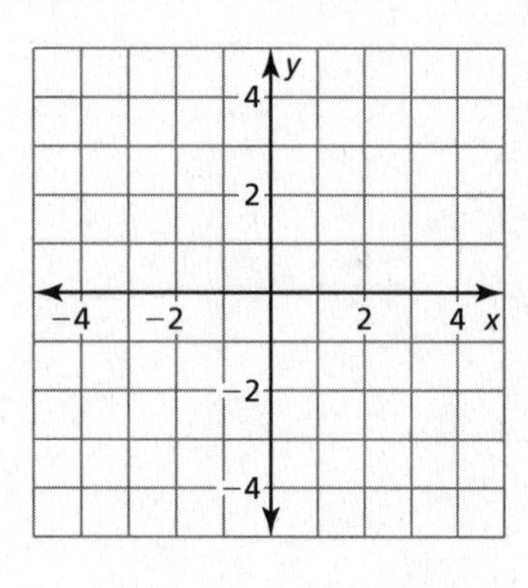

Name____ Date__________

5.6 Notetaking with Vocabulary (continued)

13. $y < 3x + 1$

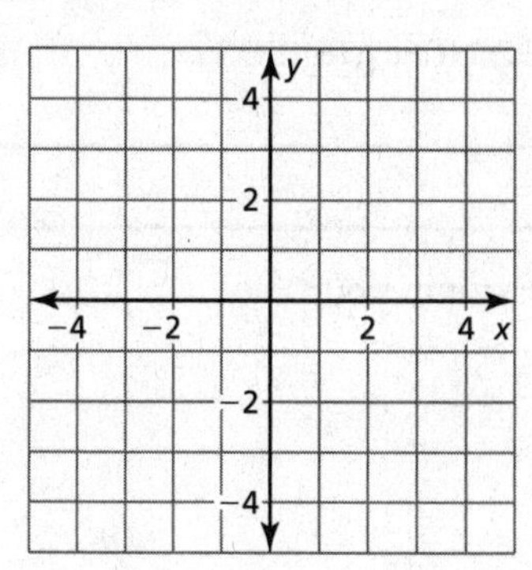

14. $y \geq -x + 1$

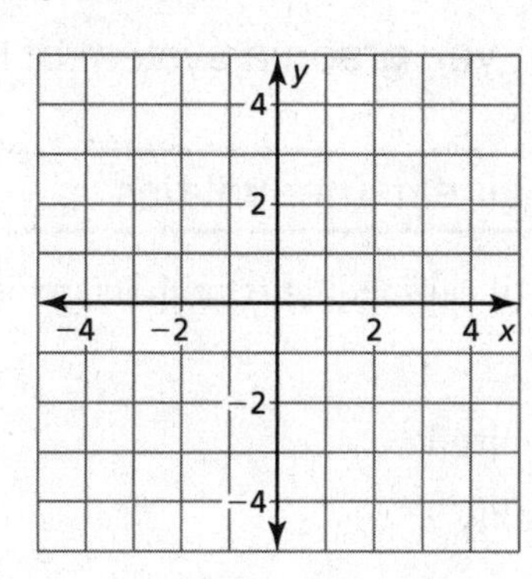

15. $x - y < 2$

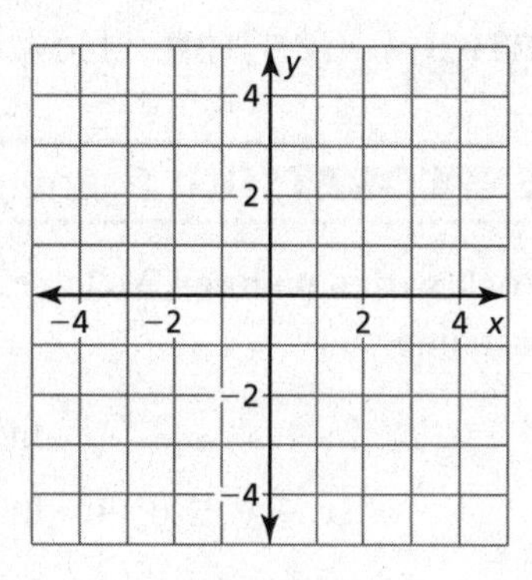

16. $x + y \geq -3$

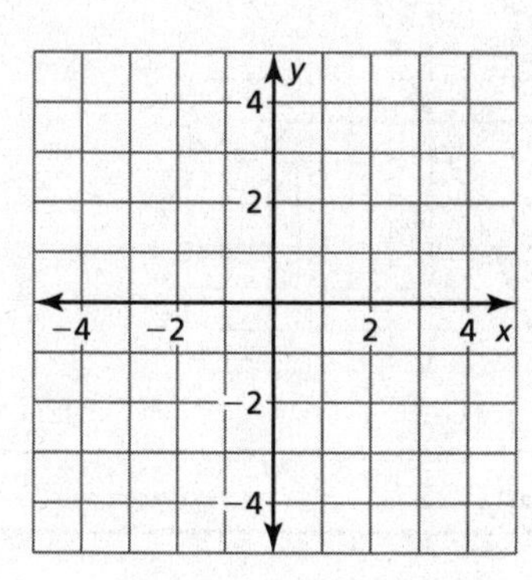

17. $x + 2y < 4$

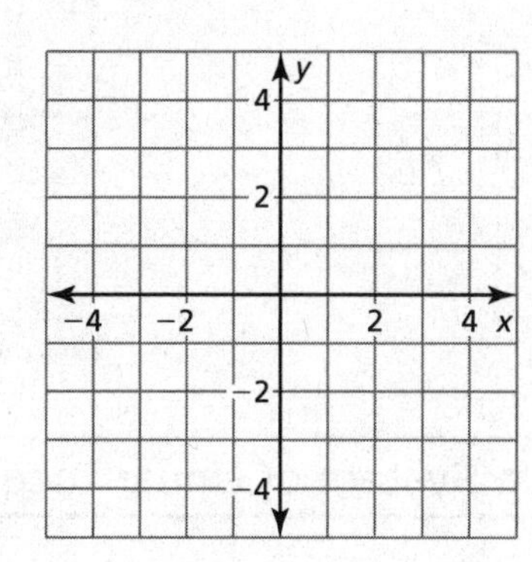

18. $-2x + 3y > 6$

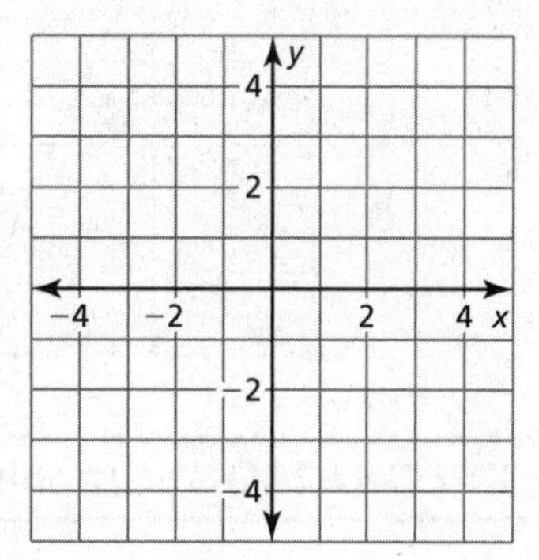

19. An online store sells digital cameras and cell phones. The store makes a \$100 profit on the sale of each digital camera x and a \$50 profit on the sale of each cell phone y. The store wants to make a profit of at least \$300 from its sales of digital cameras and cell phones. Write and graph an inequality that represents how many digital cameras and cell phones they must sell. Identify and interpret two solutions of the inequality.

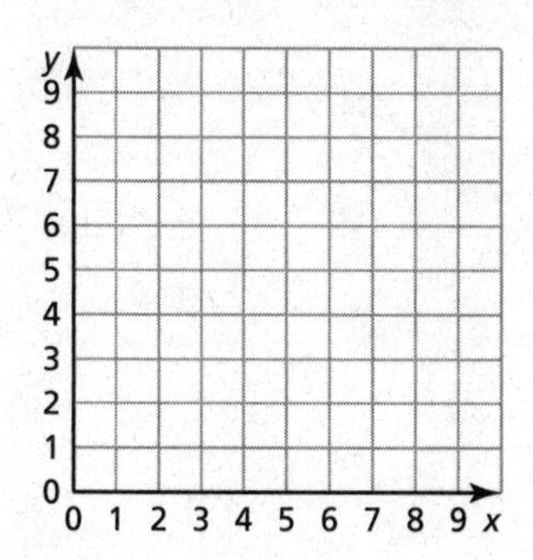

Name ______________________________ Date __________

5.7 Systems of Linear Inequalities
For use with Exploration 5.7

Essential Question How can you graph a system of linear inequalities?

1 EXPLORATION: Graphing Linear Inequalities

Work with a partner. Match each linear inequality with its graph. Explain your reasoning.

$2x + y \leq 4$ Inequality 1

$2x - y \leq 0$ Inequality 2

A.

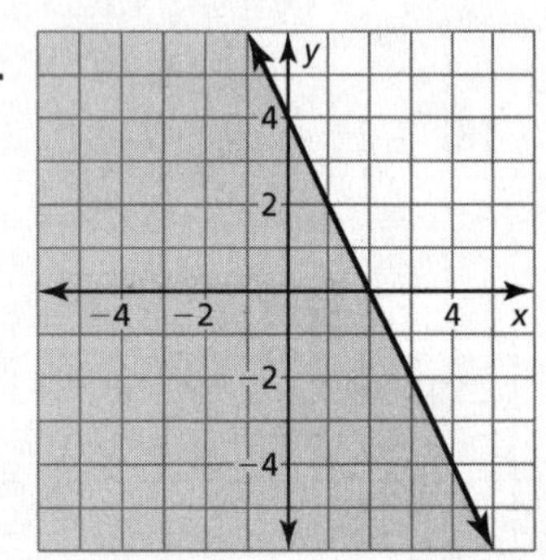

B.

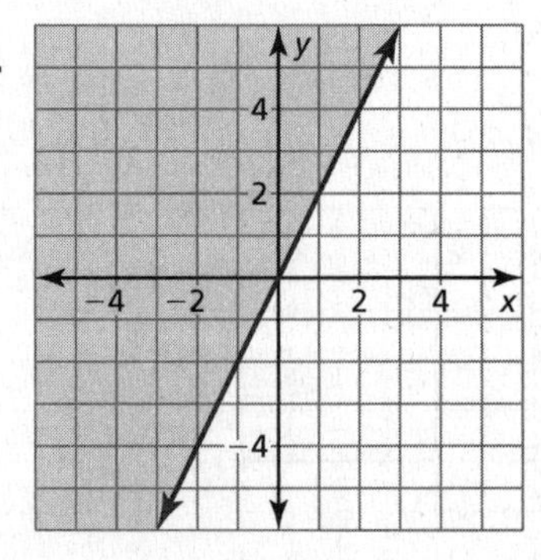

2 EXPLORATION: Graphing a System of Linear Inequalities

Go to *BigIdeasMath.com* for an interactive tool to investigate this exploration.

Work with a partner. Consider the linear inequalities given in Exploration 1.

$2x + y \leq 4$ Inequality 1

$2x - y \leq 0$ Inequality 2

a. Use two different colors to graph the inequalities in the same coordinate plane. What is the result?

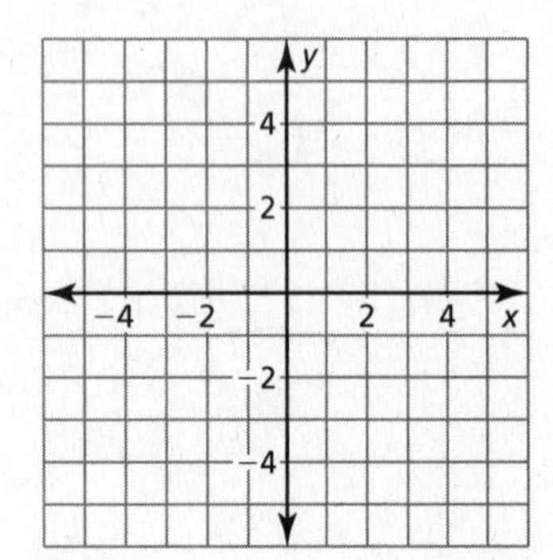

2 EXPLORATION: Graphing a System of Linear Inequalities (continued)

b. Describe each of the shaded regions of the graph. What does the unshaded region represent?

Communicate Your Answer

3. How can you graph a system of linear inequalities?

4. When graphing a system of linear inequalities, which region represents the solution of the system?

5. Do you think all systems of linear inequalities have a solution? Explain your reasoning.

6. Write a system of linear inequalities represented by the graph.

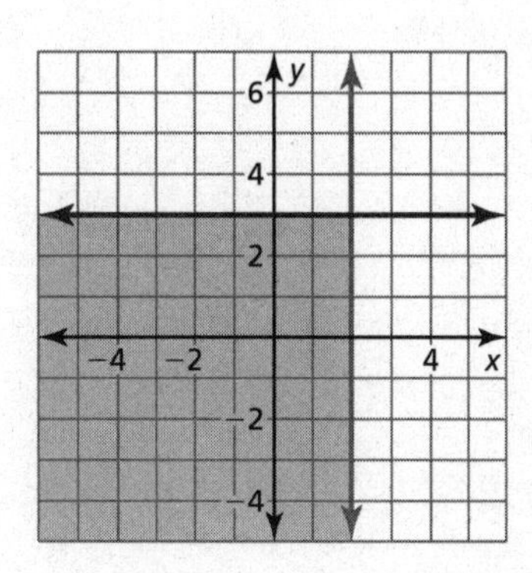

Name ______________________________ Date __________

5.7 Notetaking with Vocabulary

For use after Lesson 5.7

In your own words, write the meaning of each vocabulary term.

system of linear inequalities

solution of a system of linear inequalities

graph of a system of linear inequalities

Core Concepts

Graphing a System of Linear Inequalities

Step 1 Graph each inequality in the same coordinate plane.

Step 2 Find the intersection of the half-planes that are solutions of the inequalities. This intersection is the graph of the system.

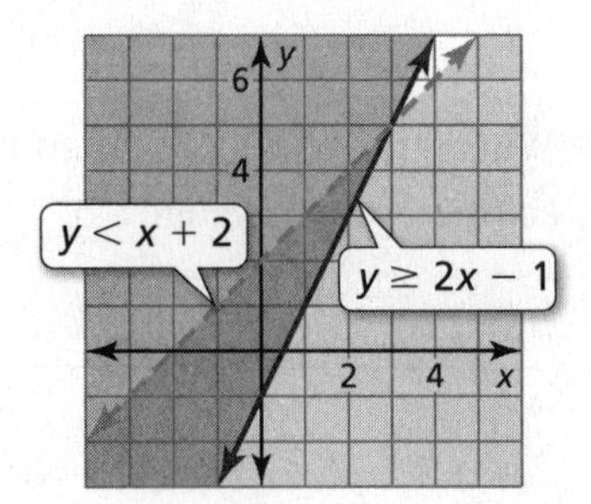

Notes:

Name____ Date__________

Extra Practice

In Exercises 1–4, tell whether the ordered pair is a solution of the system of linear inequalities.

1. $(0, 0)$; $y > 2$
$y < x - 2$

2. $(-1, 1)$; $y < 3$
$y > x - 4$

3. $(2, 3)$; $y \geq x + 4$
$y \leq 2x + 4$

4. $(0, 4)$; $y \leq -x + 4$
$y \geq 5x - 3$

In Exercises 5–8, graph the system of linear inequalities.

5. $y > -2$
$y \leq 3x$

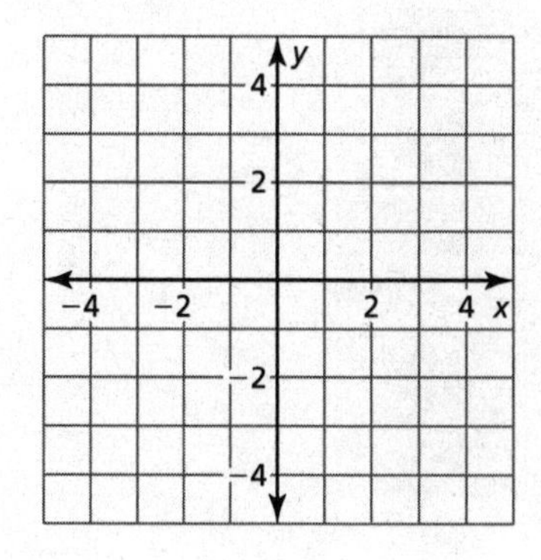

6. $y < 3$
$x < 2$

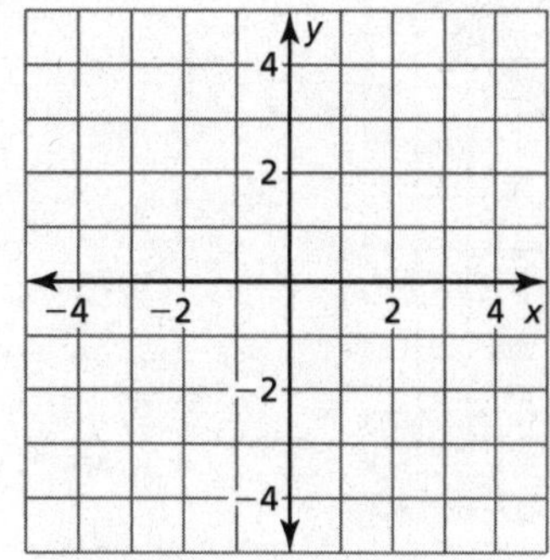

5.7 Notetaking with Vocabulary (continued)

7. $y \geq x - 2$

$y < -x + 2$

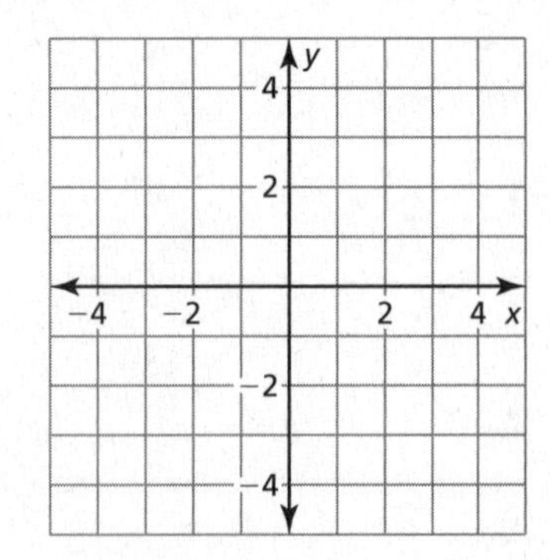

8. $2x + 3y < 6$

$y - 1 \geq -2x$

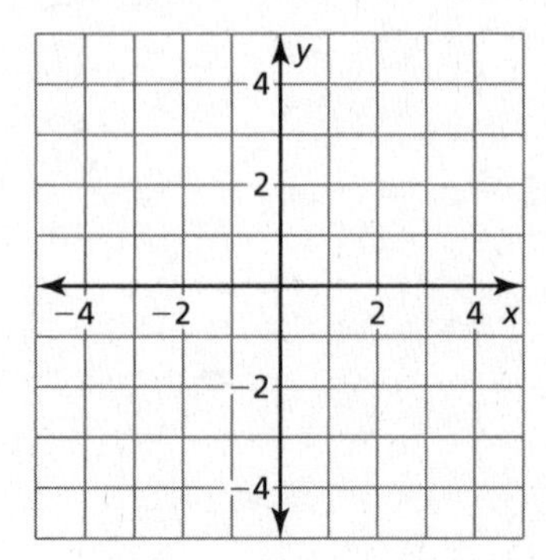

In Exercises 9–12, write a system of linear inequalities represented by the graph.

9.

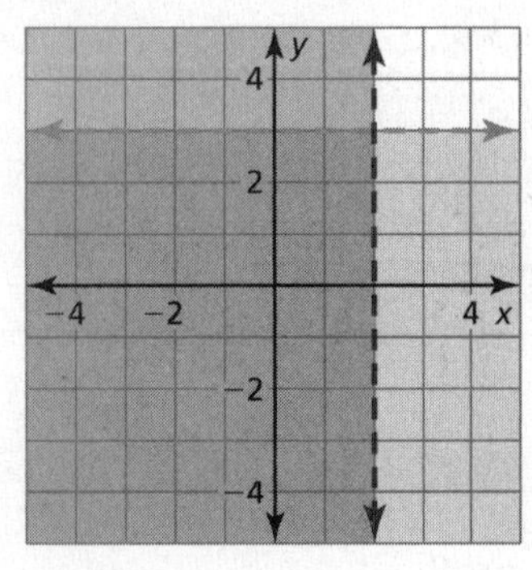

10.

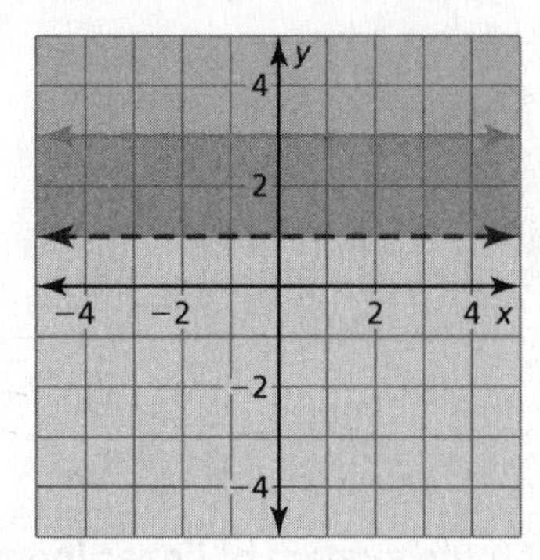

11.

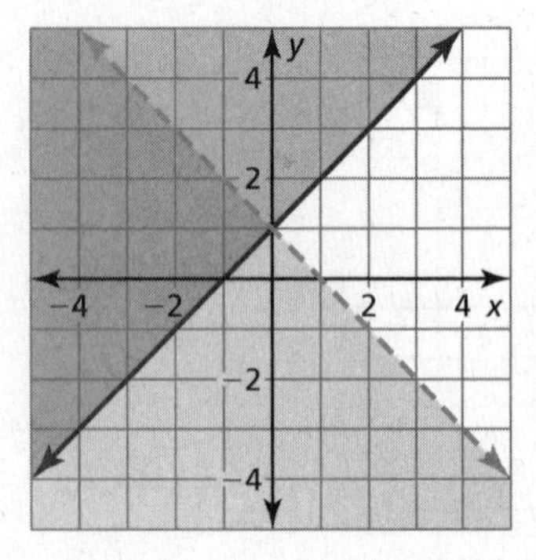

12.

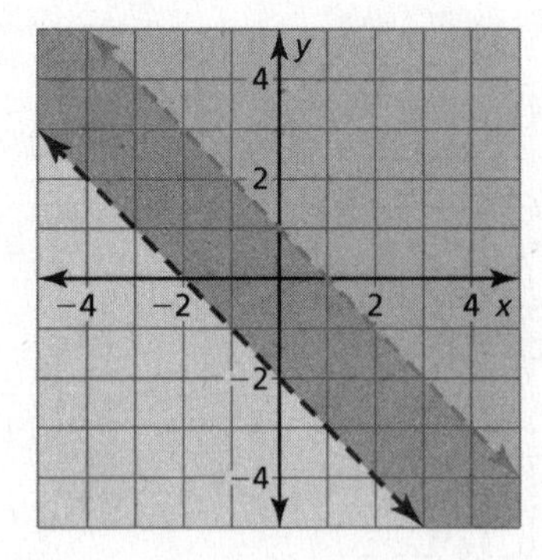

Name____ Date__________

Chapter 6 Maintaining Mathematical Proficiency

Evaluate the expression.

1. $(14 + 20 - 6) \div 4 - 6^2$ **2.** $(8 + 4)^2 + (13 - 10 \div 5)$ **3.** $8 \div 4 \bullet 19 + 18 + 13$

4. $3 \bullet 14 \bullet 11 + 4^2 + 19$ **5.** $(21 + 2)(14 - 6) + 3^2$ **6.** $7(3 \bullet 10 - 4^2) + 8$

Find the square root(s).

7. $\sqrt{36}$ **8.** $-\sqrt{49}$ **9.** $-\sqrt{225}$ **10.** $\sqrt{144}$

11. $\sqrt{169}$ **12.** $\sqrt{9}$ **13.** $-\sqrt{16}$ **14.** $\pm\sqrt{256}$

Write an equation for the *n*th term of the arithmetic sequence.

15. $1, 5, 9, 13, \ldots$ **16.** $21, 15, 9, 3, \ldots$ **17.** $-2, 1, 4, 7, \ldots$

18. $8, 6, 4, 2, \ldots$ **19.** $-10, -4, 2, 8, \ldots$ **20.** $16, 8, 0, -8, \ldots$

Name ____________________ Date ________

6.1 Properties of Exponents

For use with Exploration 6.1

Essential Question How can you write general rules involving properties of exponents?

1 EXPLORATION: Writing Rules for Properties of Exponents

Work with a partner.

a. What happens when you multiply two powers with the same base? Write the product of the two powers as a single power. Then write a *general rule* for finding the product of two powers with the same base.

i. $(2^2)(2^3) =$ ________ **ii.** $(4^1)(4^5) =$ ________

iii. $(5^3)(5^5) =$ ________ **iv.** $(x^2)(x^6) =$ ________

b. What happens when you divide two powers with the same base? Write the quotient of the two powers as a single power. Then write a *general rule* for finding the quotient of two powers with the same base.

i. $\frac{4^3}{4^2} =$ ________ **ii.** $\frac{2^5}{2^2} =$ ________

iii. $\frac{x^6}{x^3} =$ ________ **iv.** $\frac{3^4}{3^4} =$ ________

c. What happens when you find a power of a power? Write the expression as a single power. Then write a *general rule* for finding a power of a power.

i. $(2^2)^4 =$ ________ **ii.** $(7^3)^2 =$ ________

iii. $(y^3)^3 =$ ________ **iv.** $(x^4)^2 =$ ________

Name____ Date__________

1 EXPLORATION: Writing Rules for Properties of Exponents (continued)

d. What happens when you find a power of a product? Write the expression as the product of two powers. Then write a *general rule* for finding a power of a product.

i. $(2 \bullet 5)^2 =$ ________ **ii.** $(5 \bullet 4)^3 =$ ________

iii. $(6a)^2 =$ ________ **iv.** $(3x)^2 =$ ________

e. What happens when you find a power of a quotient? Write the expression as the quotient of two powers. Then write a *general rule* for finding a power of a quotient.

i. $\left(\frac{2}{3}\right)^2 =$ ________ **ii.** $\left(\frac{4}{3}\right)^3 =$ ________

iii. $\left(\frac{x}{2}\right)^3 =$ ________ **iv.** $\left(\frac{a}{b}\right)^4 =$ ________

Communicate Your Answer

2. How can you write general rules involving properties of exponents?

3. There are 3^3 small cubes in the cube below. Write an expression for the number of small cubes in the large cube at the right.

Name ______________________________ Date ________

6.1 Notetaking with Vocabulary

For use after Lesson 6.1

In your own words, write the meaning of each vocabulary term.

power

exponent

base

scientific notation

Core Concepts

Zero Exponent

Words For any nonzero number a, $a^0 = 1$. The power 0^0 is undefined.

Numbers $4^0 = 1$ **Algebra** $a^0 = 1$, where $a \neq 0$

Negative Exponents

Words For any integer n and any nonzero number a, a^{-n} is the reciprocal of a^n.

Numbers $4^{-2} = \frac{1}{4^2}$ **Algebra** $a^{-n} = \frac{1}{a^n}$, where $a \neq 0$

Notes:

6.1 Notetaking with Vocabulary (continued)

Product of Powers Property

Let a be a real number, and let m and n be integers.

Words To multiply powers with the same base, add their exponents.

Numbers $4^6 \bullet 4^3 = 4^{6+3} = 4^9$ **Algebra** $a^m \bullet a^n = a^{m+n}$

Quotient of Powers Property

Let a be a nonzero real number, and let m and n be integers.

Words To divide powers with the same base, subtract their exponents.

Numbers $\frac{4^6}{4^3} = 4^{6-3} = 4^3$ **Algebra** $\frac{a^m}{a^n} = a^{m-n}$, where $a \neq 0$

Power of a Power Property

Let a be a real number, and let m and n be integers.

Words To find a power of a power, multiply the exponents.

Numbers $\left(4^6\right)^3 = 4^{6 \bullet 3} = 4^{18}$ **Algebra** $\left(a^m\right)^n = a^{mn}$

Notes:

Power of a Product Property

Let a and b be real numbers, and let m be an integer.

Words To find a power of a product, find the power of each factor and multiply.

Numbers $(3 \bullet 2)^5 = 3^5 \bullet 2^5$ **Algebra** $(ab)^m = a^m b^m$

Power of a Quotient Property

Let a and b be real numbers with $b \neq 0$, and let m be an integer.

Words To find the power of a quotient, find the power of the numerator and the power of the denominator and divide.

Numbers $\left(\frac{3}{2}\right)^5 = \frac{3^5}{2^5}$ **Algebra** $\left(\frac{a}{b}\right)^m = \frac{a^m}{b^m}$, where $b \neq 0$

Notes:

Name ______________________________ Date ________

Extra Practice

In Exercises 1–8, evaluate the expression.

1. 3^0

2. $(-2)^0$

3. 3^{-4}

4. $(-4)^{-3}$

5. $\dfrac{2^{-3}}{5^0}$

6. $\dfrac{-3^{-2}}{2^{-3}}$

7. $\dfrac{4^{-1}}{-7^0}$

8. $\dfrac{3^{-1}}{(-5)^0}$

In Exercises 9–23, simplify the expression. Write your answer using only positive exponents.

9. z^0

10. a^{-8}

11. $6a^0b^{-2}$

12. $14m^{-4}n^0$

13. $\dfrac{3^{-2}r^{-3}}{s^0}$

14. $\dfrac{2^3a^{-3}}{8^{-1}b^{-5}c^0}$

15. $\dfrac{3^5}{3^3}$

16. $\dfrac{(-2)^7}{(-2)^5}$

17. $(-5)^3 \bullet (-5)^3$

18. $(q^5)^3$

19. $(a^{-4})^2$

20. $\dfrac{c^4 \bullet c^3}{c^6}$

21. $(-4d)^4$

22. $(-3f)^{-3}$

23. $\left(\dfrac{4}{x}\right)^{-3}$

24. A rectangular prism has length x, width $\dfrac{x}{2}$, and height $\dfrac{x}{3}$. Which of the expressions represent the volume of the prism? Select all that apply.

A. $6^{-1}x^3$

B. $6^{-1}x^{-3}$

C. $(6x^{-3})^{-1}$

D. $2^{-1} \bullet 3^{-1} \bullet x^3$

Name____ Date__________

6.2 Radicals and Rational Exponents
For use with Exploration 6.2

Essential Question How can you write and evaluate an *n*th root of a number?

Recall that you cube a number as follows.

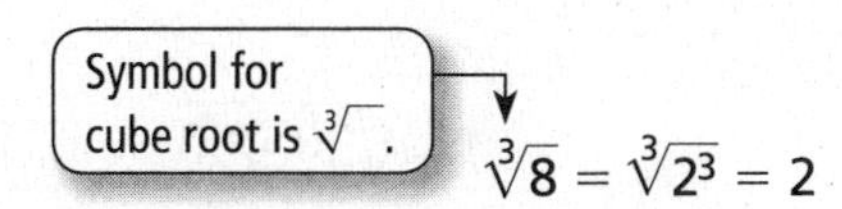

$2^3 = 2 \cdot 2 \cdot 2 = 8$ 2 cubed is 8.

To "undo" cubing a number, take the cube root of the number.

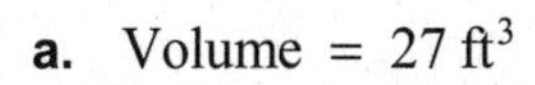

$\sqrt[3]{8} = \sqrt[3]{2^3} = 2$ The cube root of 8 is 2.

1 EXPLORATION: Finding Cube Roots

Work with a partner. Use a cube root symbol to write the side length of each cube. Then find the cube root. Check your answers by multiplying. Which cube is the largest? Which two cubes are the same size? Explain your reasoning.

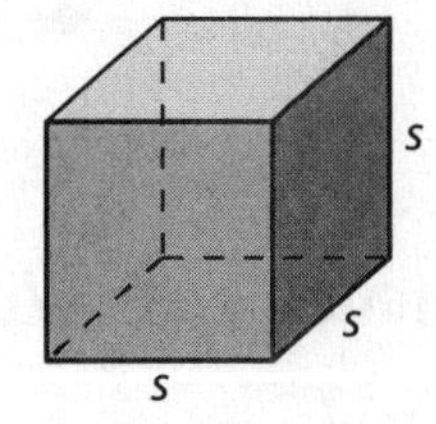

a. Volume $= 27\ \text{ft}^3$ **b.** Volume $= 125\ \text{cm}^3$ **c.** Volume $= 3375\ \text{in.}^3$

d. Volume $= 3.375\ \text{m}^3$ **e.** Volume $= 1\ \text{yd}^3$ **f.** Volume $= \frac{125}{8}\ \text{mm}^3$

2 EXPLORATION: Estimating *n*th Roots

Work with a partner. Estimate each positive *n*th root. Then match each *n*th root with the point on the number line. Justify your answers.

a. $\sqrt[4]{25}$ **b.** $\sqrt{0.5}$ **c.** $\sqrt[5]{2.5}$

d. $\sqrt[3]{65}$ **e.** $\sqrt[3]{55}$ **f.** $\sqrt[6]{20,000}$

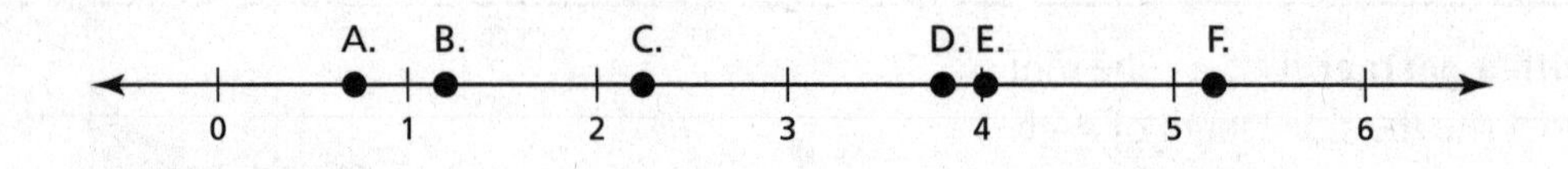

Communicate Your Answer

3. How can you write and evaluate an *n*th root of a number?

4. The body mass *m* (in kilograms) of a dinosaur that walked on two feet can be modeled by

$$m = (0.00016)C^{2.73}$$

where *C* is the circumference (in millimeters) of the dinosaur's femur. The mass of a *Tyrannosaurus rex* was 4000 kilograms. Use a calculator to approximate the circumference of its femur.

Name______________________________ Date__________

6.2 Notetaking with Vocabulary
For use after Lesson 6.2

In your own words, write the meaning of each vocabulary term.

*n*th root of *a*

radical

index of a radical

Core Concepts

Real *n*th Roots of *a*

Let n be an integer greater than 1, and let a be a real number.

- If n is odd, then a has one real nth root: $\sqrt[n]{a} = a^{1/n}$
- If n is even and $a > 0$, then a has two real nth roots: $\pm\sqrt[n]{a} = \pm a^{1/n}$
- If n is even and $a = 0$, then a has one real nth root: $\sqrt[n]{0} = 0$
- If n is even and $a < 0$, then a has no real nth roots.

Notes:

Name  Date ________

6.2 Notetaking with Vocabulary (continued)

Rational Exponents

Let $a^{1/n}$ be an nth root of a, and let m be a positive integer.

Algebra $a^{m/n} = \left(a^{1/n}\right)^m = \left(\sqrt[n]{a}\right)^m$

Numbers $27^{2/3} = \left(27^{1/3}\right)^2 = \left(\sqrt[3]{27}\right)^2$

Notes:

Extra Practice

In Exercises 1–6, find the indicated real *n*th root(s) of *a*.

1. $n = 2, a = 64$

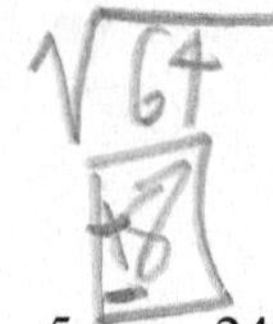

2. $n = 3, a = 27$

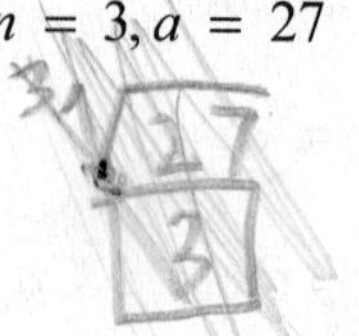

3. $n = 4, a = 256$

4. $n = 5, a = 243$

5. $n = 8, a = 256$

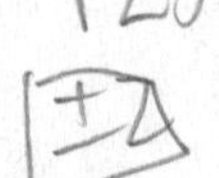

6. $n = 4, a = 10{,}000$

In Exercises 7–12, evaluate the expression.

7. $\sqrt[4]{625}$

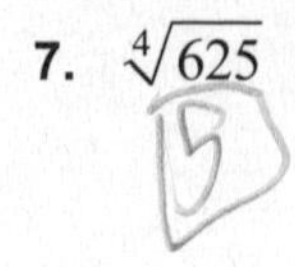

8. $\sqrt[3]{-512}$

9. $\sqrt[3]{-216}$

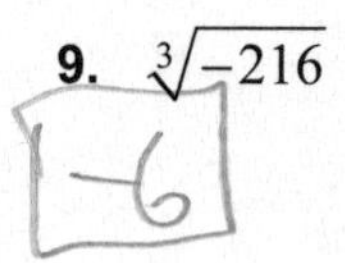

10. $\sqrt[5]{-243}$

11. $729^{1/6}$

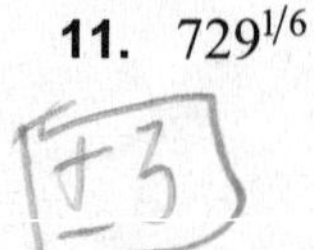

12. $(-81)^{1/2}$

Name Anit V. Date

6.2 Notetaking with Vocabulary (continued)

In Exercises 13–15, rewrite the expression in rational exponent form.

13. $\left(\sqrt[5]{4}\right)^3$ **14.** $\left(\sqrt[3]{-8}\right)^2$ **15.** $\left(\sqrt[4]{15}\right)^7$

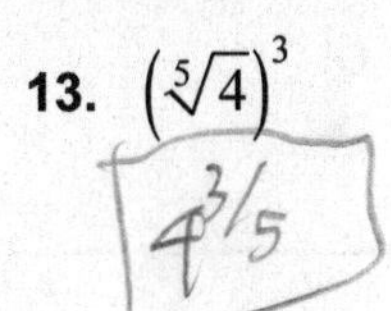

In Exercises 16–18, rewrite the expression in radical form.

16. $(-3)^{2/5}$ **17.** $6^{3/2}$ **18.** $12^{3/4}$

In Exercises 19–24, evaluate the expression.

19. $32^{2/5}$ **20.** $(-64)^{3/2}$ **21.** $343^{2/3}$

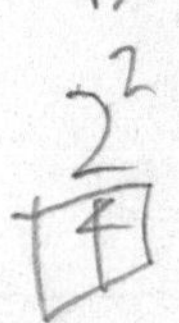

22. $256^{7/8}$ **23.** $-729^{5/6}$ **24.** $(-625)^{3/4}$

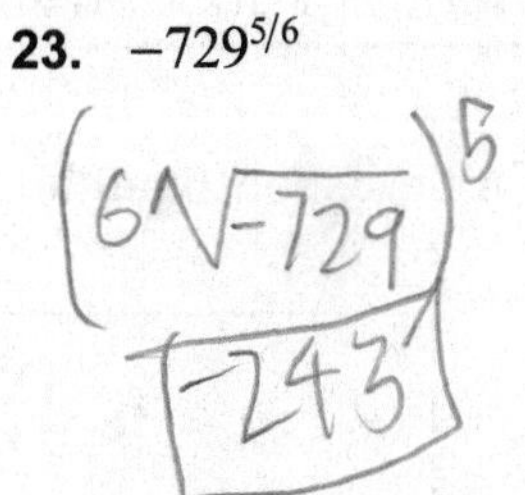

25. The radius r of a sphere is given by the equation

$$r = \left(\frac{A}{4\pi}\right)^{1/2}$$

where A is the surface area of the sphere. The surface area of a sphere is 1493 square meters. Find the radius of the sphere to the nearest tenth of a meter. Use 3.14 for π.

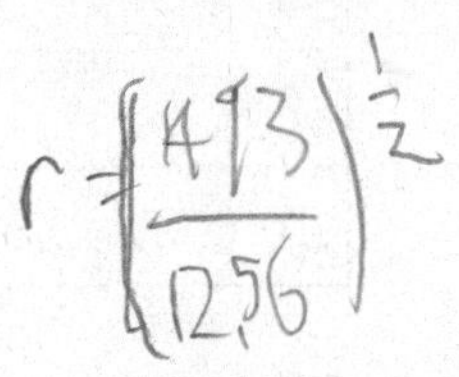

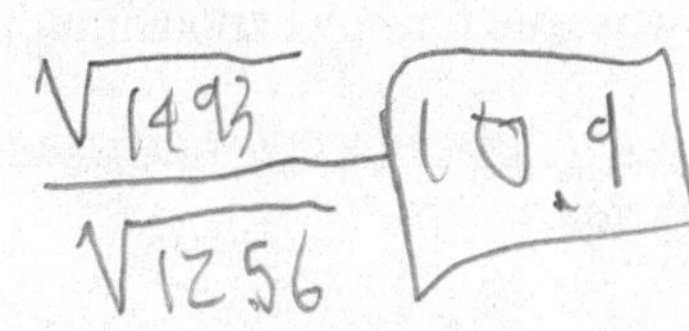

Name ______________________________ Date __________

6.3 Exponential Functions

For use with Exploration 6.3

Essential Question What are some of the characteristics of the graph of an exponential function?

1 EXPLORATION: Exploring an Exponential Function

Work with a partner. Complete each table for the *exponential function* $y = 16(2)^x$. In each table, what do you notice about the values of x? What do you notice about the values of y?

x	$y = 16(2)^x$
0	
1	
2	
3	
4	
5	

x	$y = 16(2)^x$
0	
2	
4	
6	
8	
10	

2 EXPLORATION: Exploring an Exponential Function

Work with a partner. Repeat Exploration 1 for the exponential function $y = 16\left(\frac{1}{2}\right)^x$.

x	$y = 16\left(\frac{1}{2}\right)^x$
0	
1	
2	
3	
4	
5	

x	$y = 16\left(\frac{1}{2}\right)^x$
0	
2	
4	
6	
8	
10	

Do you think the statement below is true for *any* exponential function? Justify your answer.

"*As the independent variable x changes by a constant amount, the dependent variable y is multiplied by a constant factor.*"

6.3 Exponential Functions (continued)

3 EXPLORATION: Graphing Exponential Functions

Go to *BigIdeasMath.com* for an interactive tool to investigate this exploration.

Work with a partner. Sketch the graphs of the functions given in Explorations 1 and 2. How are the graphs similar? How are they different?

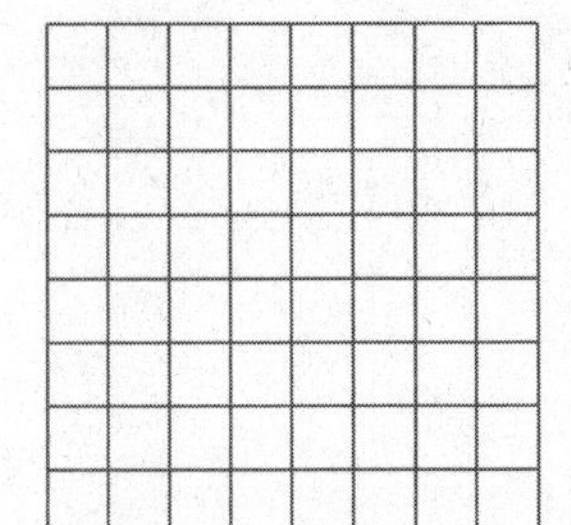

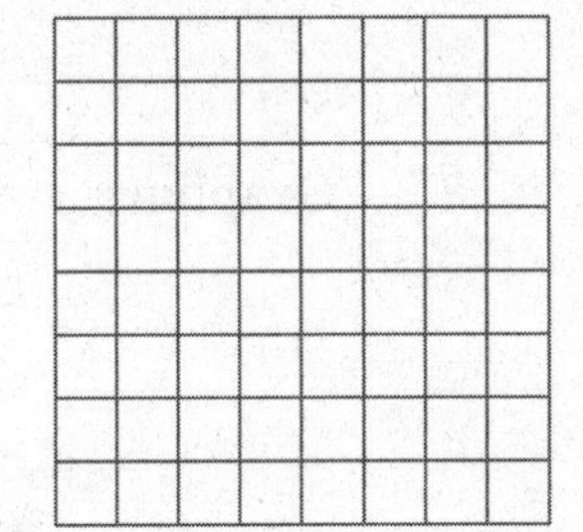

Communicate Your Answer

4. What are some of the characteristics of the graph of an exponential function?

5. Sketch the graph of each exponential function. Does each graph have the characteristics you described in Question 4? Explain your reasoning.

a. $y = 2^x$

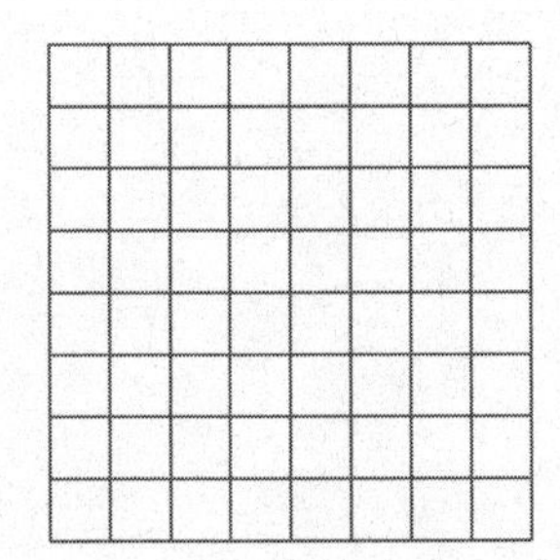

b. $y = 2(3)^x$

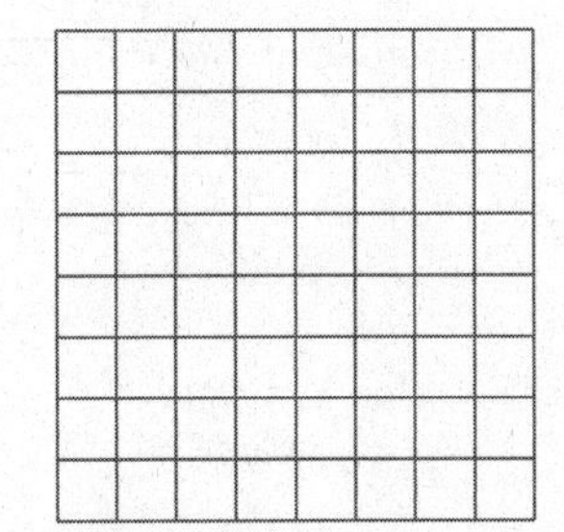

c. $y = 3(1.5)^x$

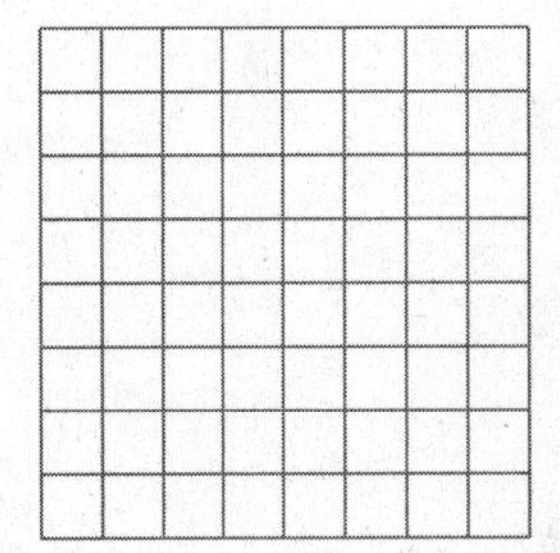

d. $y = \left(\frac{1}{2}\right)^x$

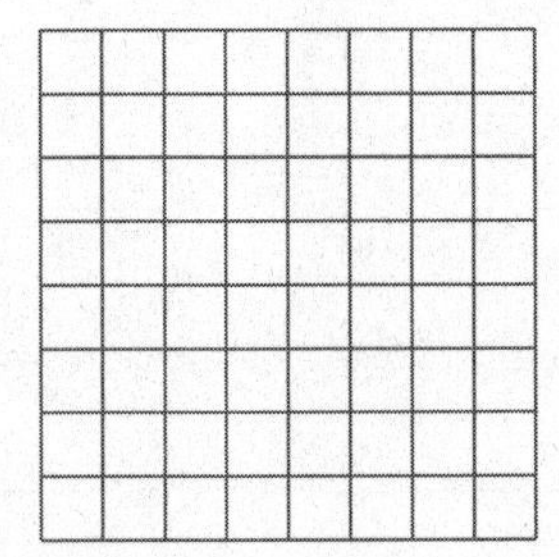

e. $y = 3\left(\frac{1}{2}\right)^x$

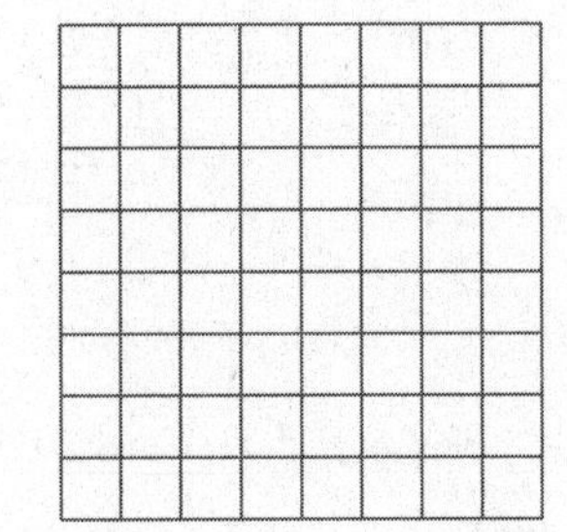

f. $y = 2\left(\frac{3}{4}\right)^x$

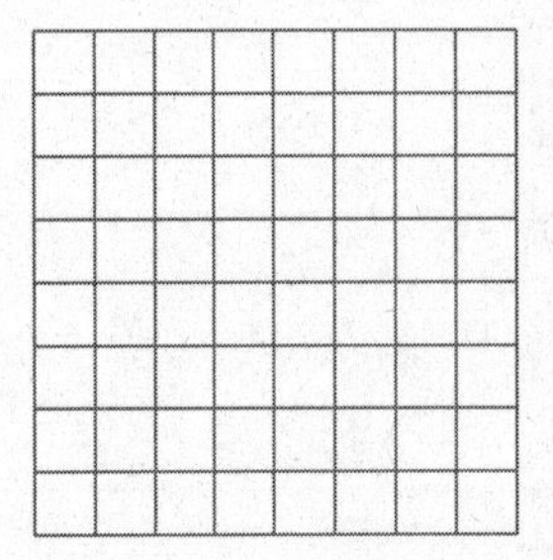

Name ______________________________ Date __________

6.3 Notetaking with Vocabulary

For use after Lesson 6.3

In your own words, write the meaning of each vocabulary term.

exponential function

Core Concepts

Graphing $y = ab^x$ When $b > 1$

y
$a > 0$
$(0, a)$
$(0, a)$
x
$a < 0$

Graphing $y = ab^x$ When $0 < b < 1$

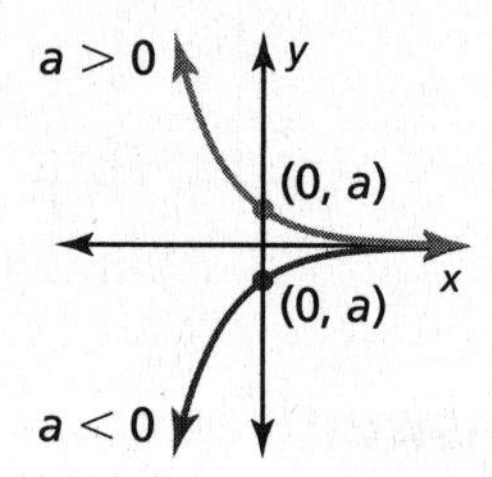

Notes:

Extra Practice

In Exercises 1–4, determine whether the table represents a *linear* or an *exponential* function. Explain.

1.

x	y
1	8
2	4
3	2
4	1

2.

x	y
1	3
2	7
3	11
4	15

3.

x	y
−1	12
0	9
1	6
2	3

4.

x	y
−1	0.125
0	0.5
1	2
2	8

In Exercises 5–7, evaluate the function for the given value of *x*.

5. $y = 3^x; x = 5$

6. $y = \left(\frac{1}{4}\right)^x; x = 3$

7. $y = 3(4)^x; x = 4$

In Exercises 8 and 9, graph the function. Compare the graph to the graph of the parent function. Describe the domain and range of *f*.

8. $f(x) = -2^x$

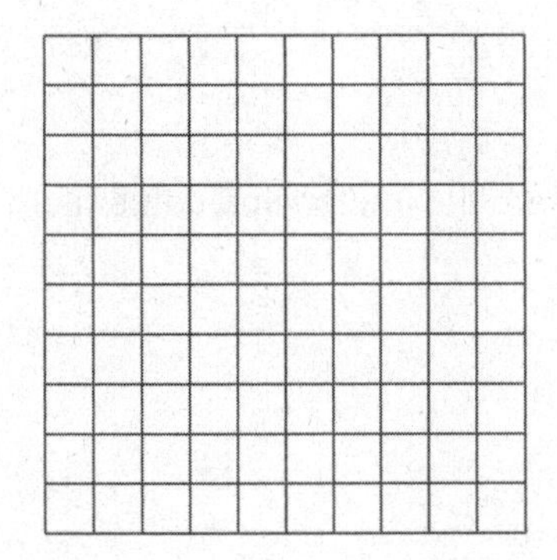

9. $f(x) = 2\left(\frac{1}{4}\right)^x$

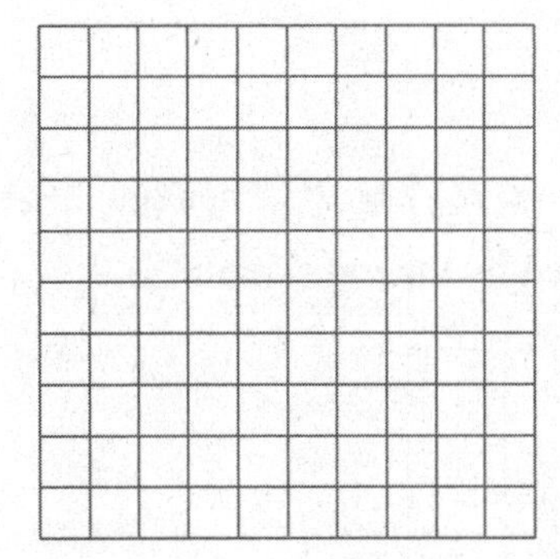

Name ________________________________ Date __________

6.3 Notetaking with Vocabulary (continued)

In Exercises 10 and 11, graph the function. Describe the domain and range.

10. $f(x) = 4^x - 2$

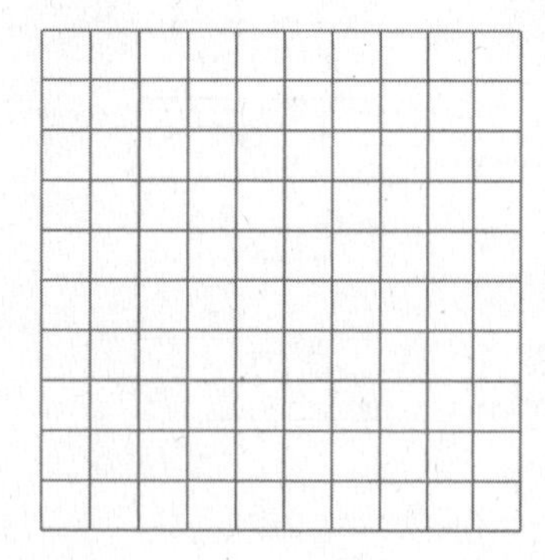

11. $f(x) = 4\left(\frac{1}{2}\right)^{x+1}$

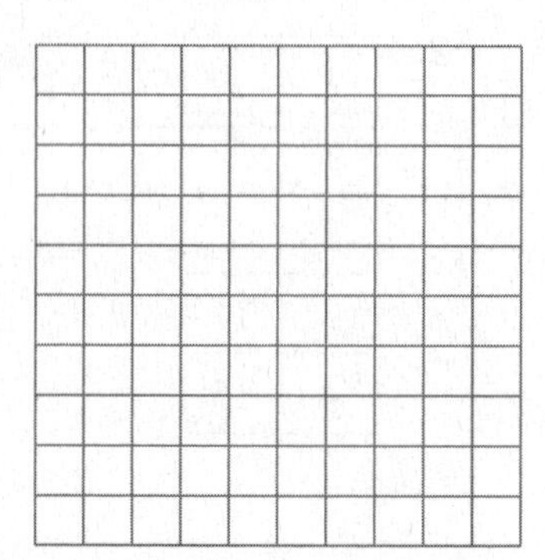

In Exercises 12 and 13, write an exponential function represented by the table or graph.

12.

x	0	1	2	3
f(x)	3	18	108	648

13.

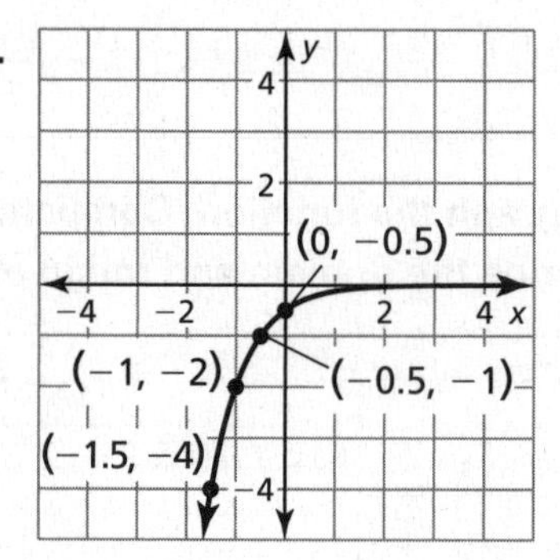

14. Graph the function $f(x) = 2^x$. Then graph $g(x) = 2^x + 3$. How are the y-intercept, domain, and range affected by the translation?

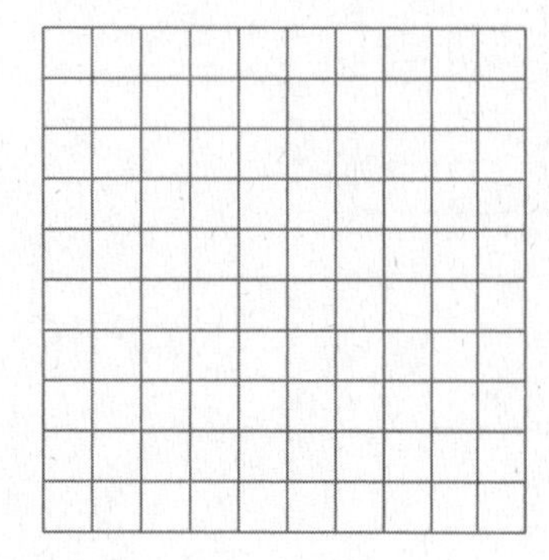

Name______________________________ Date__________

6.4 Exponential Growth and Decay

For use with Exploration 6.4

Essential Question What are some of the characteristics of exponential growth and exponential decay functions?

1 EXPLORATION: Predicting a Future Event

Work with a partner. It is estimated, that in 1782, there were about 100,000 nesting pairs of bald eagles in the United States. By the 1960s, this number had dropped to about 500 nesting pairs. In 1967, the bald eagle was declared an endangered species in the United States. With protection, the nesting pair population began to increase. Finally, in 2007, the bald eagle was removed from the list of endangered and threatened species.

Describe the pattern shown in the graph. Is it exponential growth? Assume the pattern continues. When will the population return to that of the late 1700s? Explain your reasoning.

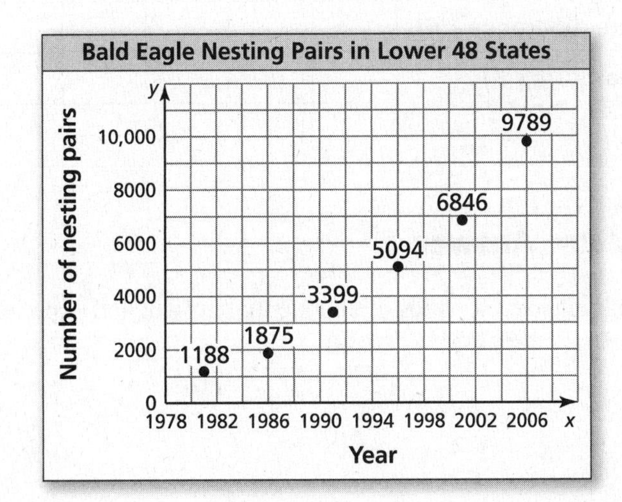

Name ____ Date __________

2 EXPLORATION: Describing a Decay Pattern

Work with a partner. A forensic pathologist was called to estimate the time of death of a person. At midnight, the body temperature was 80.5°F and the room temperature was a constant 60°F. One hour later, the body temperature was 78.5°F.

a. By what percent did the difference between the body temperature and the room temperature drop during the hour?

b. Assume that the original body temperature was 98.6°F. Use the percent decrease found in part (a) to make a table showing the decreases in body temperature. Use the table to estimate the time of death.

Time (h)								
Temperature difference (°F)								
Body temperature (°F)								

Communicate Your Answer

3. What are some of the characteristics of exponential growth and exponential decay functions?

4. Use the Internet or some other reference to find an example of each type of function. Your examples should be different than those given in Explorations 1 and 2.

a. exponential growth

b. exponential decay

Name____ Date__________

6.4 Notetaking with Vocabulary
For use after Lesson 6.4

In your own words, write the meaning of each vocabulary term.

exponential growth

exponential growth function

exponential decay

exponential decay function

compound interest

Core Concepts

Exponential Growth Functions

A function of the form $y = a(1 + r)^t$, where $a > 0$ and $r > 0$, is an **exponential growth function**.

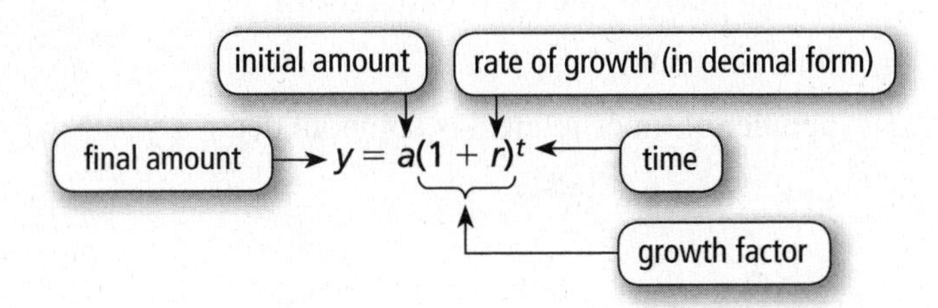

Notes:

Exponential Decay Functions

A function of the form $y = a(1 - r)^t$, where $a > 0$ and $0 < r < 1$, is an **exponential decay function**.

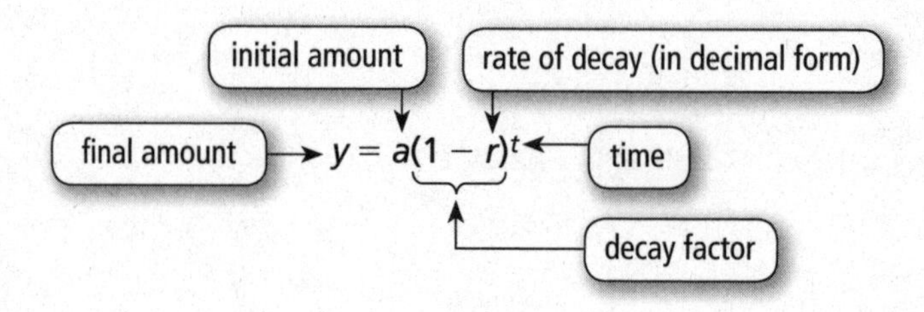

Notes:

Compound Interest

Compound interest is the interest earned on the principal *and* on previously earned interest. The balance *y* of an account earning compound interest is

$$y = P\left(1 + \frac{r}{n}\right)^{nt}.$$

P = principal (initial amount)

r = annual interest rate (in decimal form)

t = time (in years)

n = number of times interest is compounded per year

Notes:

Name______________________________ Date__________

Extra Practice

1. In 2005, there were 100 rabbits in Polygon Park. The population increased by 11% each year.

a. Write an exponential growth function that represents the population t years after 2005.

b. What will the population be in 2025? Round your answer to the nearest whole number.

In Exercises 2–5, determine whether the table represents an *exponential growth function*, an *exponential decay function*, or *neither*. Explain.

2.

x	y
0	20
1	30
2	45
3	67.5

3.

x	y
–1	160
0	40
1	10
2	2.5

4.

x	y
1	32
2	22
3	12
4	2

5.

x	y
–1	4
0	10
1	25
2	62.5

In Exercises 6–8, determine whether each function represents *exponential growth* or *exponential decay*. Identify the percent rate of change.

6. $y = 4(0.95)^t$

7. $y = 500(1.08)^t$

8. $w(t) = \left(\frac{3}{4}\right)^t$

In Exercises 9 and 10, write a function that represents the balance after t years.

9. \$3000 deposit that earns 6% annual interest compounded quarterly.

10. \$5000 deposit that earns 7.2% annual interest compounded monthly.

Name ______________________________ Date __________

6.5 Solving Exponential Equations

For use with Exploration 6.5

Essential Question How can you solve an exponential equation graphically?

1 EXPLORATION: Solving an Exponential Equation Graphically

Go to *BigIdeasMath.com* for an interactive tool to investigate this exploration.

Work with a partner. Use a graphing calculator to solve the exponential equation $2.5^{x-3} = 6.25$ graphically. Describe your process and explain how you determined the solution.

2 EXPLORATION: The Number of Solutions of an Exponential Equation

Go to *BigIdeasMath.com* for an interactive tool to investigate this exploration.

Work with a partner.

a. Use a graphing calculator to graph the equation $y = 2^x$.

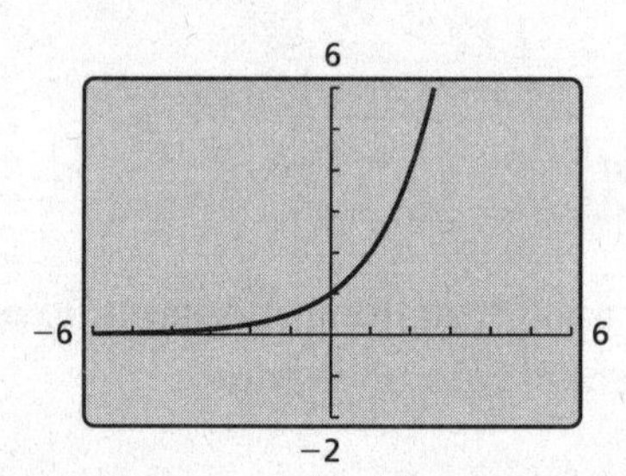

b. In the same viewing window, graph a linear equation (if possible) that does not intersect the graph of $y = 2^x$.

c. In the same viewing window, graph a linear equation (if possible) that intersects the graph of $y = 2^x$ in more than one point.

d. Is it possible for an exponential equation to have no solution? more than one solution? Explain your reasoning.

Name_____________________________ Date__________

3 EXPLORATION: Solving Exponential Equations Graphically

Go to *BigIdeasMath.com* for an interactive tool to investigate this exploration.

Work with a partner. Use a graphing calculator to solve each equation.

a. $2^x = \frac{1}{2}$ **b.** $2^{x+1} = 0$ **c.** $2^x = \sqrt{2}$

d. $3^x = 9$ **e.** $3^{x-1} = 0$ **f.** $4^{2x} = 2$

g. $2^{x/2} = \frac{1}{4}$ **h.** $3^{x+2} = \frac{1}{9}$ **i.** $2^{x-2} = \frac{3}{2}x - 2$

Communicate Your Answer

4. How can you solve an exponential equation graphically?

5. A population of 30 mice is expected to double each year. The number p of mice in the population each year is given by $p = 30(2^n)$. In how many years will there be 960 mice in the population?

Name ______________________________ Date __________

6.5 Notetaking with Vocabulary

For use after Lesson 6.5

In your own words, write the meaning of each vocabulary term.

exponential equation

Core Concepts

Property of Equality for Exponential Equations

Words Two powers with the *same positive base b*, where $b \neq 1$, are equal if and only if their exponents are equal.

Numbers If $2^x = 2^5$, then $x = 5$. If $x = 5$, then $2^x = 2^5$.

Algebra If $b > 0$ and $b \neq 1$, then $b^x = b^y$ if and only if $x = y$.

Notes:

Name______________________________ Date__________

Extra Practice

In Exercises 1–15, solve the equation. Check your solution.

1. $3^{4x} = 3^{12}$

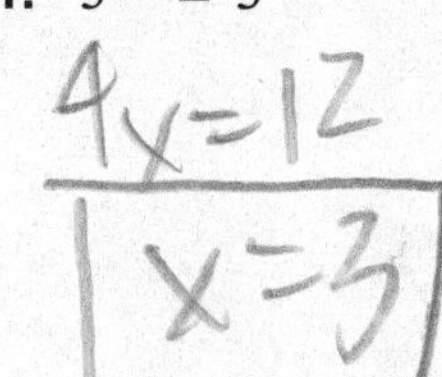

2. $8^{x+5} = 8^{20}$

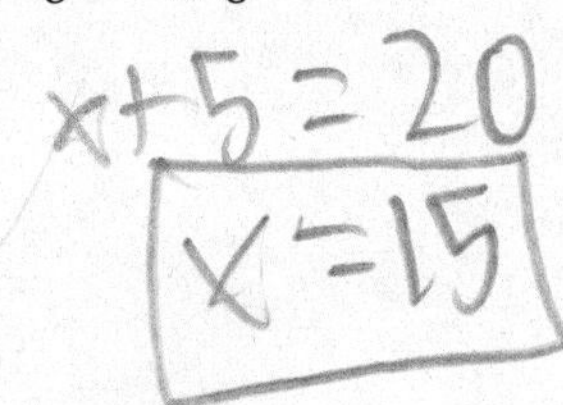

3. $6^{4x-5} = 6^{2x}$

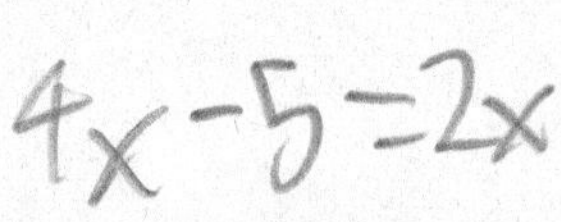

4. $5^{6x-3} = 5^{-3+4x}$

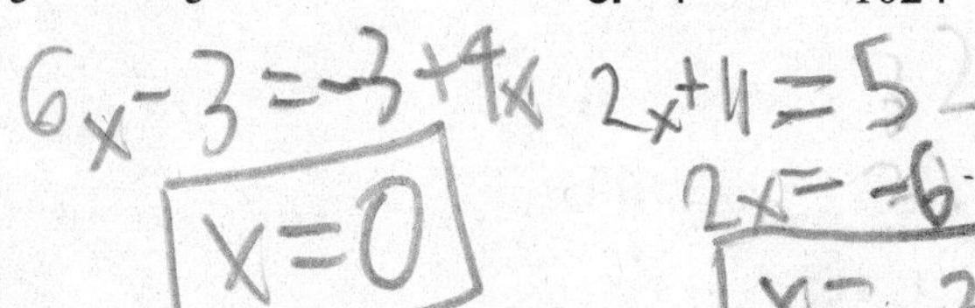

5. $4^{2x+11} = 1024$

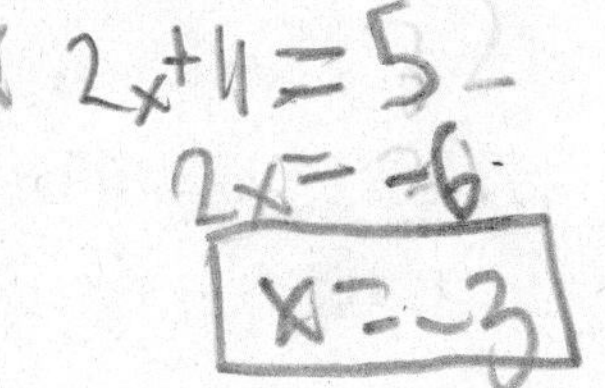

6. $8^{3-2x} = 512$

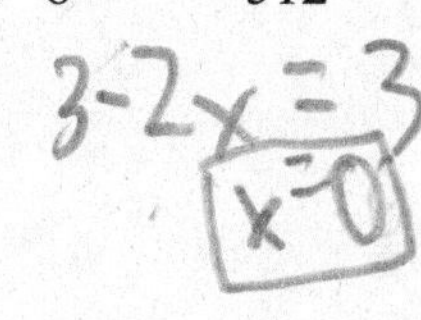

7. $4^{7-x} = 256$

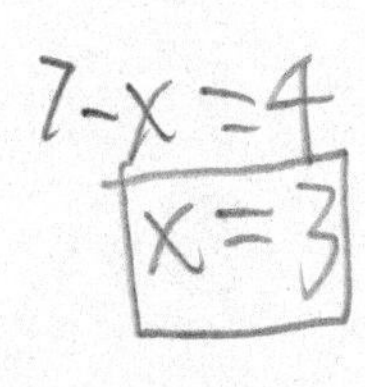

8. $49^{x-2} = 343$

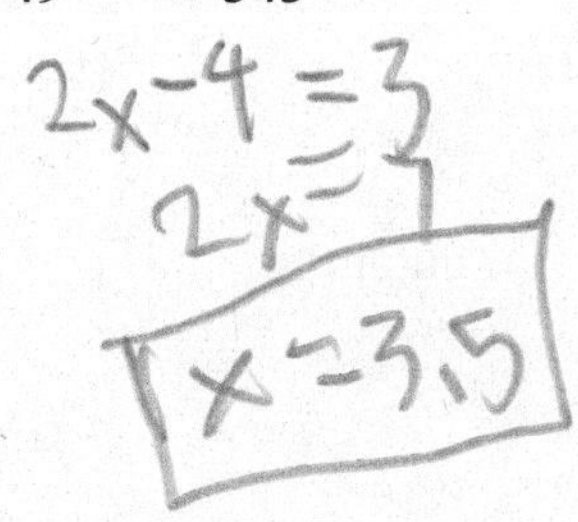

9. $36^{6x-1} = 6^{5x}$

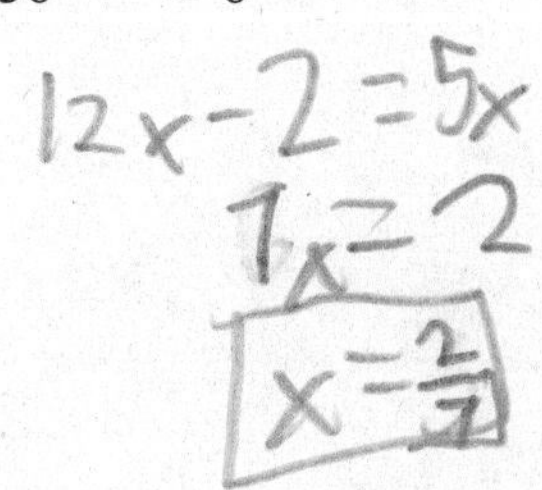

10. $9^{x-4} = 81^{3x}$

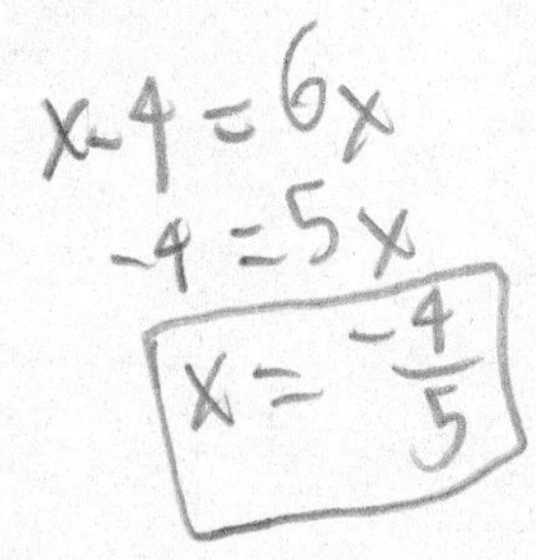

11. $64^{x+1} = 512^{x}$

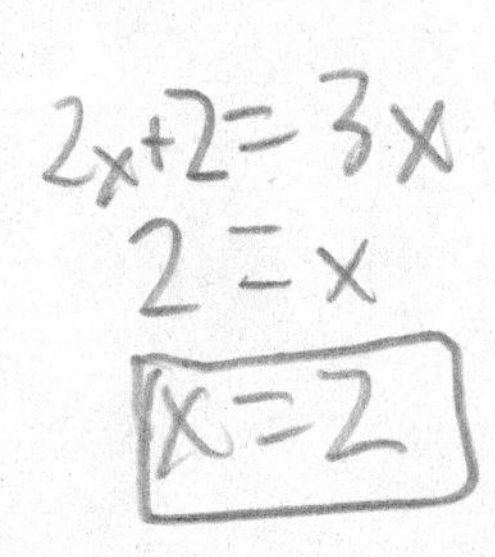

12. $6^{2x} = 36^{2x+1}$

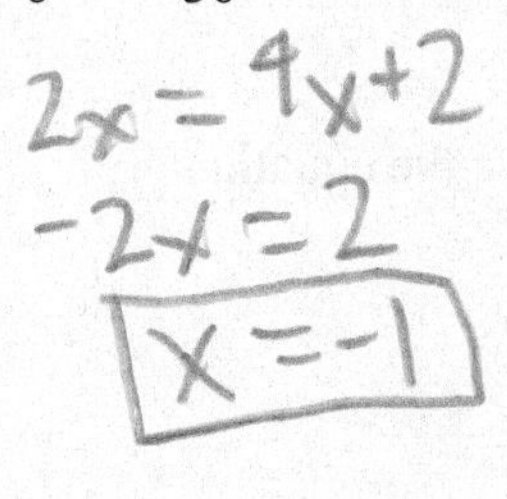

Name ______________________________ Date ________

6.5 **Notetaking with Vocabulary (continued)**

13. $\left(\frac{1}{7}\right)^{x} = 2401$

$-x = 4$

$x = -4$

14. $\frac{1}{512} = 2^{3x-1}$

$-9 = 3x - 1$

$3x = -8$

$x = -\frac{8}{3}$

15. $25^{2-2x} = \left(\frac{1}{625}\right)^{x+1}$

$4 - 4x = -4x - 4$

No solution

In Exercises 16–21, use a graphing calculator to solve the equation.

16. $3^{x+3} = -9$

No solution

17. $\left(\frac{1}{4}\right)^{-x-1} = 18$

$x \approx 1.08$

18. $3^{x} = -2^{-x+1}$

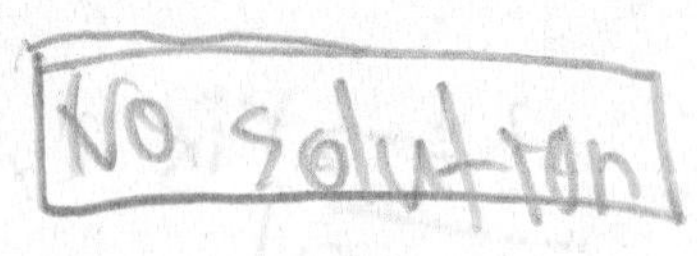

No solution

19. $2^{x+2} = 5^{x-3}$

$x \approx 6.78$

20. $7^{-x+1} = -4^{x-1}$

No solution

21. $\frac{1}{4}x + 1 = \left(\frac{2}{3}\right)^{2x-1}$

$x \approx 0.39$

22. You deposit \$1000 in a savings account that earns 5% annual interest compounded yearly.

a. Write an exponential equation to determine when the balance of the account will be \$1500.

b. Solve the equation.

Name______________________________ Date__________

6.6 Geometric Sequences

For use with Exploration 6.6

Essential Question How can you use a geometric sequence to describe a pattern?

In a **geometric sequence**, the ratio between each pair of consecutive terms is the same. This ratio is called the **common ratio**.

1 EXPLORATION: Describing Calculator Patterns

Work with a partner. Enter the keystrokes on a calculator and record the results in the table. Describe the pattern.

a. Step 1 [2] [=]

Step 2 [×] [2] [=]

Step 3 [×] [2] [=]

Step 4 [×] [2] [=]

Step 5 [×] [2] [=]

Step	1	2	3	4	5
Calculator display					

b. Step 1 [6] [4] [=]

Step 2 [×] [.] [5] [=]

Step 3 [×] [.] [5] [=]

Step 4 [×] [.] [5] [=]

Step 5 [×] [.] [5] [=]

Step	1	2	3	4	5
Calculator display					

c. Use a calculator to make your own sequence. Start with any number and multiply by 3 each time. Record your results in the table.

Step	1	2	3	4	5
Calculator display					

d. Part (a) involves a geometric sequence with a common ratio of 2. What is the common ratio in part (b)? part (c)?

2 EXPLORATION: Folding a Sheet of Paper

Work with a partner. A sheet of paper is about 0.1 millimeter thick.

a. How thick will it be when you fold it in half once? twice? three times?

b. What is the greatest number of times you can fold a piece of paper in half? How thick is the result?

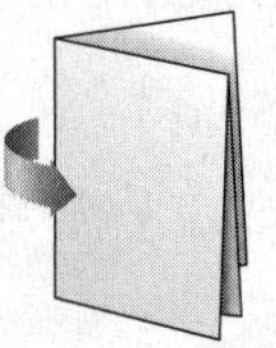

c. Do you agree with the statement below? Explain your reasoning.

"If it were possible to fold the paper in half 15 times, it would be taller than you."

Communicate Your Answer

3. How can you use a geometric sequence to describe a pattern?

4. Give an example of a geometric sequence from real life other than paper folding.

Name____ Date__________

6.6 Notetaking with Vocabulary

For use after Lesson 6.6

In your own words, write the meaning of each vocabulary term.

geometric sequence

common ratio

Core Concepts

Geometric Sequence

In a **geometric sequence**, the ratio between each pair of consecutive terms is the same. This ratio is called the **common ratio**. Each term is found by multiplying the previous term by the common ratio.

1, 5, 25, 125, . . . Terms of a geometric sequence

$\times 5$ $\times 5$ $\times 5$ ← common ratio

Notes:

Equation for a Geometric Sequence

Let a_n be the nth term of a geometric sequence with first term a_1 and common ratio r. The nth term is given by

$$a_n = a_1 r^{n-1}.$$

Notes:

Name ______________________________ Date __________

6.6 **Notetaking with Vocabulary (continued)**

Extra Practice

In Exercises 1–6, determine whether the sequence is *arithmetic*, *geometric*, or *neither*. Explain your reasoning.

1. $1, -4, 16, -64, \ldots$

2. $3, 7, 11, 15, \ldots$

3. $2, 4, 8, 32, \ldots$

4. $12, 9, 7, 5, \ldots$

5. $6, 18, 54, 162, \ldots$

6. $11, 19, 27, 35, \ldots$

In Exercises 7–9, write the next three terms of the geometric sequence.

7. $7, 21, 63, 189, \ldots$

8. $576, 288, 144, 72, \ldots$

9. $5, -10, 20, -40, \ldots$

In Exercises 10–12, write the next three terms of the geometric sequence. Then graph the sequence.

10. $12, 6, 3, \frac{3}{2}, \ldots$

11. $3, 12, 48, 192, \ldots$

12. $0.008, 0.04, 0.2, 1, \ldots$

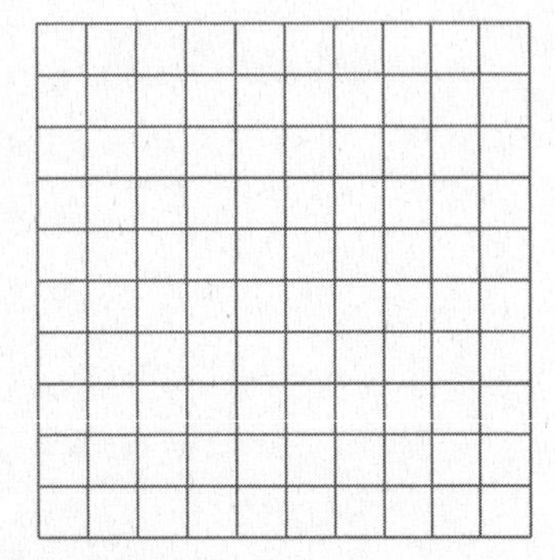

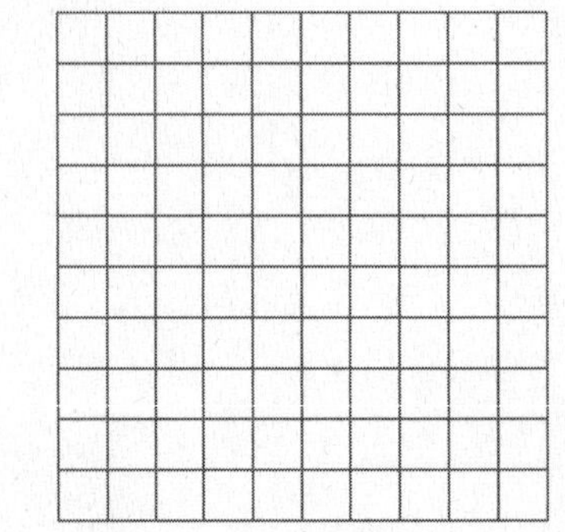

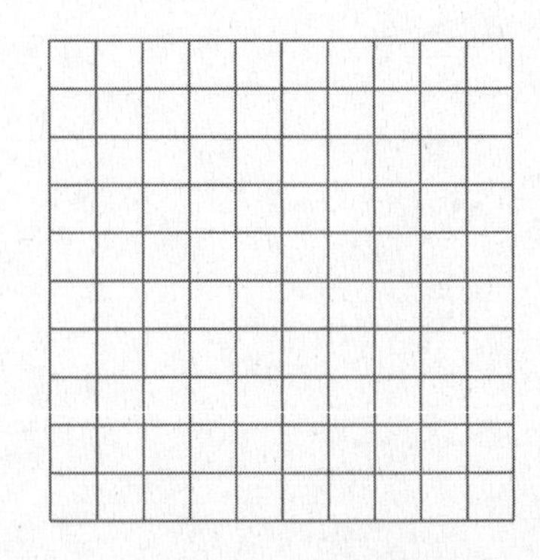

6.6 Notetaking with Vocabulary (continued)

In Exercises 13–20, write an equation for the *n*th term of the geometric sequence. Then find a_6.

13. 6561, 2187, 729, 243, ...

14. 8, −24, 72, −216, ...

15. 3, 15, 75, 375, ...

16.

n	1	2	3	4
a_n	2916	972	324	108

17.

n	1	2	3	4
a_n	11	44	176	704

18.

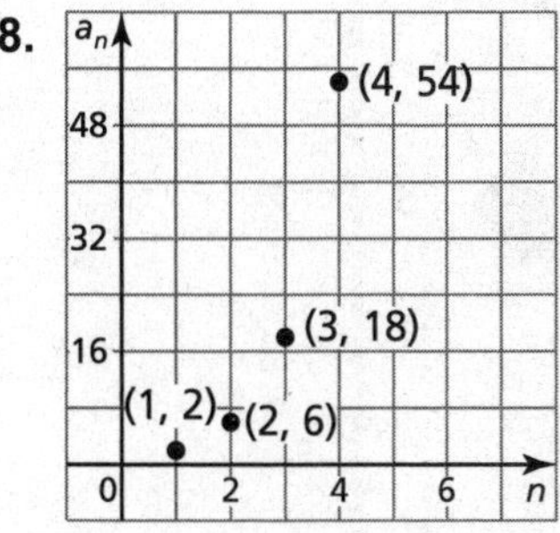

19.

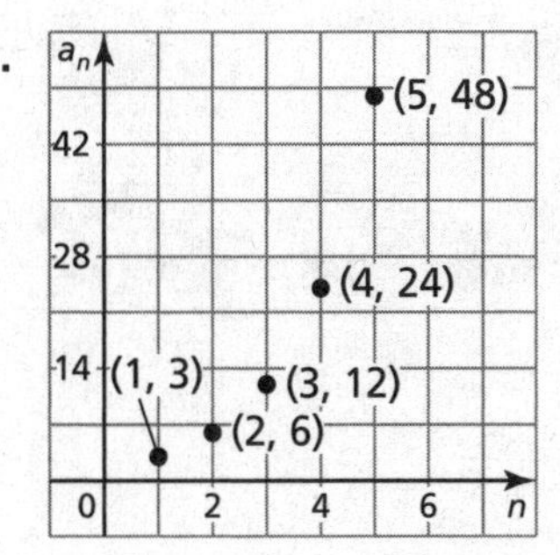

20.

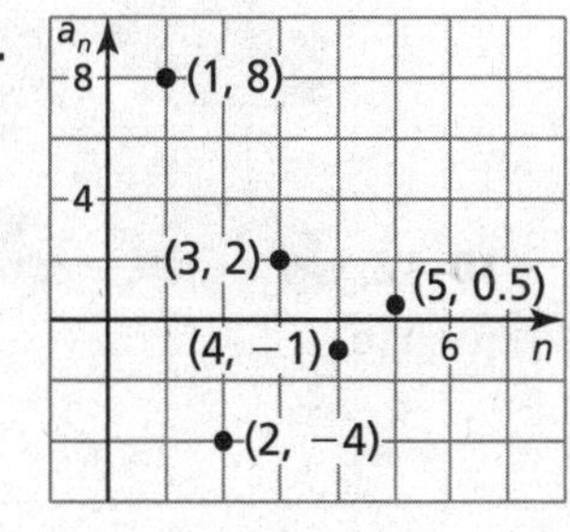

Name ____ Date __________

6.7 Recursively Defined Sequences

For use with Exploration 6.7

Essential Question How can you define a sequence recursively?

A **recursive rule** gives the beginning term(s) of a sequence and a *recursive equation* that tells how a_n is related to one or more preceding terms

1 EXPLORATION: Describing a Pattern

Work with a partner. Consider a hypothetical population of rabbits. Start with one breeding pair. After each month, each breeding pair produces another breeding pair. The total number of rabbits each month follows the exponential pattern 2, 4, 8, 16, 32, …. Now suppose that in the first month after each pair is born, the pair is too young to reproduce. Each pair produces another pair after it is 2 months old. Find the total number of pairs in months 6, 7, and 8.

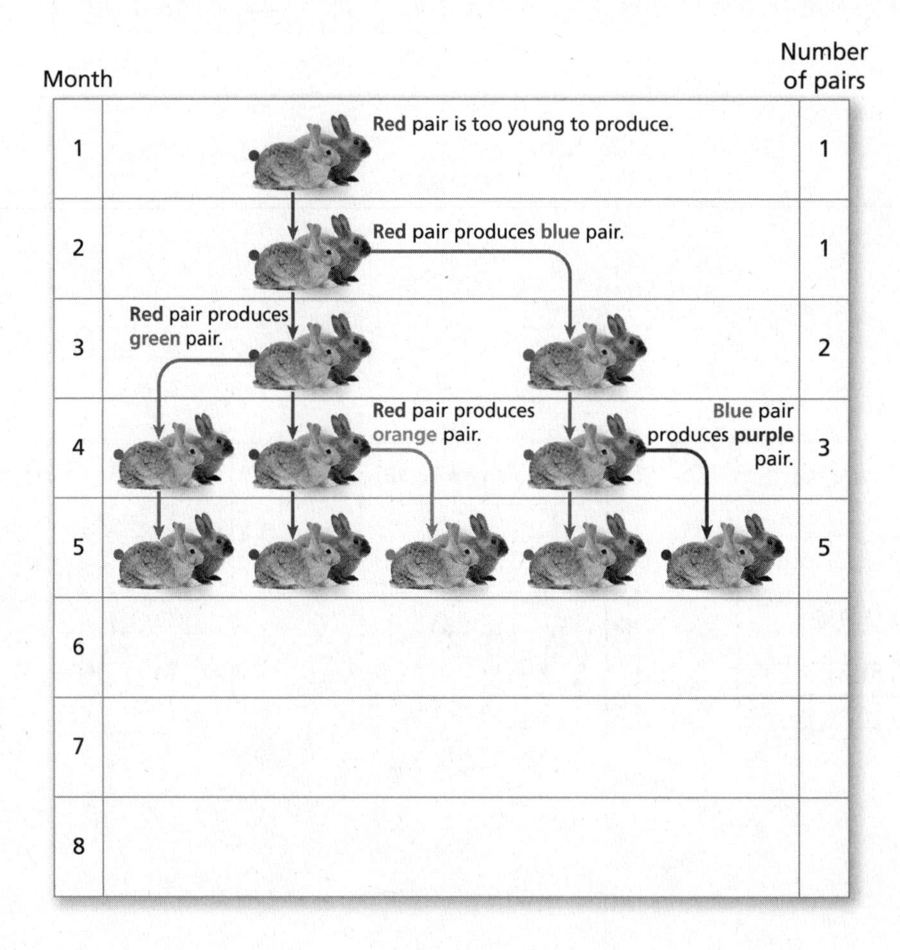

Name____ Date__________

2 EXPLORATION: Using a Recursive Equation

Work with a partner. Consider the following recursive equation.

$$a_n = a_{n-1} + a_{n-2}$$

Each term in the sequence is the sum of the two preceding terms.

Complete the table. Compare the results with the sequence of the number of pairs in Exploration 1.

a_1	a_2	a_3	a_4	a_5	a_6	a_7	a_8
1	1						

Communicate Your Answer

3. How can you define a sequence recursively?

4. Use the Internet or some other reference to determine the mathematician who first described the sequences in Explorations 1 and 2.

Name ______________________________ Date __________

6.7 Notetaking with Vocabulary

For use after Lesson 6.7

In your own words, write the meaning of each vocabulary term.

explicit rule

recursive rule

Core Concepts

Recursive Equation for an Arithmetic Sequence

$a_n = a_{n-1} + d$, where d is the common difference

Recursive Equation for a Geometric Sequence

$a_n = r \bullet a_{n-1}$, where r is the common ratio

Notes:

Name____ Date__________

Extra Practice

In Exercises 1–6, write the first six terms of the sequence. Then graph the sequence.

1. $a_1 = -2; a_n = -2a_{n-1}$ **2.** $a_1 = -4; a_n = a_{n-1} + 3$ **3.** $a_1 = 4; a_n = 1.5a_{n-1}$

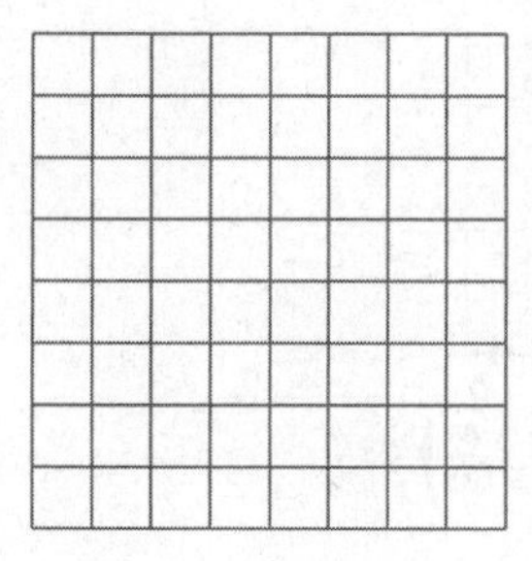

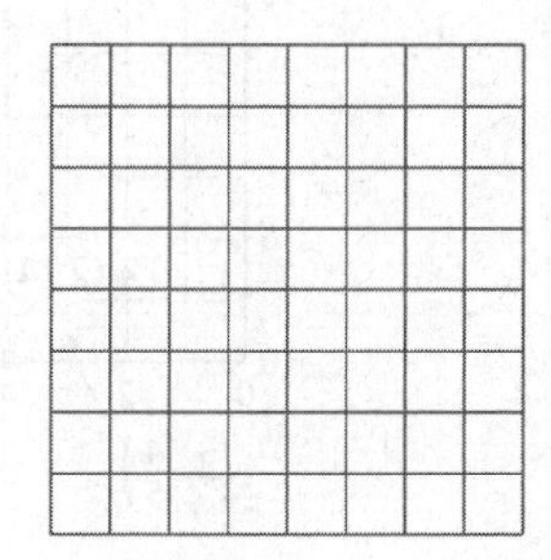

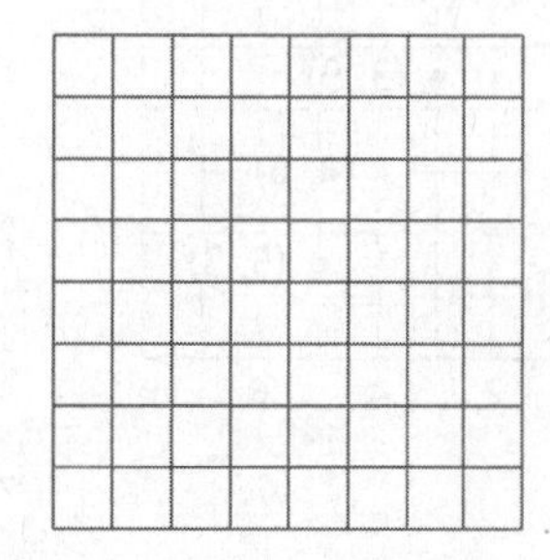

4. $a_1 = 14; a_n = a_{n-1} - 4$ **5.** $a_1 = -\frac{1}{2}; a_n = -2a_{n-1}$ **6.** $a_1 = -3; a_n = a_{n-1} + 2$

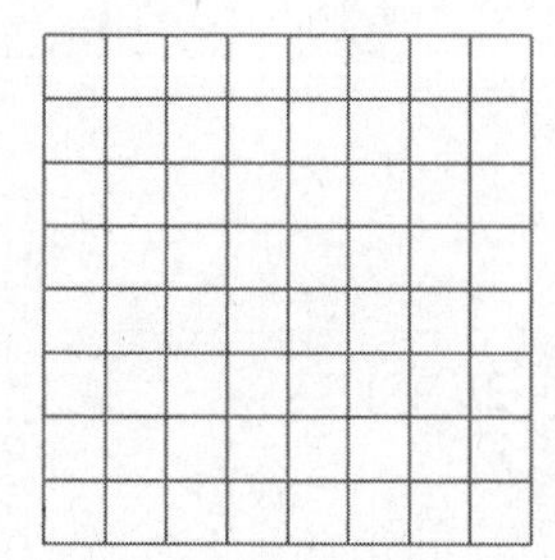

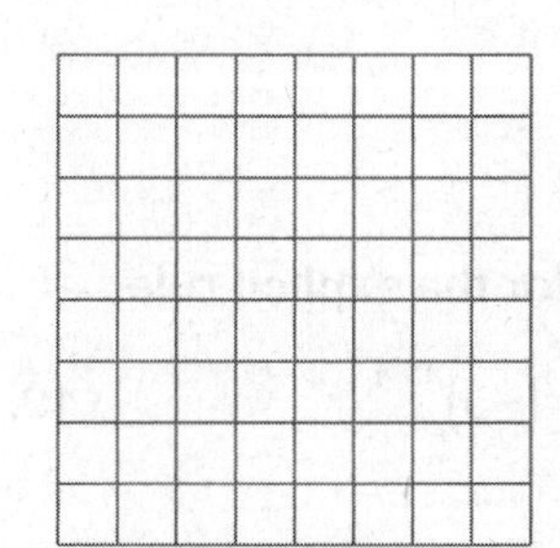

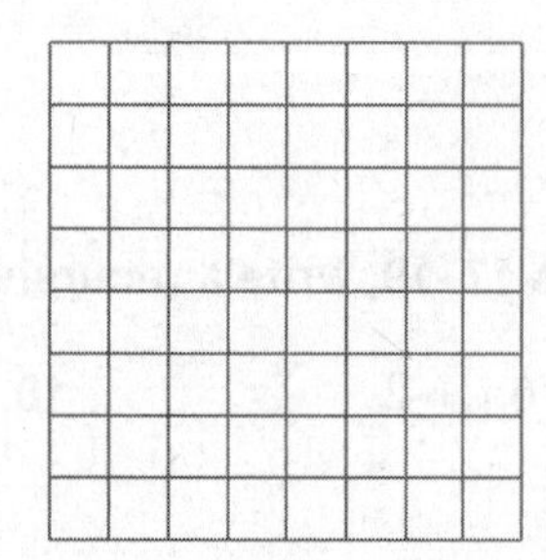

In Exercises 7 and 8, write a recursive rule for the sequence.

7.

n	1	2	3	4
a_n	324	108	36	12

8.

n	1	2	3	4
a_n	9	14	19	24

6.7 Notetaking with Vocabulary (continued)

In Exercises 9–13, write a recursive rule for the sequence.

9. $3125, 625, 125, 25, \ldots$ **10.** $8, -24, 72, -216, \ldots$ **11.** $7, 13, 19, 25, \ldots$

12.

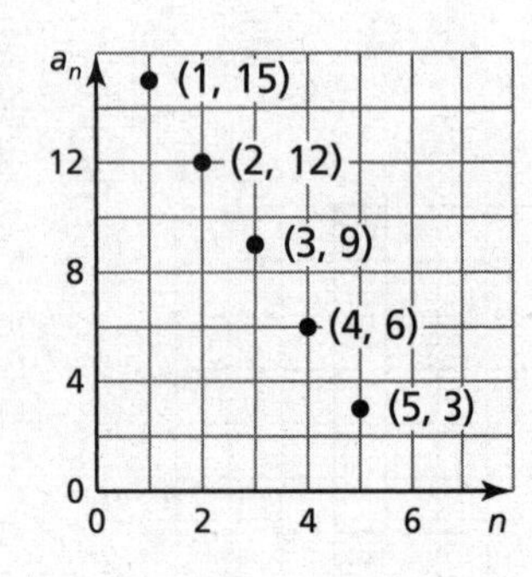

13.

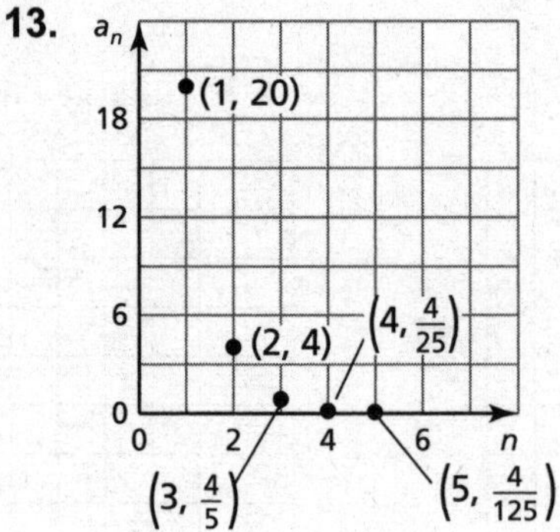

In Exercises 14–16, write an explicit rule for the recursive rule.

14. $a_1 = 4; a_n = 3a_{n-1}$ **15.** $a_1 = 6; a_n = a_{n-1} + 11$ **16.** $a_1 = -1; a_n = 5a_{n-1}$

In Exercises 17–19, write a recursive rule for the explicit rule.

17. $a_n = 6n + 2$ **18.** $a_n = (-3)^{n-1}$ **19.** $a_n = -2n + 1$

In Exercises 20–22, write a recursive rule for the sequence. Then write the next two terms of the sequence.

20. $2, 4, 6, 10, 16, 26, \ldots$ **21.** $1, 3, -2, 5, -7, 12, \ldots$ **22.** $1, 2, 2, 4, 8, 32, \ldots$

Name______________________________ Date__________

Chapter 7 Maintaining Mathematical Proficiency

Simplify the expression.

1. $5x - 6 + 3x$

2. $3t + 7 - 3t - 4$

3. $8s - 4 + 4s - 6 - 5s$

4. $9m + 3 + m - 3 + 5m$

5. $-4 - 3p - 7 - 3p - 4$

6. $12(z - 1) + 4$

7. $-6(x + 2) - 4$

8. $3(h + 4) - 3(h - 4)$

9. $7(z + 4) - 3(z + 2) - 2(z - 3)$

Find the greatest common factor.

10. 24, 32

11. 30, 55

12. 48, 84

13. 28, 72

14. 42, 60

15. 35, 99

16. Explain how to find the greatest common factor of 42, 70, and 84.

Name ______________________________ Date __________

7.1 Adding and Subtracting Polynomials

For use with Exploration 7.1

Essential Question How can you add and subtract polynomials?

1 EXPLORATION: Adding Polynomials

Go to *BigIdeasMath.com* for an interactive tool to investigate this exploration.

Work with a partner. Write the expression modeled by the algebra tiles in each step.

Step 1

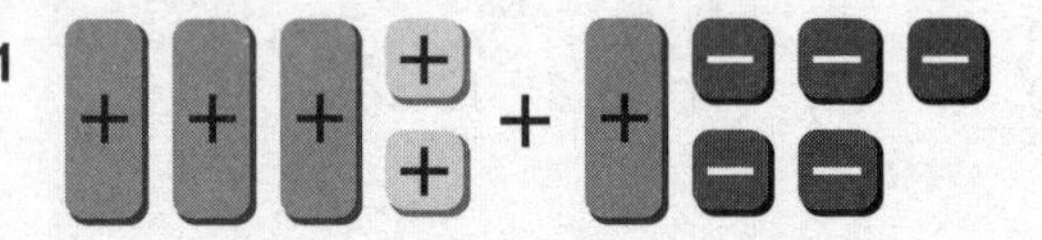

$(3x + 2) + (x - 5)$

Step 2

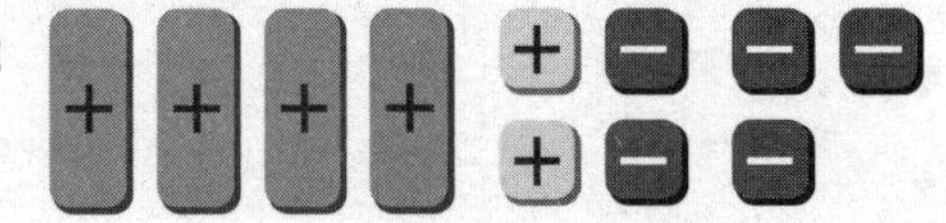

____

Step 3

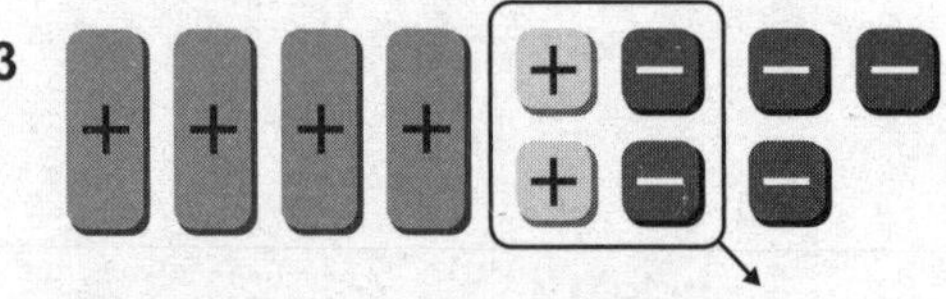

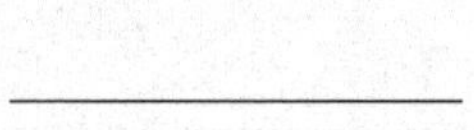

____

Step 4

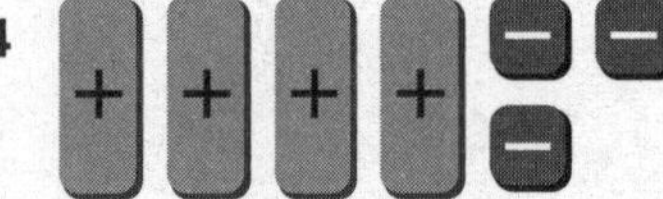

____

2 EXPLORATION: Subtracting Polynomials

Go to *BigIdeasMath.com* for an interactive tool to investigate this exploration.

Work with a partner. Write the expression modeled by the algebra tiles in each step.

Step 1

$(x^2 + 2x + 2) - (x - 1)$

Step 2

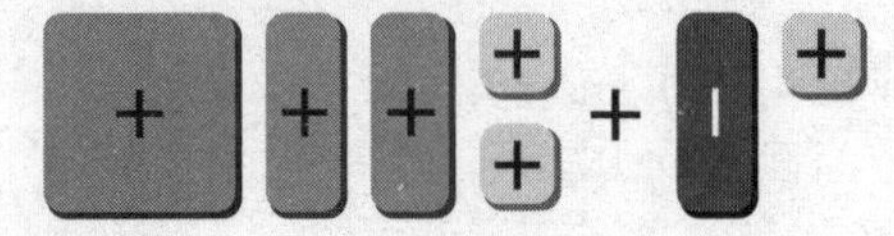

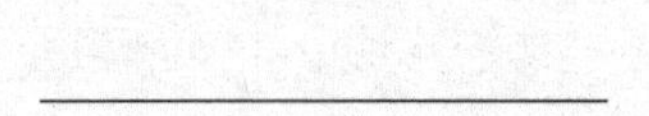

____

Step 3

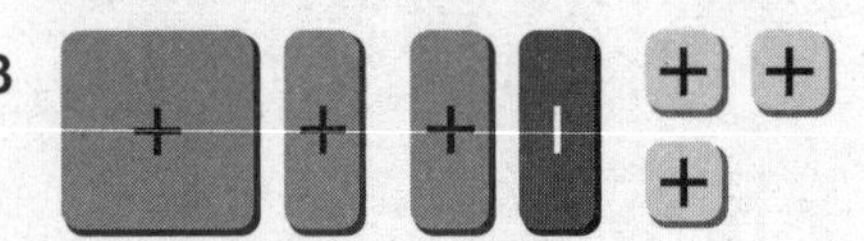

____

Name____ Date__________

2 EXPLORATION: Subtracting Polynomials (continued)

Step 4

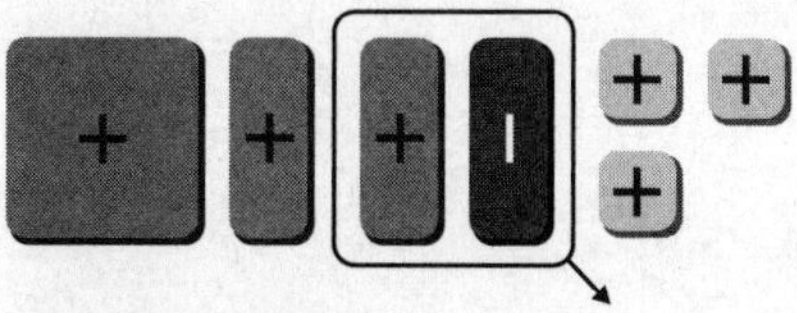

____

Step 5

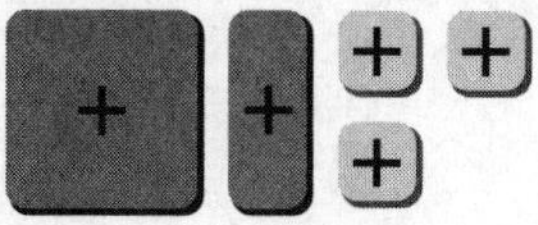

____

Communicate Your Answer

3. How can you add and subtract polynomials?

4. Use your methods in Question 3 to find each sum or difference.

a. $\left(x^2 + 2x - 1\right) + \left(2x^2 - 2x + 1\right)$

b. $(4x + 3) + (x - 2)$

c. $\left(x^2 + 2\right) - \left(3x^2 + 2x + 5\right)$

d. $(2x - 3x) - \left(x^2 - 2x + 4\right)$

Name ____ Date __________

7.1 Notetaking with Vocabulary
For use after Lesson 7.1

In your own words, write the meaning of each vocabulary term.

monomial

degree of a monomial

polynomial

binomial

trinomial

degree of a polynomial

standard form

leading coefficient

closed

Notes:

7.1 Notetaking with Vocabulary (continued)

Core Concepts

Polynomials

A **polynomial** is a monomial or a sum of monomials. Each monomial is called a *term* of the polynomial. A polynomial with two terms is a **binomial**. A polynomial with three terms is a **trinomial**.

Binomial	Trinomial
$5x + 2$	$x^2 + 5x + 2$

The **degree of a polynomial** is the greatest degree of its terms. A polynomial in one variable is in **standard form** when the exponents of the terms decrease from left to right. When you write a polynomial in standard form, the coefficient of the first term is the **leading coefficient**.

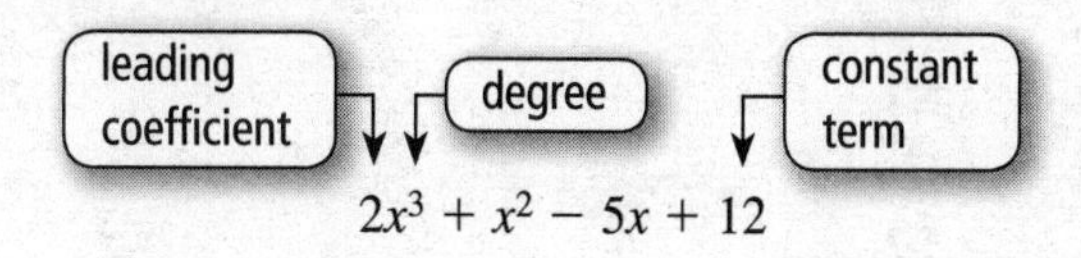

Notes:

Extra Practice

In Exercises 1–8, find the degree of the monomial.

1. $-6s$ **2.** w **3.** 8 **4.** $-2abc$

5. $7x^2y$ **6.** $4r^2s^3t$ **7.** $10mn^3$ **8.** $\frac{2}{3}$

Name ______________________________ Date __________

7.1 Notetaking with Vocabulary (continued)

In Exercises 9–12, write the polynomial in standard form. Identify the degree and leading coefficient of the polynomial. Then classify the polynomial by the number of terms.

9. $x + 3x^2 + 5$ **10.** $\sqrt{5}\,y$ **11.** $3x^5 + 6x^8$ **12.** $f^2 - 2f + f^4$

In Exercises 13–16, find the sum.

13. $(-4x + 9) + (6x - 14)$ **14.** $(-3a - 2) + (7a + 5)$

15. $(x^2 + 3x + 5) + (-x^2 + 6x - 4)$ **16.** $(t^2 + 3t^3 - 3) + (2t^2 + 7t - 2t^3)$

In Exercises 17–20, find the difference.

17. $(g - 4) - (3g - 6)$ **18.** $(-5h - 2) - (7h + 6)$

19. $(-x^2 - 5) - (-3x^2 - x - 8)$ **20.** $(k^2 + 6k^3 - 4) - (5k^3 + 7k - 3k^2)$

Name ________________________________ Date __________

7.2 Multiplying Polynomials
For use with Exploration 7.2

Essential Question How can you multiply two polynomials?

1 EXPLORATION: Multiplying Monomials Using Algebra Tiles

Work with a partner. Write each product. Explain your reasoning.

a. [+] • [+] = ________

b. [+] • [−] = ________

c. [−] • [−] = ________

d. [+] • [+] = ________

e. [+] • [−] = ________

f. [−] • [+] = ________

g. [−] • [−] = ________

h. [+] • [+] = ________

i. [+] • [−] = ________

j. [−] • [−] = ________

2 EXPLORATION: Multiplying Binomials Using Algebra Tiles

Go to *BigIdeasMath.com* for an interactive tool to investigate this exploration.

Work with a partner. Write the product of two binomials modeled by each rectangular array of algebra tiles. In parts (c) and (d), first draw the rectangular array of algebra tiles that models each product.

a. $(x + 3)(x - 2) =$ ____________

b. $(2x - 1)(2x + 1) =$ ____________

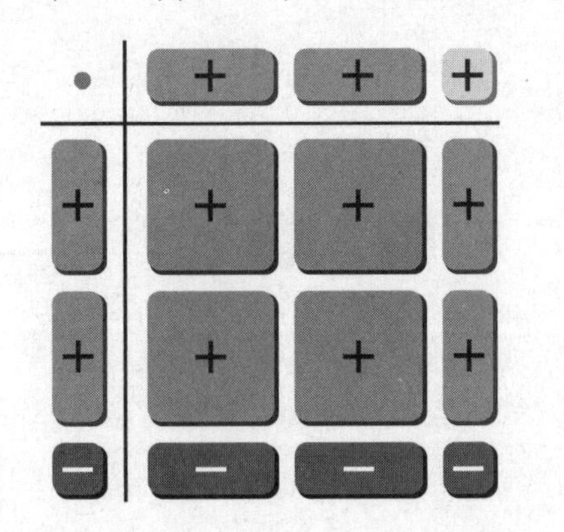

c. $(x + 2)(2x - 1) =$ ____________

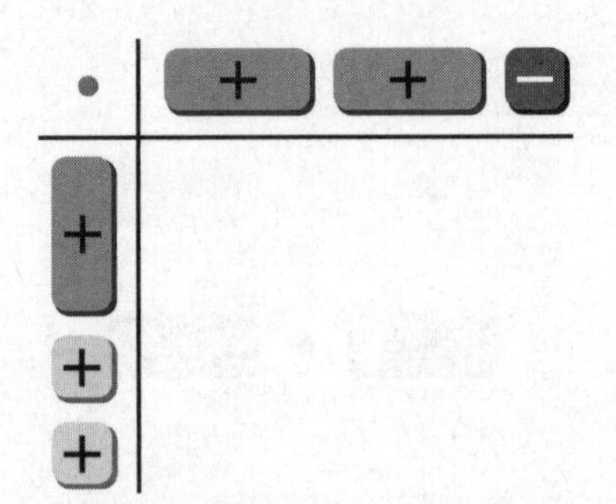

d. $(-x - 2)(x - 3) =$ ____________

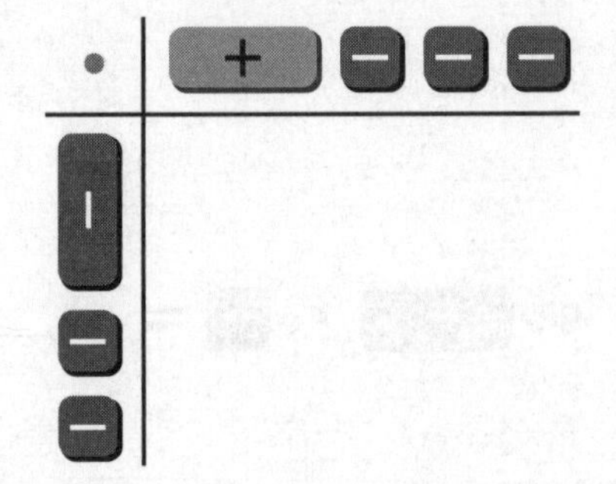

Communicate Your Answer

3. How can you multiply two polynomials?

4. Give another example of multiplying two binomials using algebra tiles that is similar to those in Exploration 2.

Name______________________________ Date__________

7.2 Notetaking with Vocabulary

For use after Lesson 7.2

In your own words, write the meaning of each vocabulary term.

FOIL Method

Core Concepts

FOIL Method

To multiply two binomials using the FOIL Method, find the sum of the products of the

First terms,	$(x + 1)(x + 2)$	➡	$x(x) = x^2$
Outer terms,	$(x + 1)(x + 2)$	➡	$x(2) = 2x$
Inner terms, and	$(x + 1)(x + 2)$	➡	$1(x) = x$
Last terms.	$(x + 1)(x + 2)$	➡	$1(2) = 2$

$$(x + 1)(x + 2) = x^2 + 2x + x + 2 = x^2 + 3x + 2$$

Notes:

Name ______________________________ Date ________

Extra Practice

In Exercises 1–6, use the Distributive Property to find the product.

1. $(x-2)(x-1)$

2. $(b-3)(b+2)$

3. $(g+2)(g+4)$

4. $(a-1)(2a+5)$

5. $(3n-4)(n+1)$

6. $(r+3)(3r+2)$

In Exercises 7–12, use a table to find the product.

7. $(x-3)(x-2)$

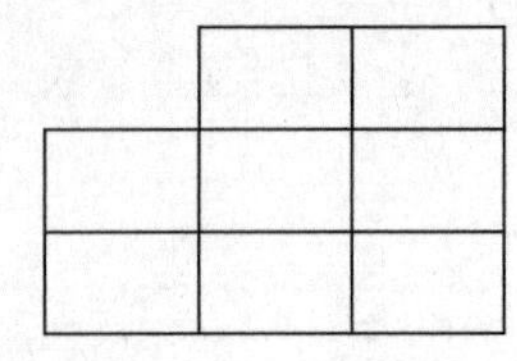

8. $(y+1)(y-6)$

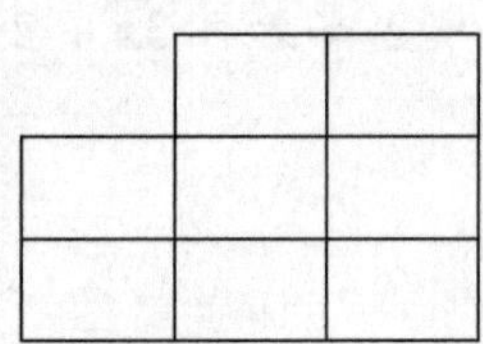

9. $(q+3)(q+7)$

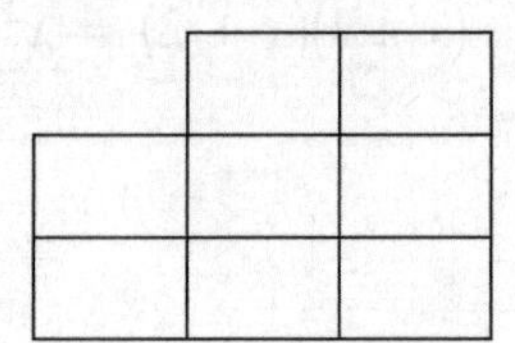

10. $(2w-5)(w-3)$

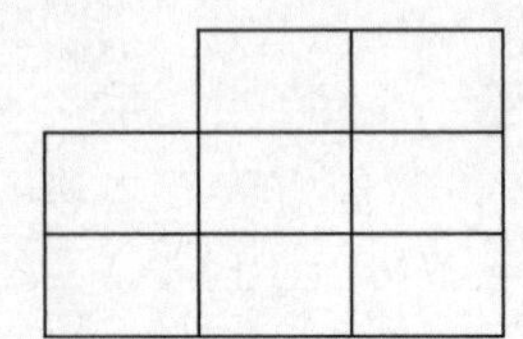

11. $(6h-2)(-3-2h)$

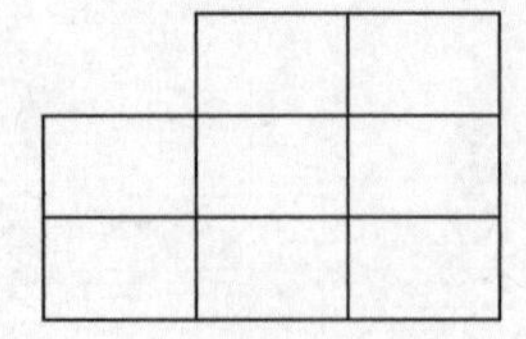

12. $(-3+4j)(3j+4)$

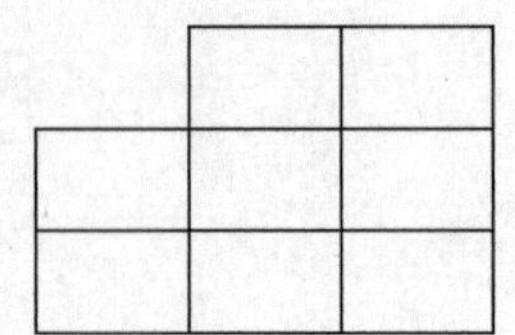

Name____________________ Date__________

7.2 Notetaking with Vocabulary (continued)

In Exercises 13–18, use the FOIL Method to find the product.

13. $(x + 2)(x - 3)$

14. $(z + 3)(z + 2)$

15. $(h - 2)(h + 4)$

16. $(2m - 1)(m + 2)$

17. $(4n - 1)(3n + 4)$

18. $(-q - 1)(q + 1)$

In Exercises 19–24, find the product.

19. $(x - 2)(x^2 + x - 1)$

20. $(2 - a)(3a^2 + 3a - 5)$

21. $(h + 1)(h^2 - h - 1)$

22. $(d + 3)(d^2 - 4d + 1)$

23. $(3n^2 + 2n - 5)(2n + 1)$

24. $(2p^2 + p - 3)(3p - 1)$

Name ____ Date __________

7.3 Special Products of Polynomials

For use with Exploration 7.3

Essential Question What are the patterns in the special products $(a + b)(a - b)$, $(a + b)^2$, and $(a - b)^2$?

1 EXPLORATION: Finding a Sum and Difference Pattern

Work with a partner. Write the product of two binomials modeled by each rectangular array of algebra tiles.

a. $(x + 2)(x - 2) =$ ____________

b. $(2x - 1)(2x + 1) =$ ____________

2 EXPLORATION: Finding the Square of a Binomial Pattern

Go to *BigIdeasMath.com* for an interactive tool to investigate this exploration.

Work with a partner. Draw the rectangular array of algebra tiles that models each product of two binomials. Write the product.

a. $(x + 2)^2 =$ ____________

b. $(2x - 1)^2 =$ ____________

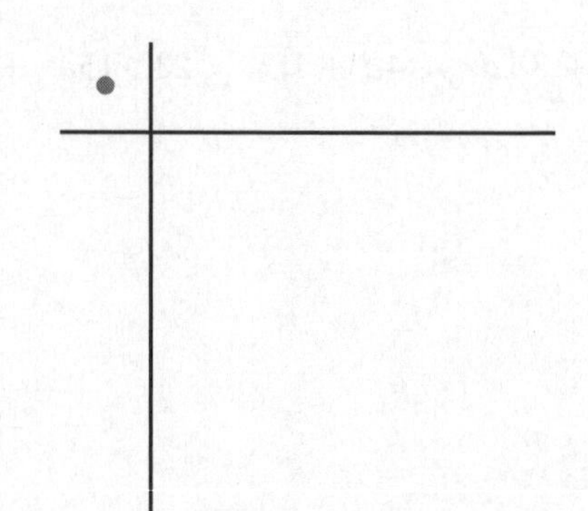

Communicate Your Answer

3. What are the patterns in the special products $(a + b)(a - b)$, $(a + b)^2$, and $(a - b)^2$?

4. Use the appropriate special product pattern to find each product. Check your answers using algebra tiles.

a. $(x + 3)(x - 3)$ **b.** $(x - 4)(x + 4)$ **c.** $(3x + 1)(3x - 1)$

d. $(x + 3)^2$ **e.** $(x - 2)^2$ **f.** $(3x + 1)^2$

Name ______________________ Date __________

7.3 Notetaking with Vocabulary

For use after Lesson 7.3

In your own words, write the meaning of each vocabulary term.

binomial

Core Concepts

Square of a Binomial Pattern

Algebra

$(a + b)^2 = a^2 + 2ab + b^2$

$(a - b)^2 = a^2 - 2ab + b^2$

Example

$$(x + 5)^2 = (x)^2 + 2(x)(5) + (5)^2$$
$$= x^2 + 10x + 25$$

$$(2x - 3)^2 = (2x)^2 - 2(2x)(3) + (3)^2$$
$$= 4x^2 - 12x + 9$$

Notes:

Sum and Difference Pattern

Algebra

$(a + b)(a - b) = a^2 - b^2$

Example

$(x + 3)(x - 3) = x^2 - 9$

Notes:

Name______________________________ Date__________

Extra Practice

In Exercises 1–18, find the product.

1. $(a + 3)^2$

2. $(b - 2)^2$

3. $(c + 4)^2$

4. $(-2x + 1)^2$

5. $(3x - 2)^2$

6. $(-4p - 3)^2$

7. $(3x + 2y)^2$

8. $(2a - 3b)^2$

9. $(-4c + 5d)^2$

10. $(x - 3)(x + 3)$

11. $(q + 5)(q - 5)$

12. $(t - 11)(t + 11)$

13. $(5a - 1)(5a + 1)$

14. $\left(\frac{1}{4}b + 1\right)\left(\frac{1}{4}b - 1\right)$

15. $\left(\frac{1}{2}c + \frac{1}{3}\right)\left(\frac{1}{2}c - \frac{1}{3}\right)$

16. $(-m + 2n)(-m - 2n)$

17. $(-3j - 2k)(-3j + 2k)$

18. $\left(6a + \frac{1}{2}b\right)\left(-6a + \frac{1}{2}b\right)$

In Exercises 19–24, use special product patterns to find the product.

19. $18 \bullet 22$

20. $49 \bullet 51$

21. $19\frac{3}{5} \bullet 20\frac{2}{5}$

22. $(31)^2$

23. $(20.7)^2$

24. $(109)^2$

25. Find k so that $kx^2 - 12x + 9$ is the square of a binomial.

Name ____________________ Date __________

7.4 Solving Polynomial Equations in Factored Form

For use with Exploration 7.4

Essential Question How can you solve a polynomial equation?

1 EXPLORATION: Matching Equivalent Forms of an Equation

Work with a partner. An equation is considered to be in *factored form* when the product of the factors is equal to 0. Match each factored form of the equation with its equivalent standard form and nonstandard form.

Factored Form	Standard Form	Nonstandard Form
a. $(x-1)(x-3)=0$	**A.** $x^2-x-2=0$	**1.** $x^2-5x=-6$
b. $(x-2)(x-3)=0$	**B.** $x^2+x-2=0$	**2.** $(x-1)^2=4$
c. $(x+1)(x-2)=0$	**C.** $x^2-4x+3=0$	**3.** $x^2-x=2$
d. $(x-1)(x+2)=0$	**D.** $x^2-5x+6=0$	**4.** $x(x+1)=2$
e. $(x+1)(x-3)=0$	**E.** $x^2-2x-3=0$	**5.** $x^2-4x=-3$

2 EXPLORATION: Writing a Conjecture

Go to *BigIdeasMath.com* for an interactive tool to investigate this exploration.

Work with a partner. Substitute 1, 2, 3, 4, 5, and 6 for x in each equation and determine whether the equation is true. Organize your results in the table. Write a conjecture describing what you discovered.

	Equation	$x=1$	$x=2$	$x=3$	$x=4$	$x=5$	$x=6$
a.	$(x-1)(x-2)=0$						
b.	$(x-2)(x-3)=0$						
c.	$(x-3)(x-4)=0$						
d.	$(x-4)(x-5)=0$						
e.	$(x-5)(x-6)=0$						
f.	$(x-6)(x-1)=0$						

Name ______________________________ Date ________

3 EXPLORATION: Special Properties of 0 and 1

Work with a partner. The numbers 0 and 1 have special properties that are shared by no other numbers. For each of the following, decide whether the property is true for 0, 1, both, or neither. Explain your reasoning.

a. When you add ____ to a number *n*, you get *n*.

b. If the product of two numbers is ____, then at least one of the numbers is 0.

c. The square of ____ is equal to itself.

d. When you multiply a number *n* by ____, you get *n*.

e. When you multiply a number *n* by ____, you get 0.

f. The opposite of ____ is equal to itself.

Communicate Your Answer

4. How can you solve a polynomial equation?

5. One of the properties in Exploration 3 is called the Zero-Product Property. It is one of the most important properties in all of algebra. Which property is it? Why do you think it is called the Zero-Product Property? Explain how it is used in algebra and why it so important.

Name____ Date__________

7.4 Notetaking with Vocabulary
For use after Lesson 7.4

In your own words, write the meaning of each vocabulary term.

factored form

Zero-Product Property

roots

repeated roots

Core Concepts

Zero-Product Property

Words If the product of two real numbers is 0, then at least one of the numbers is 0.

Algebra If a and b are real numbers and $ab = 0$, then $a = 0$ or $b = 0$.

Notes:

Name ____________________ Date ________

Extra Practice

In Exercises 1–12, solve the equation.

1. $x(x + 5) = 0$

2. $a(a - 12) = 0$

3. $5p(p - 2) = 0$

4. $(c - 2)(c + 1) = 0$

5. $(2b - 6)(3b + 18) = 0$

6. $(3 - 5s)(-3 + 5s) = 0$

7. $(x - 3)^2 = 0$

8. $(3d + 7)(5d - 6) = 0$

9. $(2t + 8)(2t - 8) = 0$

10. $(w + 4)^2(w + 1) = 0$

11. $g(6 - 3g)(6 + 3g) = 0$

12. $(4 - m)\left(8 + \frac{2}{3}m\right)(-2 - 3m) = 0$

7.4 Notetaking with Vocabulary (continued)

In Exercises 13–18, factor the polynomial.

13. $6x^2 + 3x$

14. $4y^4 - 20y^3$

15. $18u^4 - 6u$

16. $7z^7 + 2z^6$

17. $24h^3 + 8h$

18. $15f^4 - 45f$

In Exercises 19–24, solve the equation.

19. $6k^2 + k = 0$

20. $35n - 49n^2 = 0$

21. $4z^2 + 52z = 0$

22. $6x^2 = -72x$

23. $22s = 11s^2$

24. $7p^2 = 21p$

25. A boy kicks a ball in the air. The height y (in feet) above the ground of the ball is modeled by the equation $y = -16x^2 + 80x$, where x is the time (in seconds) since the ball was kicked. Find the roots of the equation when $y = 0$. Explain what the roots mean in this situation.

Name ______________________________ Date ________

7.5 Factoring $x^2 + bx + c$

For use with Exploration 7.5

Essential Question How can you use algebra tiles to factor the trinomial $x^2 + bx + c$ into the product of two binomials?

1 EXPLORATION: Finding Binomial Factors

Go to *BigIdeasMath.com* for an interactive tool to investigate this exploration.

Work with a partner. Use algebra tiles to write each polynomial as the product of two binomials. Check your answer by multiplying.

Sample $x^2 + 5x + 6$

Step 1 Arrange algebra tiles that model $x^2 + 5x + 6$ into a rectangular array.

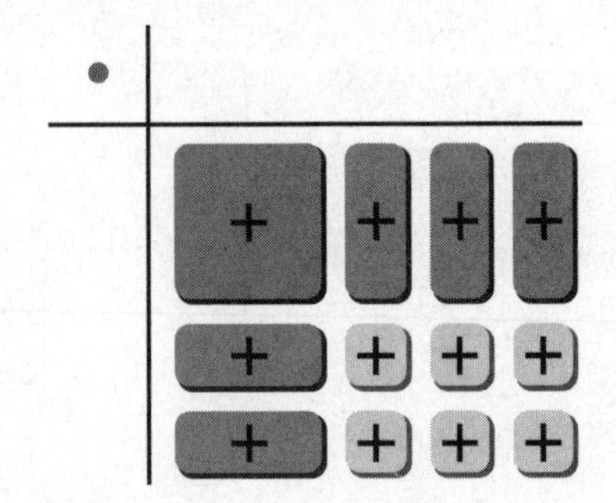

Step 2 Use additional algebra tiles to model the dimensions of the rectangle.

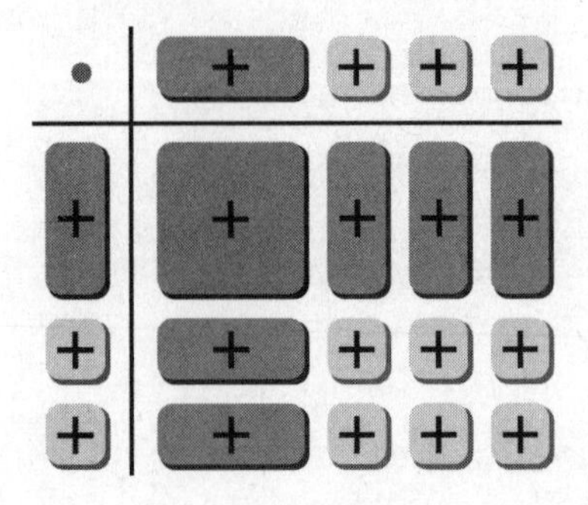

Step 3 Write the polynomial in factored form using the dimensions of the rectangle.

width — length

$$\text{Area} = x^2 + 5x + 6 = (x + 2)(x + 3)$$

a. $x^2 - 3x + 2 =$ ____________

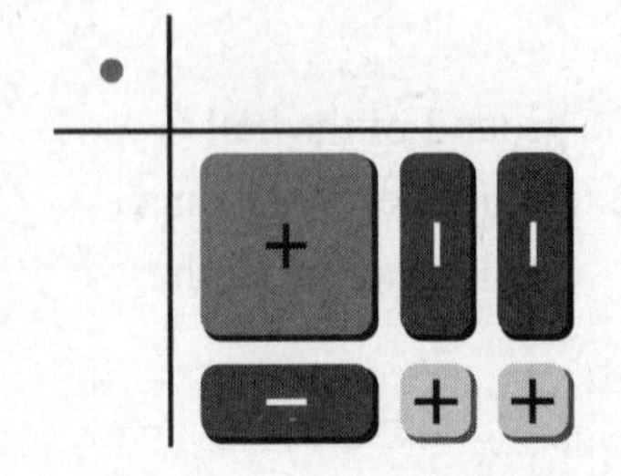

b. $x^2 + 5x + 4 =$ ____________

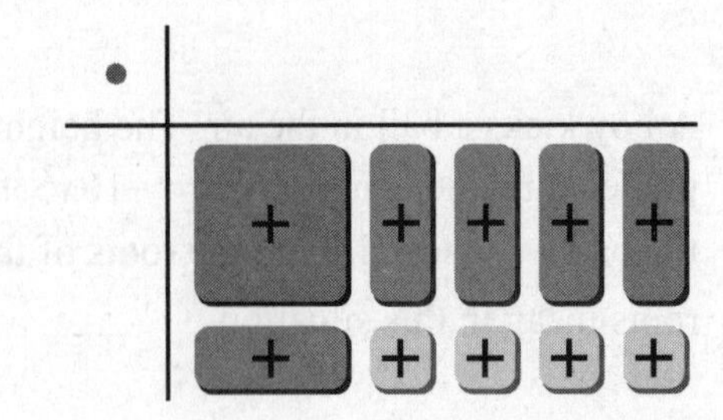

Name______________________________ Date__________

7.5 **Factoring $x^2 + bx + c$ (continued)**

1 EXPLORATION: Finding Binomial Factors (continued)

c. $x^2 - 7x + 12 =$ ____________

d. $x^2 + 7x + 12 =$ ____________

Communicate Your Answer

2. How can you use algebra tiles to factor the trinomial $x^2 + bx + c$ into the product of two binomials?

3. Describe a strategy for factoring the trinomial $x^2 + bx + c$ that does not use algebra tiles.

Name ____________________ Date ________

7.5 Notetaking with Vocabulary

For use after Lesson 7.5

In your own words, write the meaning of each vocabulary term.

polynomial

FOIL Method

Zero-Product Property

Core Concepts

Factoring $x^2 + bx + c$ When c Is Positive

Algebra $x^2 + bx + c = (x + p)(x + q)$ when $p + q = b$ and $pq = c$.

When c is positive, p and q have the same sign as b.

Examples $x^2 + 6x + 5 = (x + 1)(x + 5)$

$x^2 - 6x + 5 = (x - 1)(x - 5)$

Notes:

Factoring $x^2 + bx + c$ When c Is Negative

Algebra $x^2 + bx + c = (x + p)(x + q)$ when $p + q = b$ and $pq = c$.

When c is negative, p and q have different signs.

Example $x^2 - 4x - 5 = (x + 1)(x - 5)$

Notes:

Name____ Date__________

7.5 **Notetaking with Vocabulary (continued)**

Extra Practice

In Exercises 1–12, factor the polynomial.

1. $c^2 + 8c + 7$

2. $a^2 + 16a + 64$

3. $x^2 + 11x + 18$

4. $d^2 + 6d + 8$

5. $s^2 + 11s + 10$

6. $u^2 + 10u + 9$

7. $b^2 + 3b - 54$

8. $y^2 - y - 2$

9. $u + 3u - 18$

10. $z^2 - z - 56$

11. $h^2 + 2h - 24$

12. $f^2 - 3f - 40$

In Exercises 13–18, solve the equation.

13. $g^2 - 13g + 40 = 0$

14. $k^2 - 5k + 6 = 0$

15. $w^2 - 7w + 10 = 0$

16. $x^2 - x = 30$

17. $r^2 - 3r = -2$

18. $t^2 - 7t = 8$

19. The area of a right triangle is 16 square miles. One leg of the triangle is 4 miles longer than the other leg. Find the length of each leg.

20. You have two circular flower beds, as shown. The sum of the areas of the two flower beds is 136π square feet. Find the radius of each bed.

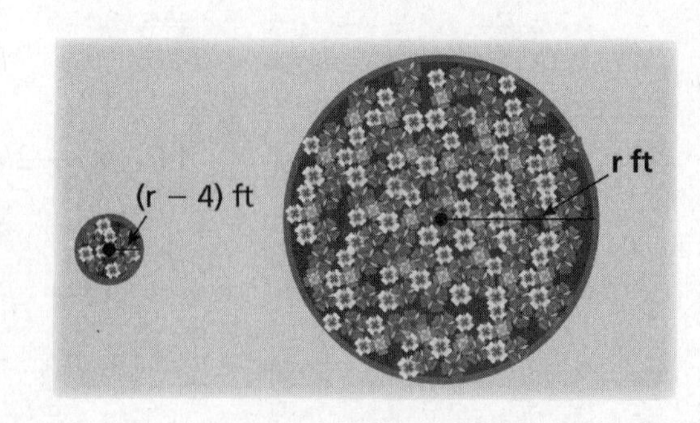

Name____ Date__________

7.6 Factoring $ax^2 + bx + c$

For use with Exploration 7.6

Essential Question How can you use algebra tiles to factor the trinomial $ax^2 + bx + c$ into the product of two binomials?

1 EXPLORATION: Finding Binomial Factors

Go to *BigIdeasMath.com* for an interactive tool to investigate this exploration.

Work with a partner. Use algebra tiles to write each polynomial as the product of two binomials. Check your answer by multiplying.

Sample $2x^2 + 5x + 2$

Step 1 Arrange algebra tiles that model $2x^2 + 5x + 2$ into a rectangular array.

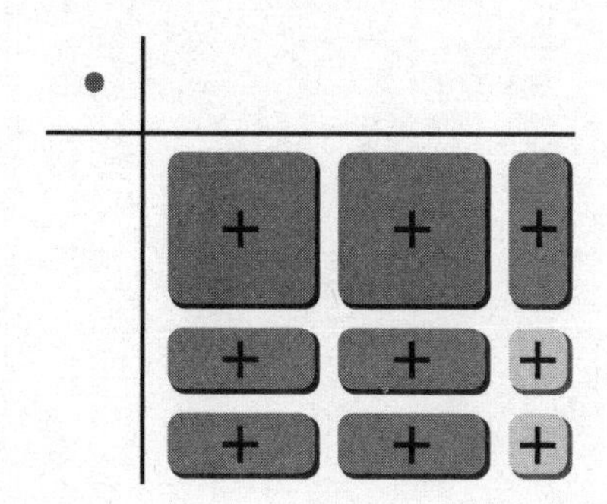

Step 2 Use additional algebra tiles to model the dimensions of the rectangle.

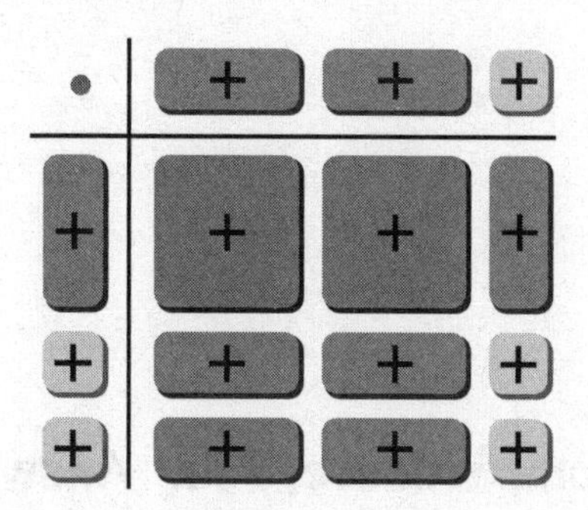

Step 3 Write the polynomial in factored form using the dimensions of the rectangle.

width, length

$$\text{Area} = 2x^2 + 5x + 2 = (x + 2)(2x + 1)$$

a. $3x^2 + 5x + 2 =$ ____________

7.6 Factoring $ax^2 + bx + c$ (continued)

1 EXPLORATION: Finding Binomial Factors (continued)

b. $4x^2 + 4x - 3 =$ ____________

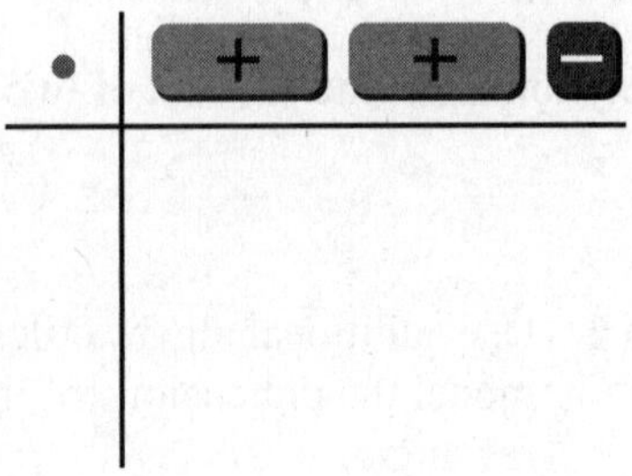

c. $2x^2 - 11x + 5 =$ ____________

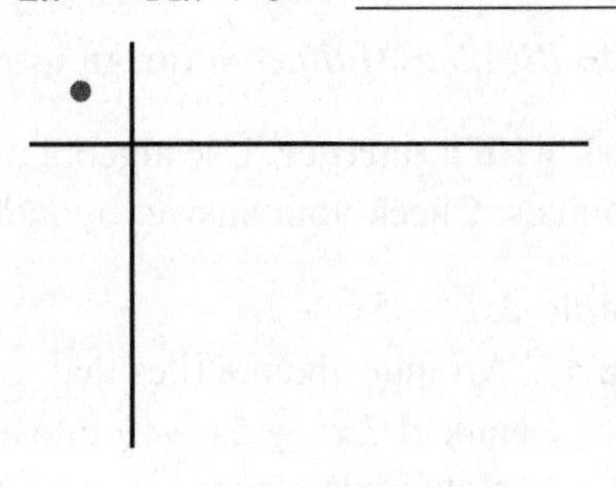

Communicate Your Answer

2. How can you use algebra tiles to factor the trinomial $ax^2 + bx + c$ into the product of two binomials?

3. Is it possible to factor the trinomial $2x^2 + 2x + 1$? Explain your reasoning.

Name____ Date__________

7.6 Notetaking with Vocabulary
For use after Lesson 7.6

In your own words, write the meaning of each vocabulary term.

polynomial

greatest common factor (GCF)

Zero-Product Property

Notes:

Name ______________________________ Date __________

7.6 Notetaking with Vocabulary (continued)

Extra Practice

In Exercises 1–18, factor the polynomial.

1. $2c^2 - 14c - 36$

2. $4a^2 + 8a - 140$

3. $3x^2 - 6x - 24$

4. $2d^2 - 2d - 60$

5. $5s^2 + 55s + 50$

6. $3q^2 + 30q + 27$

7. $12g^2 - 37g + 28$

8. $6k^2 - 11k + 4$

9. $9w^2 + 9w + 2$

10. $12a^2 + 5a - 2$

11. $15b^2 + 14b - 8$

12. $5t^2 + 12t - 9$

7.6 Notetaking with Vocabulary (continued)

13. $-12b^2 + 5b + 2$

14. $-6x^2 + x + 15$

15. $-60g^2 - 11g + 1$

16. $-2d^2 - d + 6$

17. $-3r^2 - 4r - 1$

18. $-8x^2 + 14x - 5$

19. The length of a rectangular shaped park is $(3x + 5)$ miles. The width is $(2x + 8)$ miles. The area of the park is 360 square miles. What are the dimensions of the park?

20. The sum of two numbers is 8. The sum of the squares of the two numbers is 34. What are the two numbers?

Name ______________________________ Date __________

7.7 Factoring Special Products

For use with Exploration 7.7

Essential Question How can you recognize and factor special products?

1 EXPLORATION: Factoring Special Products

Go to *BigIdeasMath.com* for an interactive tool to investigate this exploration.

Work with a partner. Use algebra tiles to write each polynomial as the product of two binomials. Check your answer by multiplying. State whether the product is a "special product" that you studied in Section 7.3.

a. $4x^2 - 1 =$ __________

b. $4x^2 - 4x + 1 =$ __________

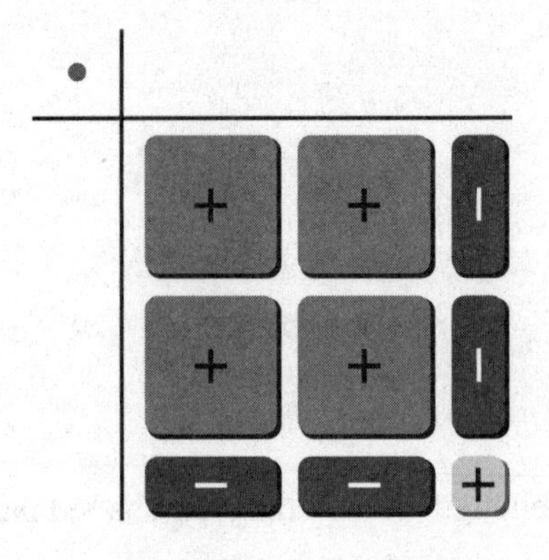

c. $4x^2 + 4x + 1 =$ __________

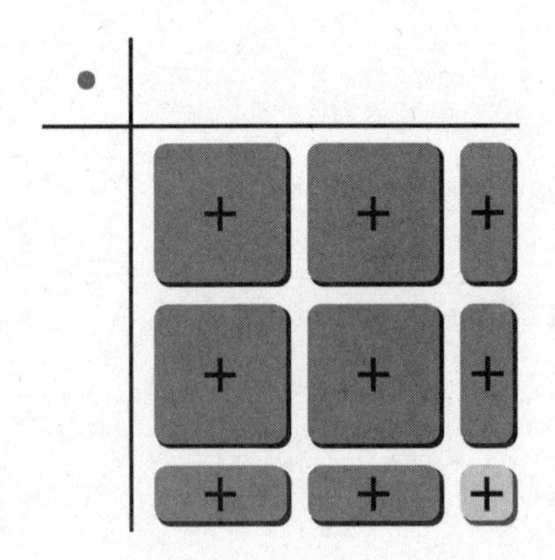

d. $4x^2 - 6x + 2 =$ __________

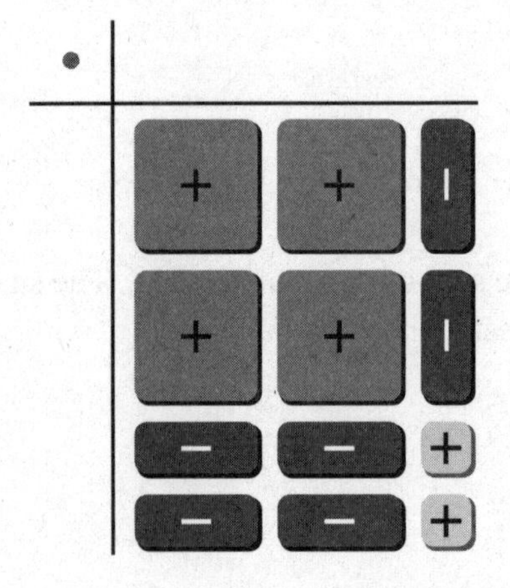

2 EXPLORATION: Factoring Special Products

Go to *BigIdeasMath.com* for an interactive tool to investigate this exploration.

Work with a partner. Use algebra tiles to complete the rectangular arrays in three different ways, so that each way represents a different special product. Write each special product in standard form and in factored form.

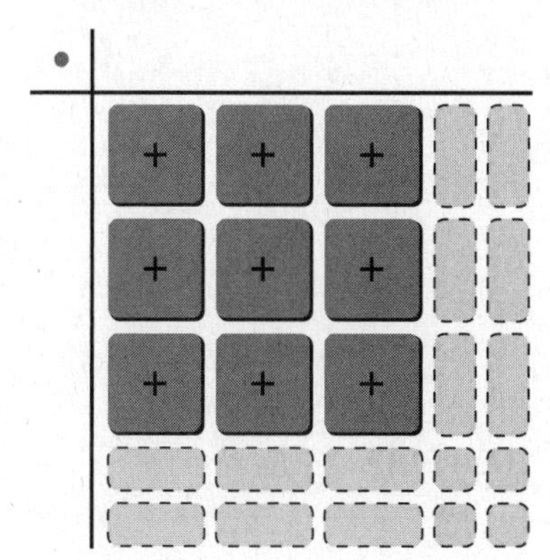

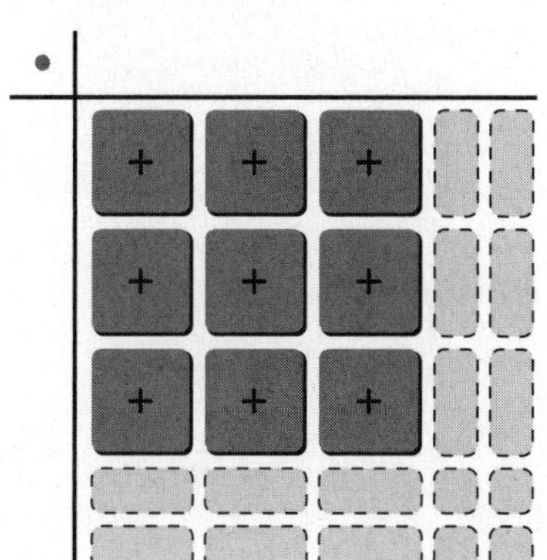

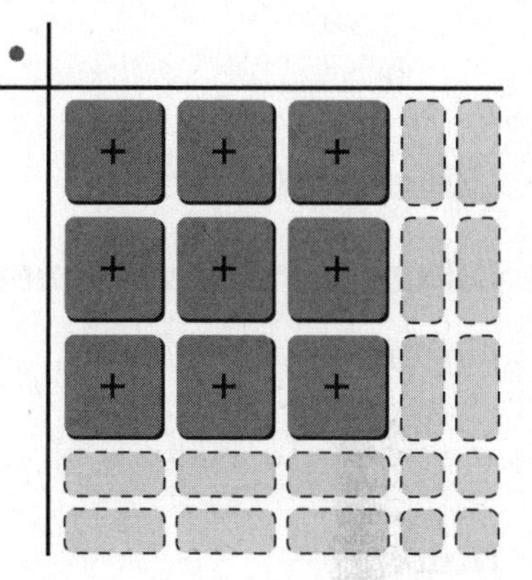

Communicate Your Answer

3. How can you recognize and factor special products? Describe a strategy for recognizing which polynomials can be factored as special products.

4. Use the strategy you described in Question 3 to factor each polynomial.

a. $25x^2 + 10x + 1$ **b.** $25x^2 - 10x + 1$ **c.** $25x^2 - 1$

Name ______________________________ Date __________

7.7 Notetaking with Vocabulary

For use after Lesson 7.7

In your own words, write the meaning of each vocabulary term.

polynomial

trinomial

Core Concepts

Difference of Two Squares Pattern

Algebra

$$a^2 - b^2 = (a + b)(a - b)$$

Example

$$x^2 - 9 = x^2 - 3^2 = (x + 3)(x - 3)$$

Notes:

Perfect Square Trinomial Pattern

Algebra

$$a^2 + 2ab + b^2 = (a + b)^2$$

$$a^2 - 2ab + b^2 = (a - b)^2$$

Example

$$x^2 + 6x + 9 = x^2 + 2(x)(3) + 3^2 = (x + 3)^2$$

$$x^2 - 6x + 9 = x^2 - 2(x)(3) + 3^2 = (x - 3)^2$$

Notes:

Name______________________________ Date__________

7.7 Notetaking with Vocabulary (continued)

Extra Practice

In Exercises 1–6, factor the polynomial.

1. $s^2 - 49$

2. $t^2 - 81$

3. $16 - x^2$

4. $4g^2 - 25$

5. $36h^2 - 121$

6. $81 - 49k^2$

In Exercises 7–12, use a special product pattern to evaluate the expression.

7. $57^2 - 53^2$

8. $38^2 - 32^2$

9. $68^2 - 64^2$

10. $45^2 - 40^2$

11. $79^2 - 71^2$

12. $86^2 - 84^2$

Name _______________ Date _______

7.7 Notetaking with Vocabulary (continued)

In Exercises 13–18, factor the polynomial.

13. $x^2 + 16x + 64$

14. $p^2 + 28p + 196$

15. $r^2 - 26r + 169$

16. $a^2 - 18a + 81$

17. $36c^2 + 84c + 49$

18. $100x^2 - 20x + 1$

In Exercises 19–24, solve the equation.

19. $x^2 - 144 = 0$

20. $9y^2 = 49$

21. $c^2 + 14c + 49 = 0$

22. $d^2 - 4d + 4 = 0$

23. $n^2 + \frac{2}{3}n = -\frac{1}{9}$

24. $-\frac{6}{5}k + \frac{9}{25} = -k^2$

25. The dimensions of a rectangular prism are $(x + 1)$ feet by $(x + 2)$ feet by 4 feet. The volume of the prism is $(24x - 1)$ cubic feet. What is the value of x?

Name____ Date__________

7.8 Factoring Polynomials Completely

For use with Exploration 7.8

Essential Question How can you factor a polynomial completely?

1 EXPLORATION: Writing a Product of Linear Factors

Work with a partner. Write the product represented by the algebra tiles. Then multiply to write the polynomial in standard form.

a. (+ +)(+ +)(− −)

b. (+ + +)(+ +)(−)

c. (+ + + +)(+)(+ +)

d. (+ +)(+ −)(+)

e. (− +)(+ +)(−)

f. (− −)(+ +)(− −)

2 EXPLORATION: Matching Standard and Factored Forms

Work with a partner. Match the standard form of the polynomial with the equivalent factored form on the next page. Explain your strategy.

a. $x^3 + x^2$ **b.** $x^3 - x$ **c.** $x^3 + x^2 - 2x$

d. $x^3 - 4x^2 + 4x$ **e.** $x^3 - 2x^2 - 3x$ **f.** $x^3 - 2x^2 + x$

g. $x^3 - 4x$ **h.** $x^3 + 2x^2$ **i.** $x^3 - x^2$

j. $x^3 - 3x^2 + 2x$ **k.** $x^3 + 2x^2 - 3x$ **l.** $x^3 - 4x^2 + 3x$

m. $x^3 - 2x^2$ **n.** $x^3 + 4x^2 + 4x$ **o.** $x^3 + 2x^2 + x$

2 EXPLORATION: Matching Standard and Factored Forms (continued)

A. $x(x + 1)(x - 1)$ **B.** $x(x - 1)^2$ **C.** $x(x + 1)^2$

D. $x(x + 2)(x - 1)$ **E.** $x(x - 1)(x - 2)$ **F.** $x(x + 2)(x - 2)$

G. $x(x - 2)^2$ **H.** $x(x + 2)^2$ **I.** $x^2(x - 1)$

J. $x^2(x + 1)$ **K.** $x^2(x - 2)$ **L.** $x^2(x + 2)$

M. $x(x + 3)(x - 1)$ **N.** $x(x + 1)(x - 3)$ **O.** $x(x - 1)(x - 3)$

Communicate Your Answer

3. How can you factor a polynomial completely?

4. Use your answer to Question 3 to factor each polynomial completely.

a. $x^3 + 4x^2 + 3x$ **b.** $x^3 - 6x^2 + 9x$ **c.** $x^3 + 6x^2 + 9x$

Name____ Date__________

7.8 Notetaking with Vocabulary

For use after Lesson 7.8

In your own words, write the meaning of each vocabulary term.

factoring by grouping

factored completely

Core Concepts

Factoring by Grouping

To factor a polynomial with four terms, group the terms into pairs. Factor the GCF out of each pair of terms. Look for and factor out the common binomial factor. This process is called **factoring by grouping**.

Notes:

Guidelines for Factoring Polynomials Completely

To factor a polynomial completely, you should try each of these steps.

1. Factor out the greatest common monomial factor. $3x^2 + 6x = 3x(x + 2)$

2. Look for a difference of two squares or a perfect square trinomial. $x^2 + 4x + 4 = (x + 2)^2$

3. Factor a trinomial of the form $ax^2 + bx + c$ into a product of binomial factors. $3x^2 - 5x - 2 = (3x + 1)(x - 2)$

4. Factor a polynomial with four terms by grouping. $x^3 + x - 4x^2 - 4 = (x^2 + 1)(x - 4)$

Notes:

Extra Practice

In Exercises 1–8, factor the polynomial by grouping.

1. $b^3 - 4b^2 + b - 4$

2. $ac + ad + bc + bd$

3. $d^2 + 2c + cd + 2d$

4. $5t^3 + 6t^2 + 5t + 6$

5. $8s^3 + s - 64s^2 - 8$

6. $12a^3 + 2a^2 - 30a - 5$

7. $4x^3 - 12x^2 - 5x + 15$

8. $21h^3 + 18h^2 - 35h - 30$

Name____ Date__________

7.8 Notetaking with Vocabulary (continued)

In Exercises 9–16, factor the polynomial completely.

9. $4c^3 - 4c$

10. $100x^4 - 25x^2$

11. $2a^2 + 3a - 2$

12. $9x^2 + 3x - 14$

13. $20p^2 + 22p - 12$

14. $12x^2 - 20x - 48$

15. $3s^3 + 2s^2 - 21s - 14$

16. $2t^4 + t^3 - 10t - 5$

In Exercises 17–22, solve the equation.

17. $3x^2 - 21x + 30 = 0$

18. $5y^2 - 5y - 30 = 0$

19. $c^4 - 81c^2 = 0$

20. $9d + 9 = d^3 + d^2$

21. $48n - 3n^2 = 0$

22. $x^3 + 3x^2 = 16x + 48$

Name ______________________________ Date __________

Chapter 8 Maintaining Mathematical Proficiency

Graph the linear equation.

1. $y = 4x - 5$

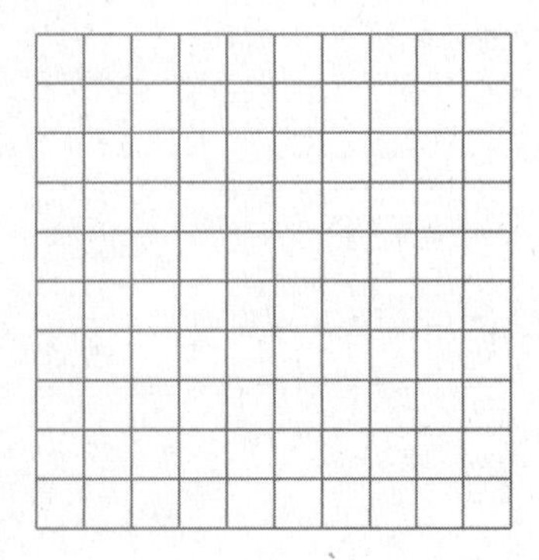

2. $y = -2x + 3$

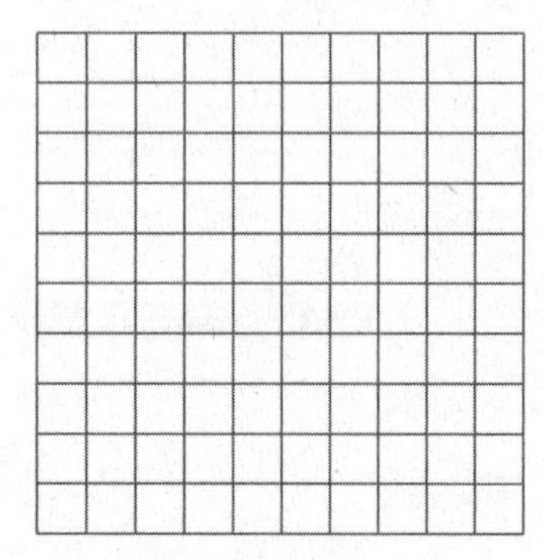

3. $y = \frac{1}{2}x + 3$

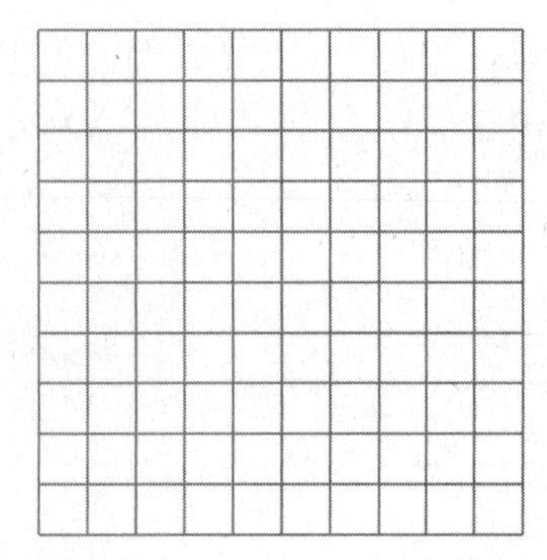

4. $y = -x + 2$

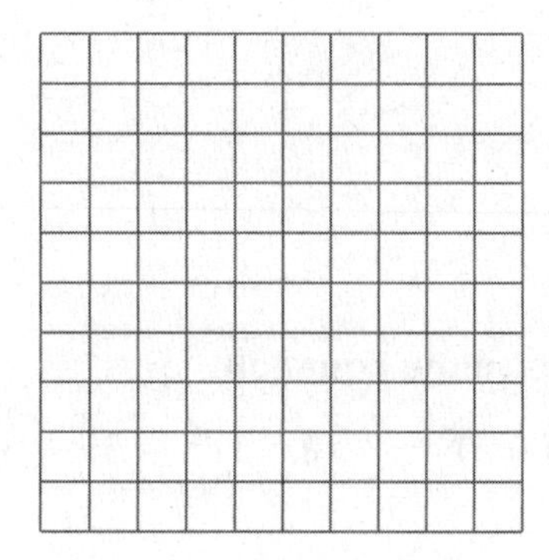

Evaluate the expression when $x = -4$.

5. $2x^2 + 8$

6. $-x^2 + 3x - 4$

7. $-3x^2 - 4$

8. $5x^2 - x + 8$

9. $4x^2 - 8x$

10. $6x^2 - 5x + 3$

11. $-2x^2 + 4x + 4$

12. $3x^2 + 2x + 2$

Name______________________________ Date__________

8.1 Graphing $f(x) = ax^2$

For use with Exploration 8.1

Essential Question What are some of the characteristics of the graph of a quadratic function of the form $f(x) = ax^2$?

1 EXPLORATION: Graphing Quadratic Functions

Go to *BigIdeasMath.com* for an interactive tool to investigate this exploration.

Work with a partner. Graph each quadratic function. Compare each graph to the graph of $f(x) = x^2$.

a. $g(x) = 3x^2$

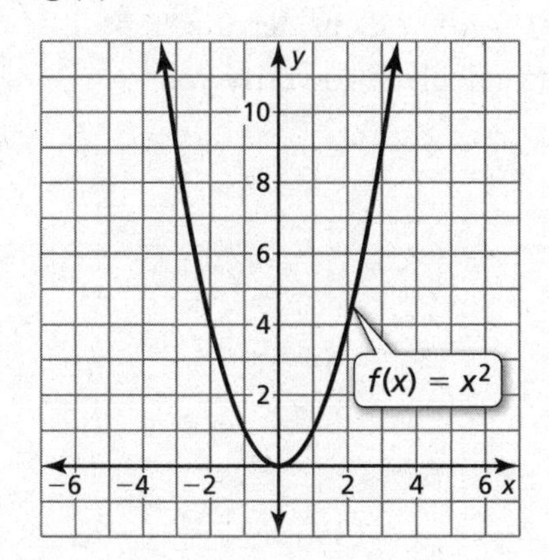

b. $g(x) = -5x^2$

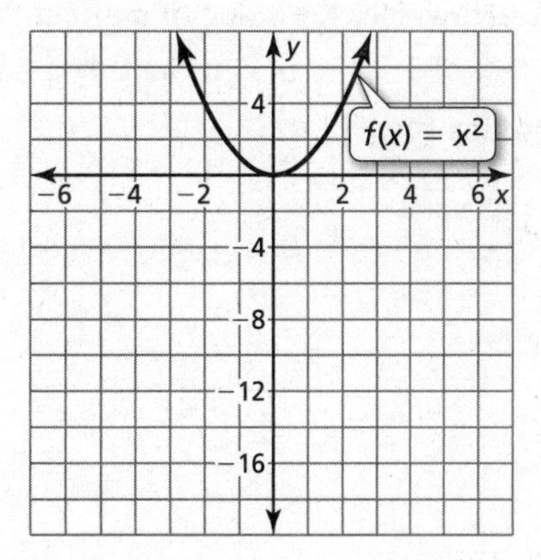

c. $g(x) = -0.2x^2$

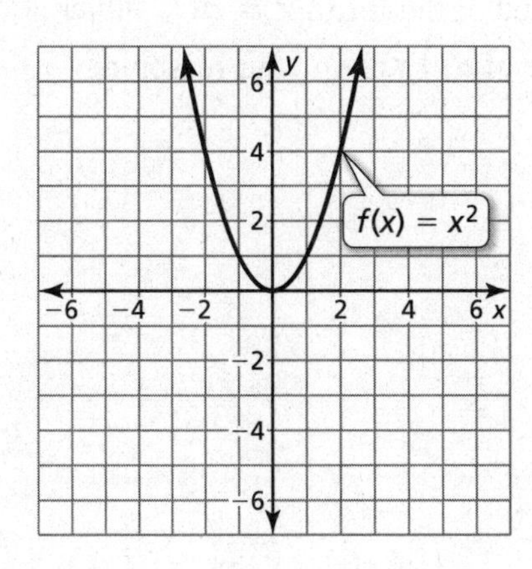

d. $g(x) = \frac{1}{10}x^2$

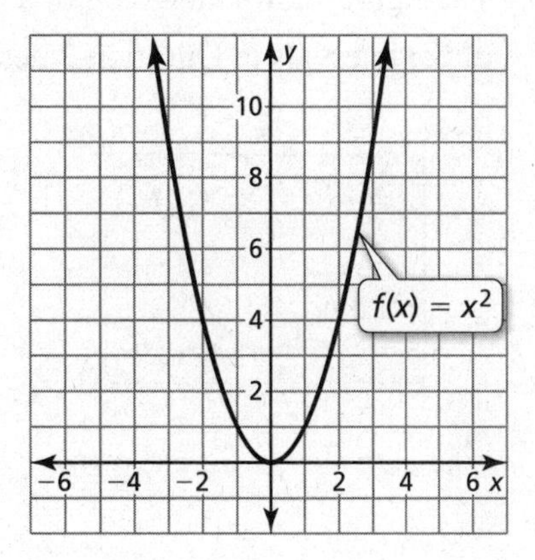

Communicate Your Answer

2. What are some of the characteristics of the graph of a quadratic function of the form $f(x) = ax^2$?

3. How does the value of a affect the graph of $f(x) = ax^2$? Consider $0 < a < 1$, $a > 1$, $-1 < a < 0$, and $a < -1$. Use a graphing calculator to verify your answers.

4. The figure shows the graph of a quadratic function of the form $y = ax^2$. Which of the intervals in Question 3 describes the value of a? Explain your reasoning.

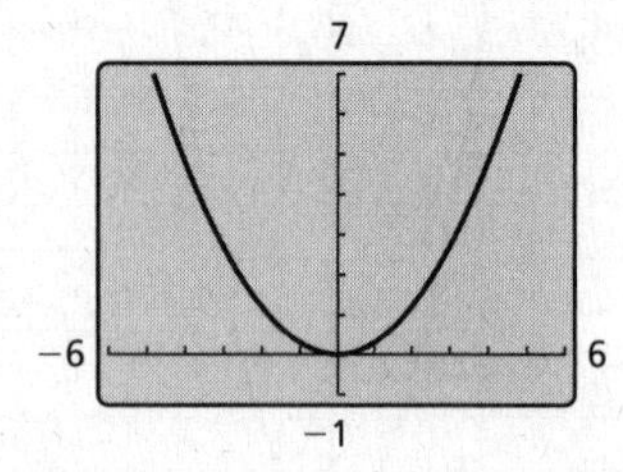

Name ______________________________ Date __________

8.1 Notetaking with Vocabulary

For use after Lesson 8.1

In your own words, write the meaning of each vocabulary term.

quadratic function

parabola

vertex

axis of symmetry

Core Concepts

Characteristics of Quadratic Functions

The *parent quadratic function* is $f(x) = x^2$. The graphs of all other quadratic functions are *transformations* of the graph of the parent quadratic function.

The lowest point on a parabola that opens up or the highest point on a parabola that opens down is the **vertex.** The vertex of the graph of $f(x) = x^2$ is (0, 0).

decreasing
increasing
axis of symmetry
vertex
y
x

The vertical line that divides the parabola into two symmetric parts is the **axis of symmetry.** The axis of symmetry passes through the vertex. For the graph of $f(x) = x^2$, the axis of symmetry is the y-axis, or $x = 0$.

Notes:

Name ______________________________ Date __________

Graphing $f(x) = ax^2$ When $a > 0$

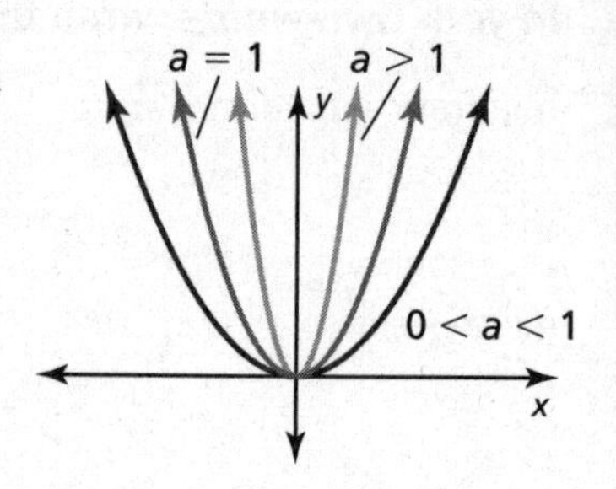

- When $0 < a < 1$, the graph of $f(x) = ax^2$ is a vertical shrink of the graph of $f(x) = x^2$.
- When $a > 1$, the graph of $f(x) = ax^2$ is a vertical stretch of the graph of $f(x) = x^2$.

Graphing $f(x) = ax^2$ When $a < 0$

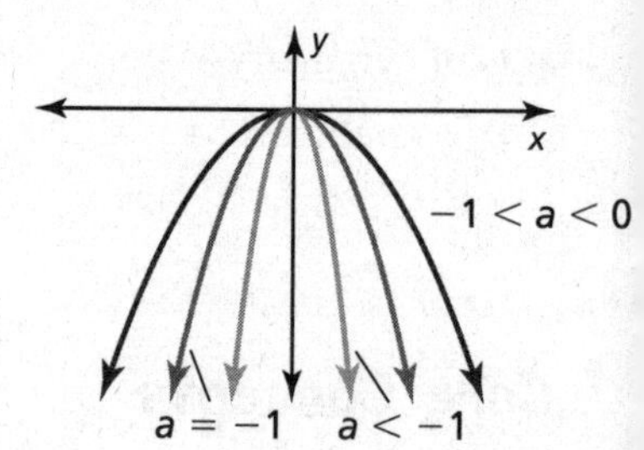

- When $-1 < a < 0$, the graph of $f(x) = ax^2$ is a vertical shrink with a reflection in the x-axis of the graph of $f(x) = x^2$.
- When $a < -1$, the graph of $f(x) = ax^2$ is a vertical stretch with a reflection in the x-axis of the graph of $f(x) = x^2$.

Notes:

Extra Practice

In Exercises 1 and 2, identify characteristics of the quadratic function and its graph.

1. 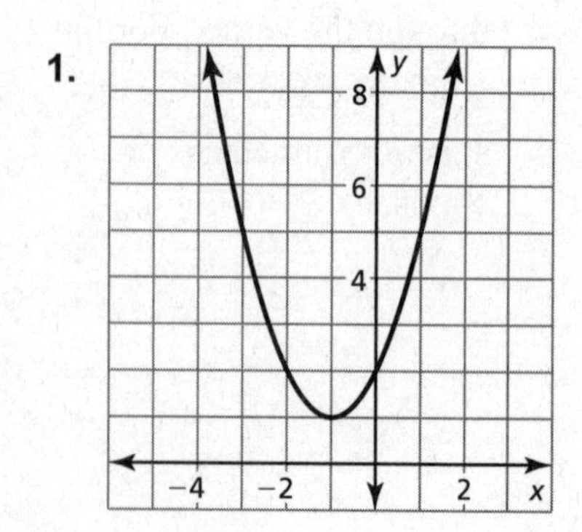

2. 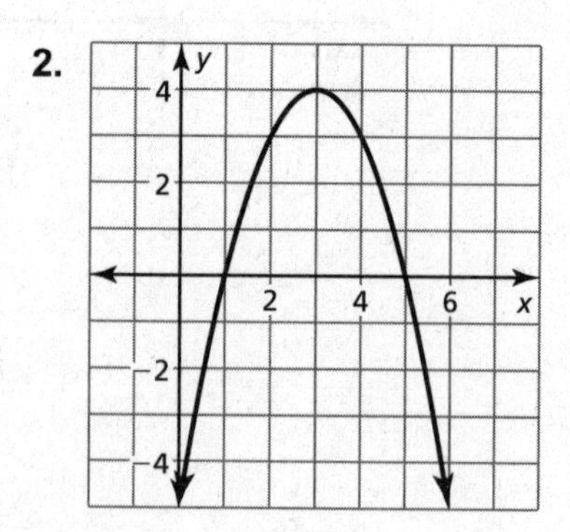

Name____ Date__________

8.1 Notetaking with Vocabulary (continued)

In Exercises 3–8, graph the function. Compare the graph to the graph of $f(x) = x^2$.

3. $g(x) = 5x^2$

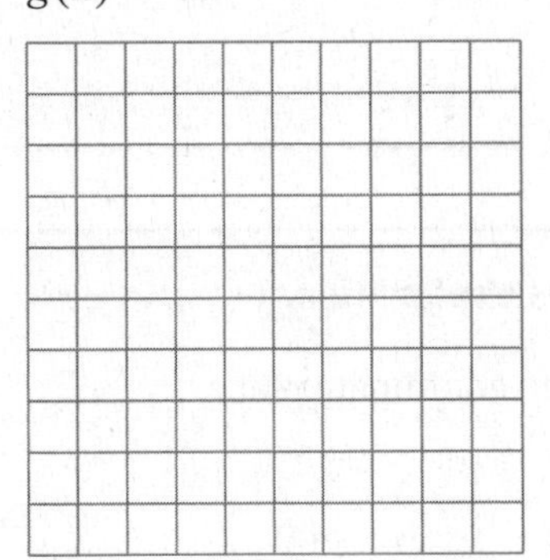

4. $m(x) = -4x^2$

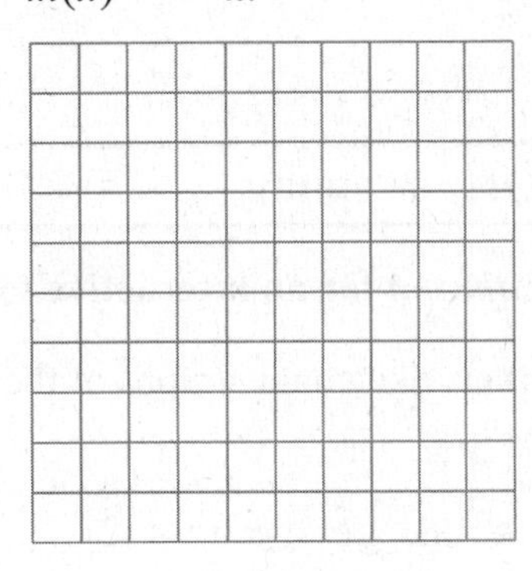

5. $k(x) = -x^2$

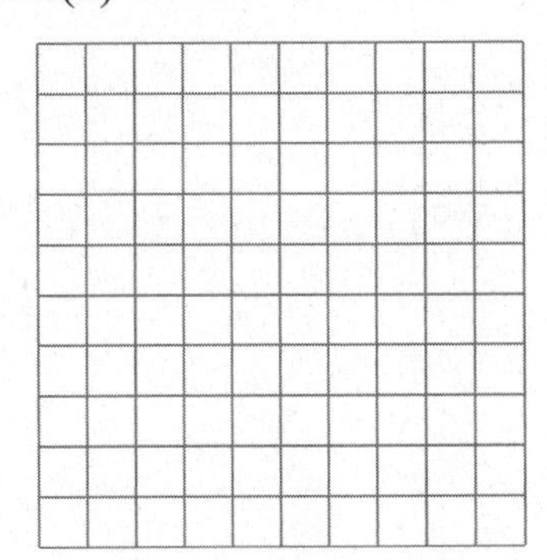

6. $l(x) = -7x^2$

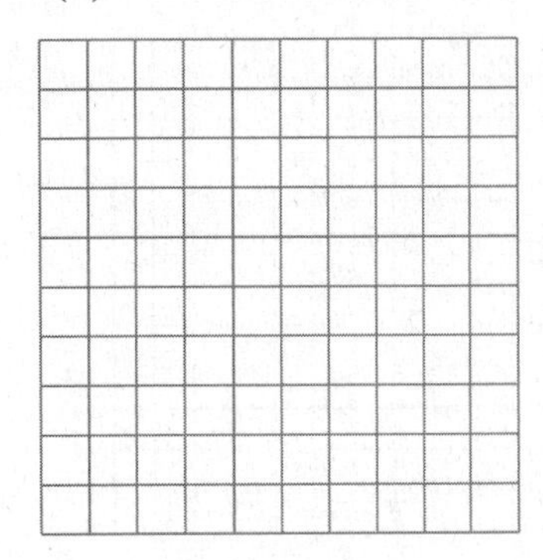

7. $n(x) = -\frac{1}{5}x^2$

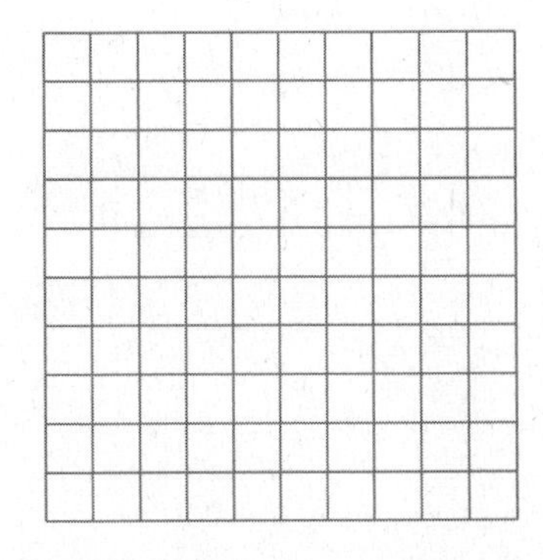

8. $p(x) = 0.6x^2$

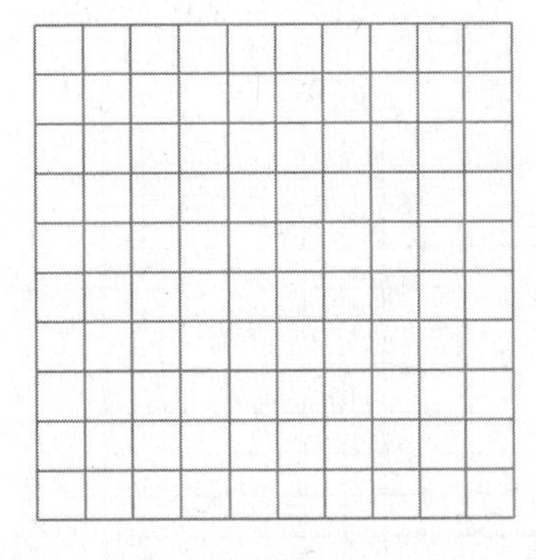

In Exercises 9 and 10, determine whether the statement is *always, sometimes,* or *never* true. Explain your reasoning.

9. The graph of $g(x) = ax^2$ is wider than the graph of $f(x) = x^2$ when $a > 0$.

10. The graph of $g(x) = ax^2$ is narrower than the graph of $f(x) = x^2$ when $|a| < 1$.

Name ______________________________ Date __________

8.2 Graphing $f(x) = ax^2 + c$

For use with Exploration 8.2

Essential Question How does the value of c affect the graph of $f(x) = ax^2 + c$?

1 EXPLORATION: Graphing $y = ax^2 + c$

Go to *BigIdeasMath.com* for an interactive tool to investigate this exploration.

Work with a partner. Sketch the graphs of the functions in the same coordinate plane. What do you notice?

a. $f(x) = x^2$ and $g(x) = x^2 + 2$

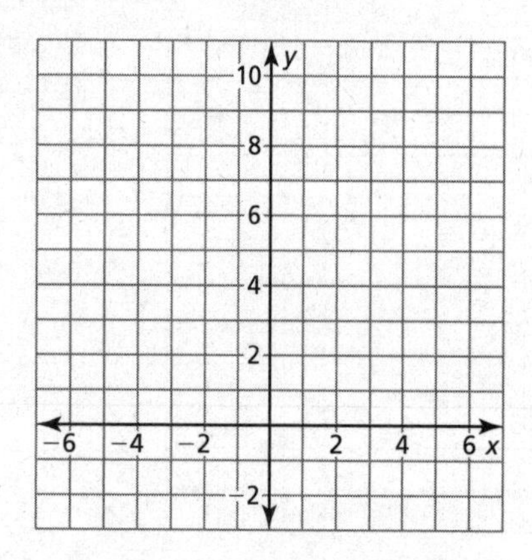

b. $f(x) = 2x^2$ and $g(x) = 2x^2 - 2$

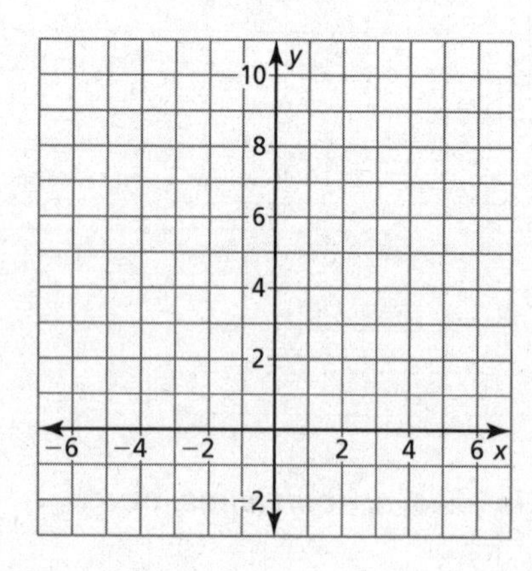

Name____ Date__________

2 EXPLORATION: Finding x-Intercepts of Graphs

Go to *BigIdeasMath.com* for an interactive tool to investigate this exploration.

Work with a partner. Graph each function. Find the x-intercepts of the graph. Explain how you found the x-intercepts.

a. $y = x^2 - 7$

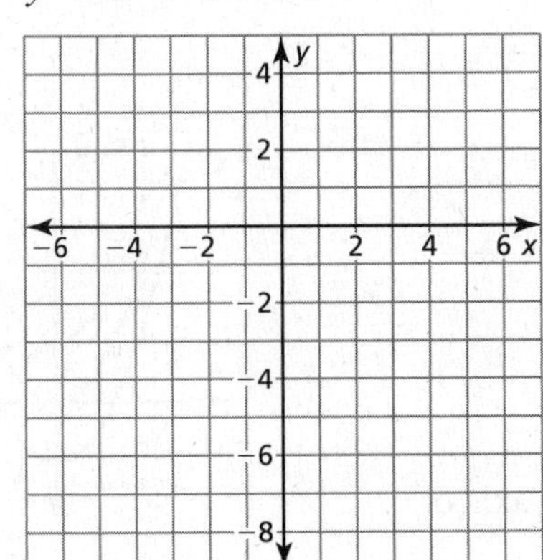

b. $y = -x^2 + 1$

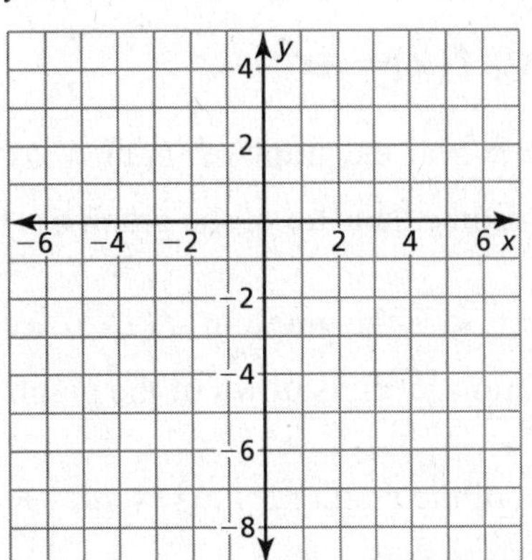

Communicate Your Answer

3. How does the value of c affect the graph of $f(x) = ax^2 + c$?

4. Use a graphing calculator to verify your answers to Question 3.

5. The figure shows the graph of a quadratic function of the form $y = ax^2 + c$. Describe possible values of a and c. Explain your reasoning.

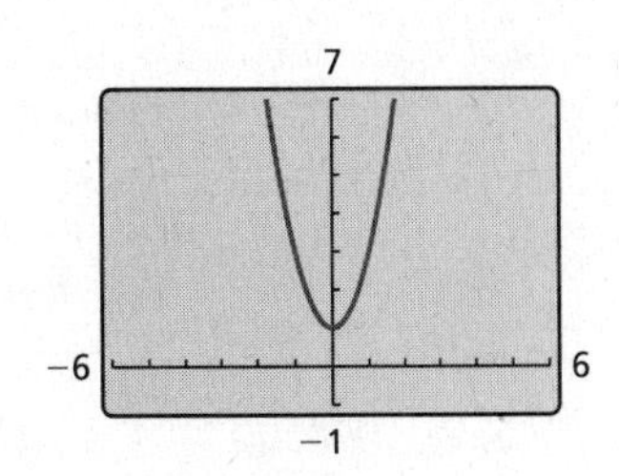

Name ______________________________ Date ________

8.2 Notetaking with Vocabulary
For use after Lesson 8.2

In your own words, write the meaning of each vocabulary term.

zero of a function

Core Concepts

Graphing $f(x) = ax^2 + c$

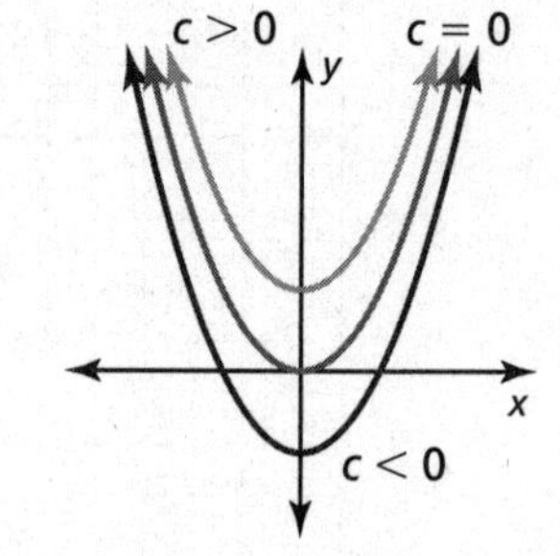

- When $c > 0$, the graph of $f(x) = ax^2 + c$ is a vertical translation c units up of the graph of $f(x) = ax^2$.
- When $c < 0$, the graph of $f(x) = ax^2 + c$ is a vertical translation $|c|$ units down of the graph of $f(x) = ax^2$.

The vertex of the graph of $f(x) = ax^2 + c$ is $(0, c)$, and the axis of symmetry is $x = 0$.

Notes:

Name____ Date__________

8.2 Notetaking with Vocabulary (continued)

Extra Practice

In Exercises 1–4, graph the function. Compare the graph to the graph of $f(x) = x^2$.

1. $g(x) = x^2 + 5$

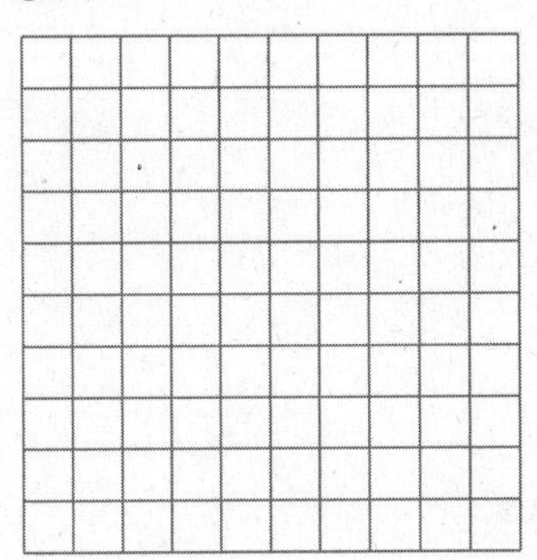

2. $m(x) = x^2 - 3$

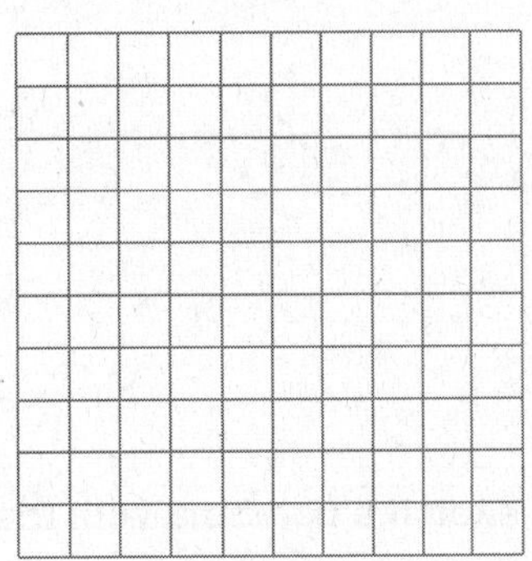

3. $n(x) = -3x^2 - 2$

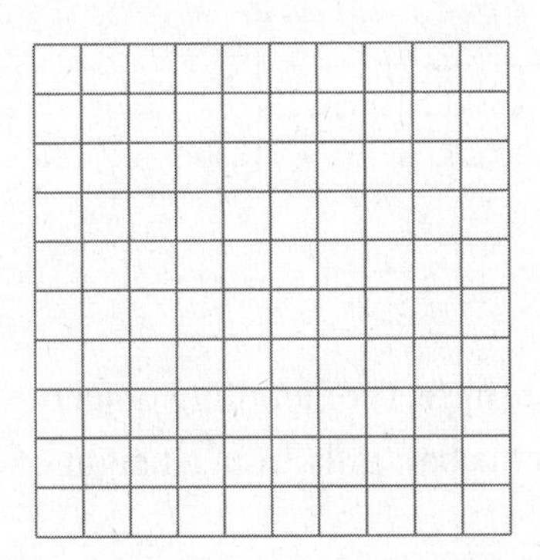

4. $q(x) = \frac{1}{2}x^2 - 4$

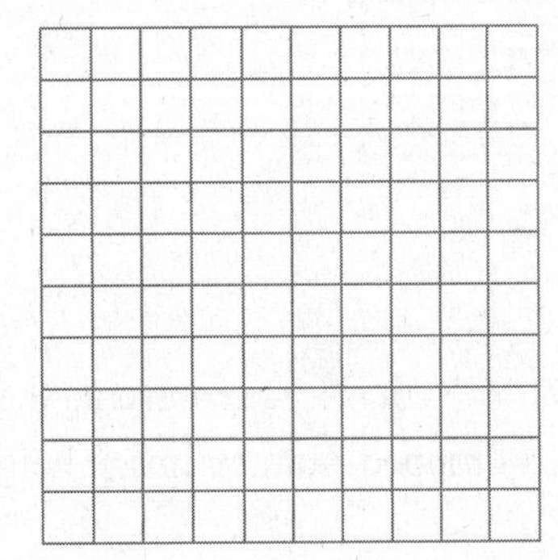

Name ______________________________ Date ________

8.2 Notetaking with Vocabulary (continued)

In Exercises 5–8, find the zeros of the function.

5. $y = -x^2 + 1$

6. $y = -4x^2 + 16$

7. $n(x) = -x^2 + 64$

8. $p(x) = -9x^2 + 1$

In Exercises 9 and 10, sketch a parabola with the given characteristics.

9. The parabola opens down, and the vertex is (0, 5).

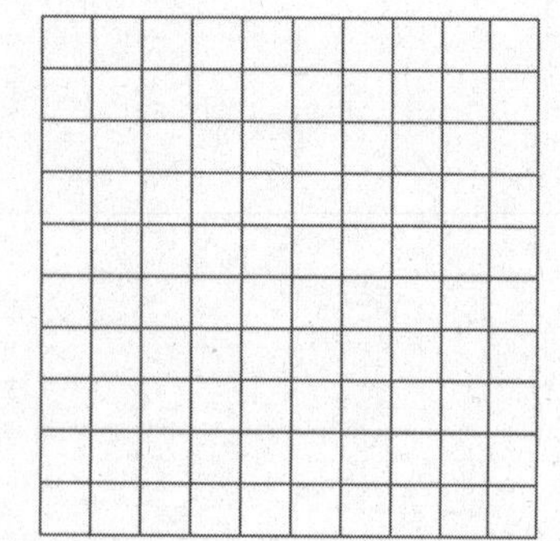

10. The lowest point on the parabola is (0, 4).

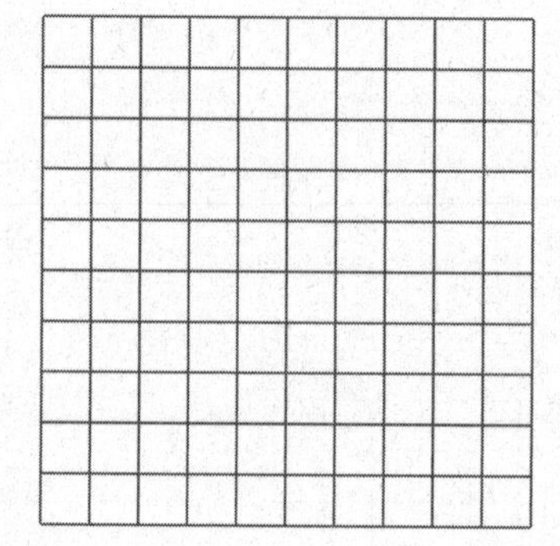

11. The function $f(t) = -16t^2 + s_0$ represents the approximate height (in feet) of a falling object t seconds after it is dropped from an initial height s_0 (in feet). A tennis ball falls from a height of 400 feet.

a. After how many seconds does the tennis ball hit the ground?

b. Suppose the initial height is decreased by 384 feet. After how many seconds does the ball hit the ground?

Name______________________________ Date__________

8.3 Graphing $f(x) = ax^2 + bx + c$

For use with Exploration 8.3

Essential Question How can you find the vertex of the graph of $f(x) = ax^2 + bx + c$?

1 EXPLORATION: Comparing x-Intercepts with the Vertex

Go to *BigIdeasMath.com* for an interactive tool to investigate this exploration.

Work with a partner.

a. Sketch the graphs of $y = 2x^2 - 8x$ and $y = 2x^2 - 8x + 6$.

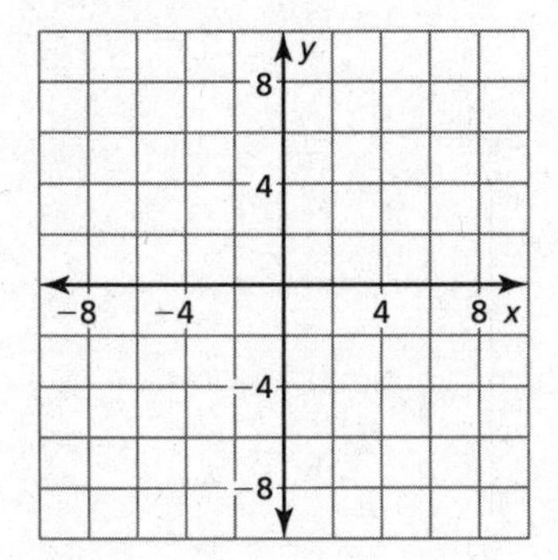

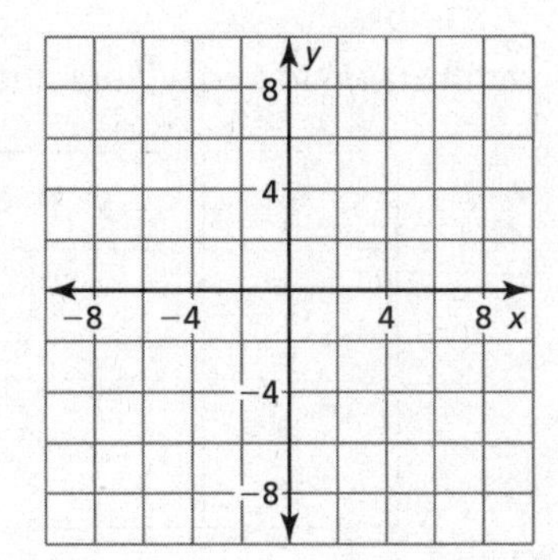

b. What do you notice about the x-coordinate of the vertex of each graph?

c. Use the graph of $y = 2x^2 - 8x$ to find its x-intercepts. Verify your answer by solving $0 = 2x^2 - 8x$.

d. Compare the value of the x-coordinate of the vertex with the values of the x-intercepts.

2 EXPLORATION: Finding x-Intercepts

Work with a partner.

a. Solve $0 = ax^2 + bx$ for x by factoring.

b. What are the x-intercepts of the graph of $y = ax^2 + bx$?

c. Complete the table to verify your answer.

x	$y = ax^2 + bx$
0	
$-\frac{b}{a}$	

3 EXPLORATION: Deductive Reasoning

Work with a partner. Complete the following logical argument.

The x-intercepts of the graph of $y = ax^2 + bx$ are 0 and $-\frac{b}{a}$.

The vertex of the graph of $y = ax^2 + bx$ occurs when $x =$ ________.

The vertices of the graphs of $y = ax^2 + bx$ and $y = ax^2 + bx + c$ have the same x-coordinate.

The vertex of the graph of $y = ax^2 + bx + c$ occurs when $x =$ ________.

Communicate Your Answer

4. How can you find the vertex of the graph of $f(x) = ax^2 + bx + c$?

5. Without graphing, find the vertex of the graph of $f(x) = x^2 - 4x + 3$. Check your result by graphing.

Name____ Date__________

8.3 Notetaking with Vocabulary

For use after Lesson 8.3

In your own words, write the meaning of each vocabulary term.

maximum value

minimum value

Core Concepts

Graphing $f(x) = ax^2 + bx + c$

- The graph opens up when $a > 0$, and the graph opens down when $a < 0$.
- The y-intercept is c.
- The x-coordinate of the vertex is $-\frac{b}{2a}$.
- The axis of symmetry is $x = -\frac{b}{2a}$.

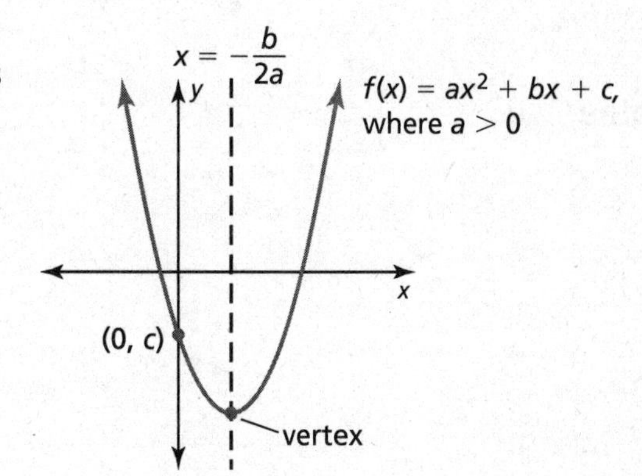

Notes:

Maximum and Minimum Values

The y-coordinate of the vertex of the graph of $f(x) = ax^2 + bx + c$ is the **maximum value** of the function when $a < 0$ or the **minimum value** of the function when $a > 0$.

$f(x) = ax^2 + bx + c, a < 0$

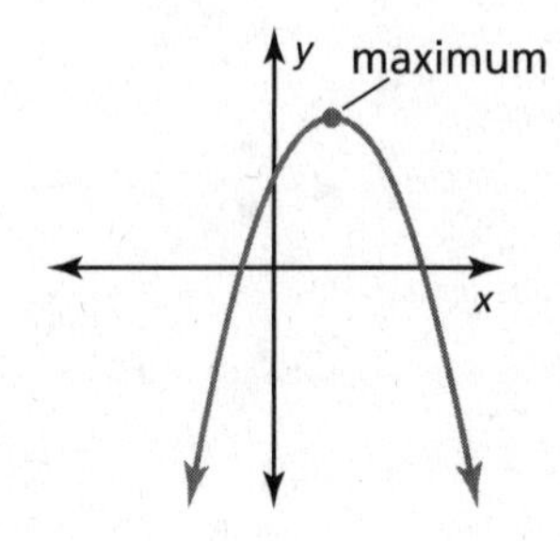

$f(x) = ax^2 + bx + c, a > 0$

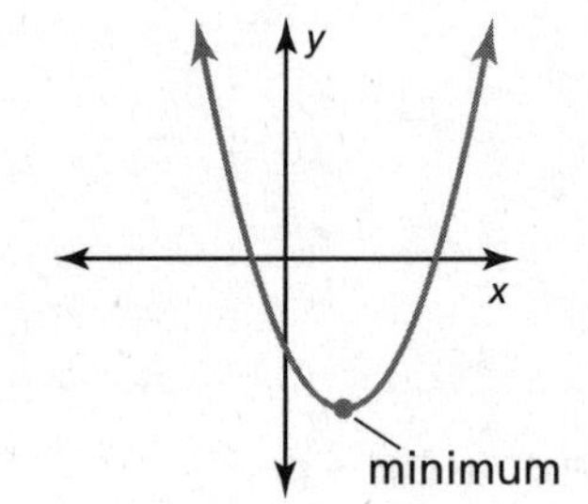

Notes:

Extra Practice

In Exercises 1–4, find (a) the axis of symmetry and (b) the vertex of the graph of the function.

1. $f(x) = x^2 - 10x + 2$

2. $y = -4x^2 + 16x$

3. $y = -2x^2 - 8x + 5$

4. $f(x) = -3x^2 + 6x + 1$

In Exercises 5–7, graph the function. Describe the domain and range.

5. $f(x) = 3x^2 + 6x + 2$

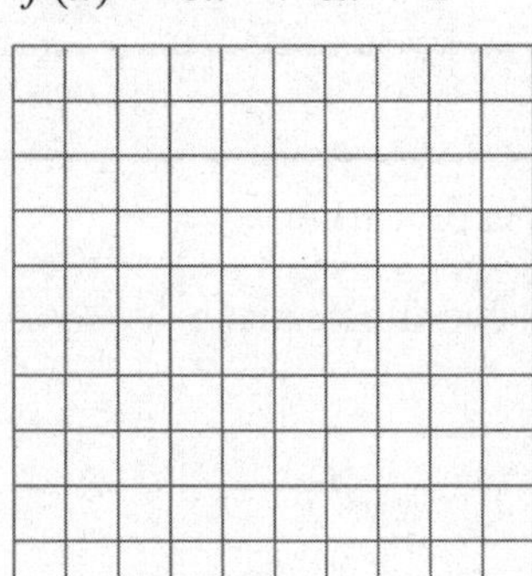

6. $y = 2x^2 - 8x - 1$

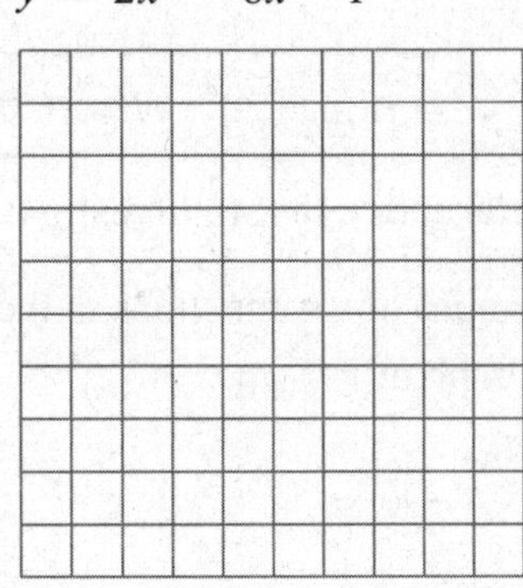

7. $y = -\frac{1}{5}x^2 - x + 5$

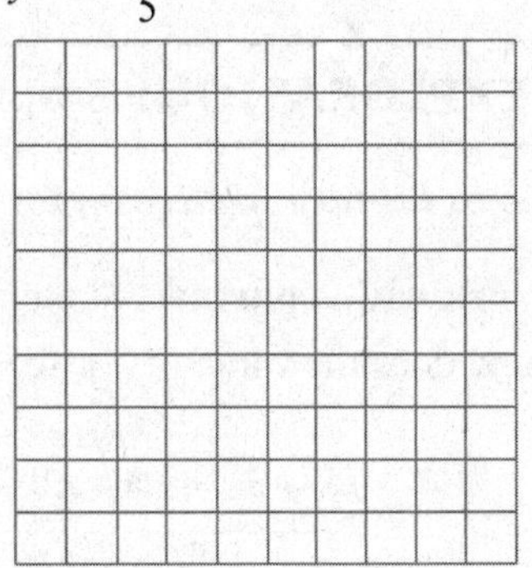

In Exercises 8–13, tell whether the function has a minimum value or a maximum value. Then find the value.

8. $y = -\frac{1}{2}x^2 - 5x + 2$

9. $y = 8x^2 + 16x - 2$

10. $y = -x^2 - 4x - 7$

11. $y = -7x^2 + 7x + 5$

12. $y = 9x^2 + 6x + 4$

13. $y = -\frac{1}{4}x^2 + x - 6$

14. The function $h = -16t^2 + 250t$ represents the height h (in feet) of a rocket t seconds after it is launched. The rocket explodes at its highest point.

a. When does the rocket explode?

b. At what height does the rocket explode?

Name ____ Date __________

8.4 Graphing $f(x) = a(x - h)^2 + k$

For use with Exploration 8.4

Essential Question How can you describe the graph of $f(x) = a(x - h)^2$?

1 EXPLORATION: Graphing $y = a(x - h)^2$ When $h > 0$

Go to *BigIdeasMath.com* for an interactive tool to investigate this exploration.

Work with a partner. Sketch the graphs of the functions in the same coordinate plane. How does the value of h affect the graph of $y = a(x - h)^2$?

a. $f(x) = x^2$ and $g(x) = (x - 2)^2$

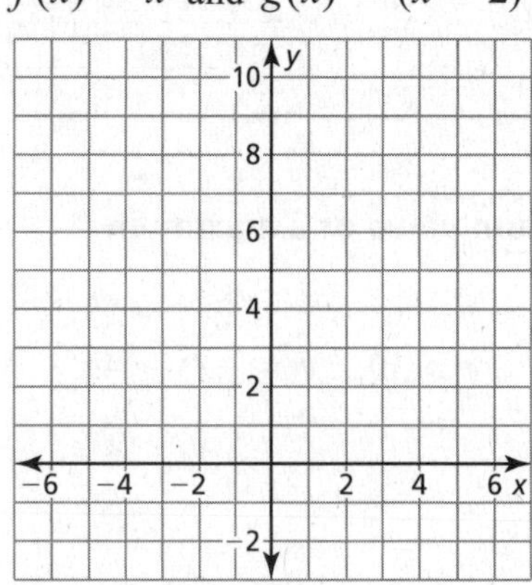

b. $f(x) = 2x^2$ and $g(x) = 2(x - 2)^2$

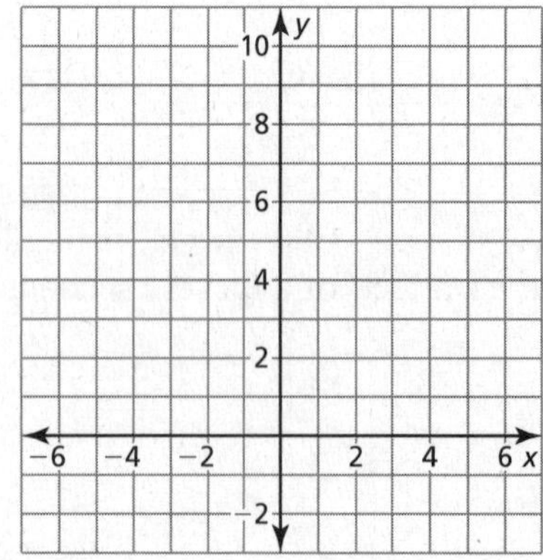

2 EXPLORATION: Graphing $y = a(x - h)^2$ When $h < 0$

Go to *BigIdeasMath.com* for an interactive tool to investigate this exploration.

Work with a partner. Sketch the graphs of the functions in the same coordinate plane. How does the value of h affect the graph of $y = a(x - h)^2$?

a. $f(x) = -x^2$ and $g(x) = -(x + 2)^2$

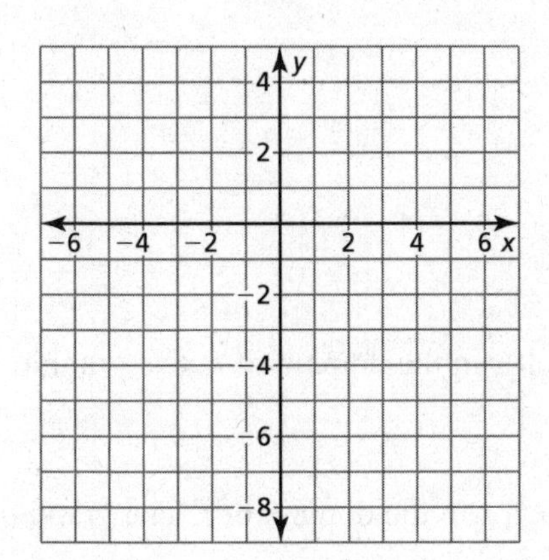

b. $f(x) = -2x^2$ and $g(x) = -2(x + 2)^2$

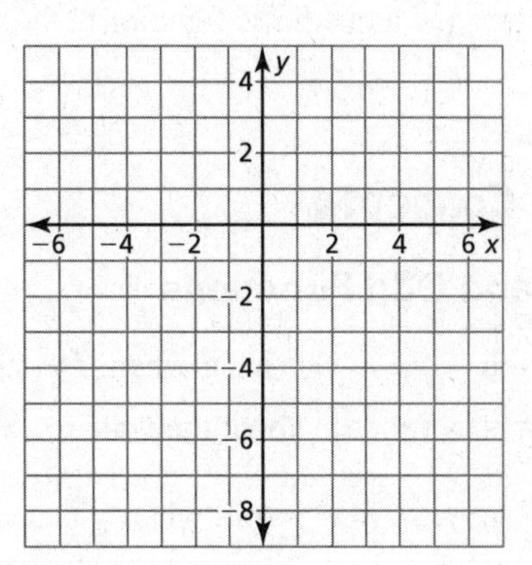

Communicate Your Answer

3. How can you describe the graph of $f(x) = a(x - h)^2$?

4. Without graphing, describe the graph of each function. Use a graphing calculator to check your answer.

a. $y = (x - 3)^2$

b. $y = (x + 3)^2$

c. $y = -(x - 3)^2$

Name ____ Date __________

8.4 Notetaking with Vocabulary

For use after Lesson 8.4

In your own words, write the meaning of each vocabulary term.

even function

odd function

vertex form (of a quadratic function)

Core Concepts

Even and Odd Functions

A function $y = f(x)$ is **even** when $f(-x) = f(x)$ for each x in the domain of f. The graph of an even function is symmetric about the y-axis.

A function $y = f(x)$ is **odd** when $f(-x) = -f(x)$ for each x in the domain of f. The graph of an odd function is symmetric about the origin. A graph is *symmetric about the origin* when it looks the same after reflections in the x-axis and then in the y-axis.

Notes:

Graphing $f(x) = a(x - h)^2$

- When $h > 0$, the graph of $f(x) = a(x - h)^2$ is a horizontal translation h units right of the graph $f(x) = ax^2$.
- When $h < 0$, the graph of $f(x) = a(x - h)^2$ is a horizontal translation $|h|$ units left of the graph of $f(x) = ax^2$.

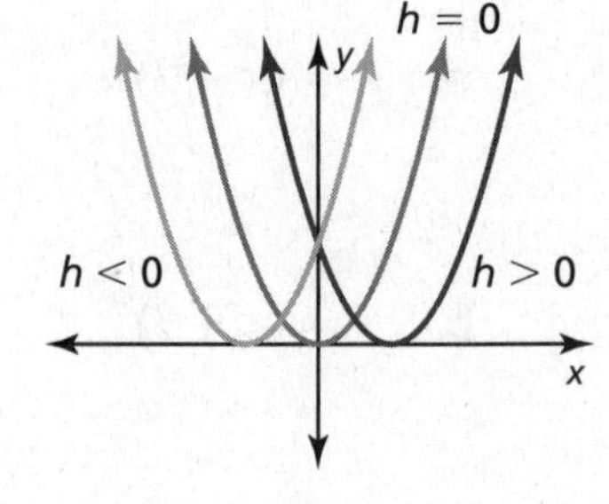

The vertex of the graph of $f(x) = a(x - h)^2$ is $(h, 0)$, and the axis of symmetry is $x = h$.

Notes:

Name____ Date__________

Graphing $f(x) = a(x - h)^2 + k$

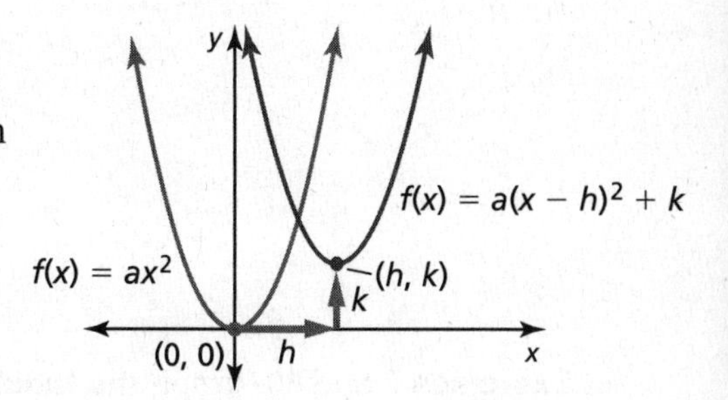

The **vertex form** of a quadratic function is $f(x) = a(x - h)^2 + k$, where $a \neq 0$. The graph of $f(x) = a(x - h)^2 + k$ is a translation h units horizontally and k units vertically of the graph of $f(x) = ax^2$.

The vertex of the graph of $f(x) = a(x - h)^2 + k$ is (h, k), and the axis of symmetry is $x = h$.

Notes:

Extra Practice

In Exercises 1–4, determine whether the function is *even, odd,* or *neither.*

1. $f(x) = 5x$

2. $f(x) = -4x^2$

3. $h(x) = \frac{1}{2}x^2$

4. $f(x) = -3x^2 + 2x + 1$

In Exercises 5–8, find the vertex and the axis of symmetry of the graph of the function.

5. $f(x) = 5(x - 2)^2$

6. $f(x) = -4(x + 8)^2$

8.4 Notetaking with Vocabulary (continued)

7. $p(x) = -\frac{1}{2}(x - 1)^2 + 4$

8. $g(x) = -(x + 1)^2 - 5$

In Exercises 9 and 10, graph the function. Compare the graph to the graph of $f(x) = x^2$.

9. $m(x) = 3(x + 2)^2$

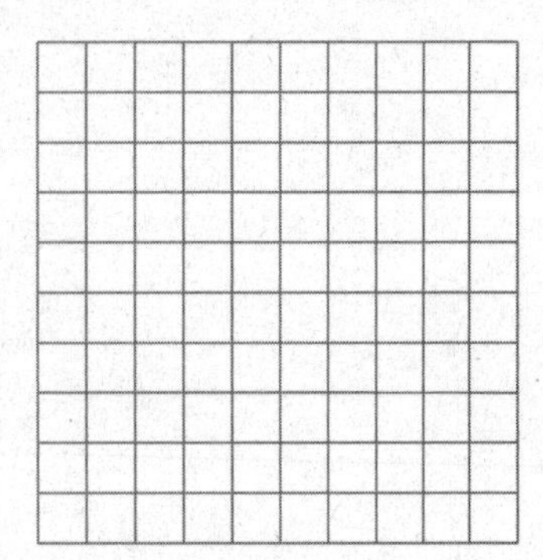

10. $g(x) = -\frac{1}{4}(x - 6)^2 + 4$

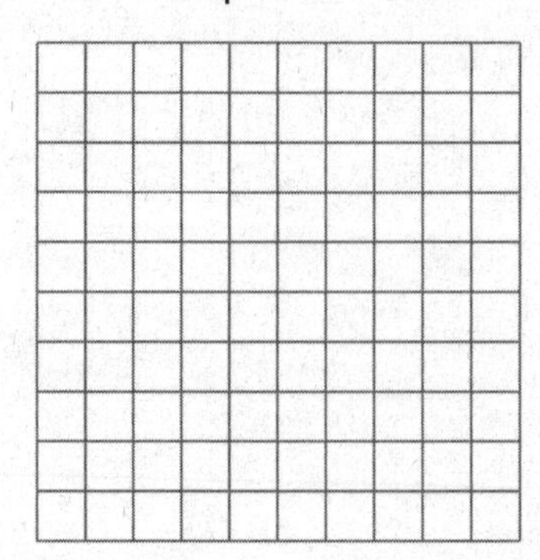

In Exercises 11 and 12, graph *g*.

11. $f(x) = 3(x + 1)^2 - 1; g(x) = f(x + 2)$

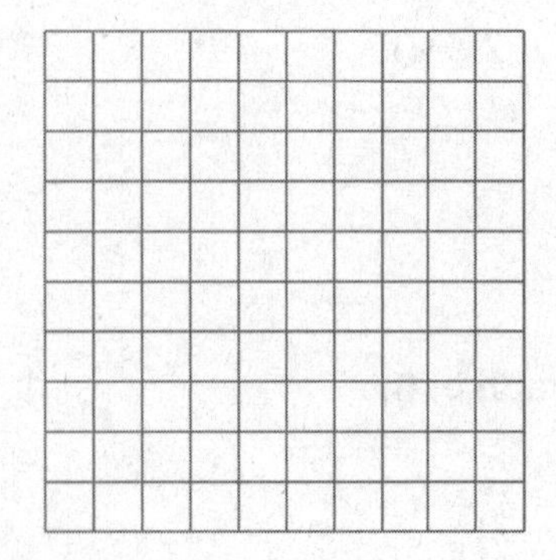

12. $f(x) = \frac{1}{2}(x - 3)^2 - 5; g(x) = -f(x)$

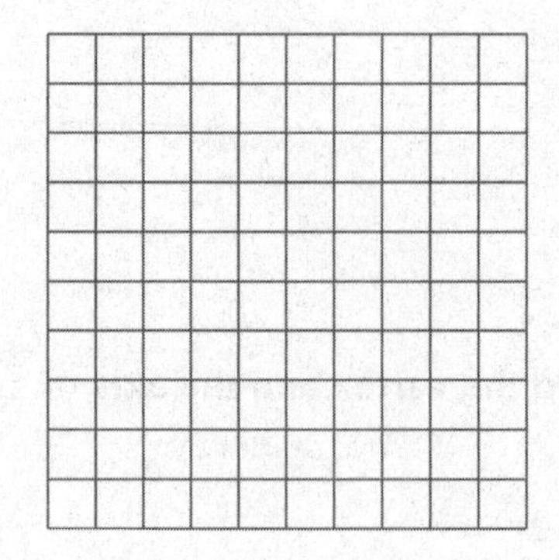

Name______________________________ Date__________

8.5 Using Intercept Form

For use with Exploration 8.5

Essential Question What are some of the characteristics of the graph of $f(x) = a(x - p)(x - q)$?

1 EXPLORATION: Using Zeros to Write Functions

Work with a partner. Each graph represents a function of the form $f(x) = (x - p)(x - q)$ or $f(x) = -(x - p)(x - q)$. Write the function represented by each graph. Explain your reasoning.

a.

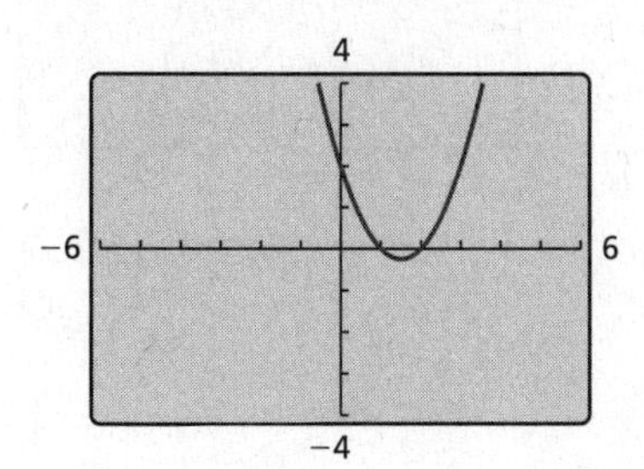

b.

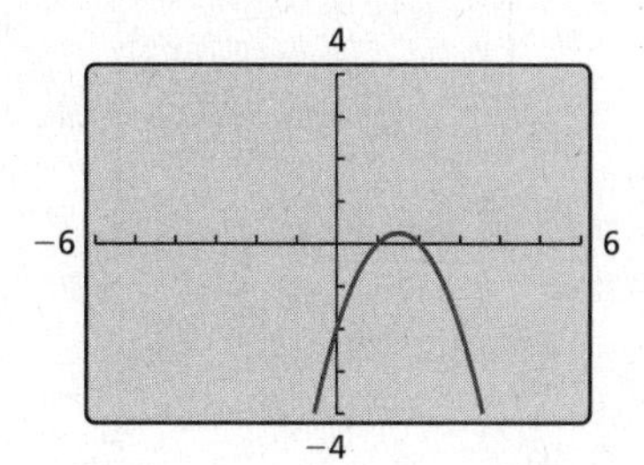

c.

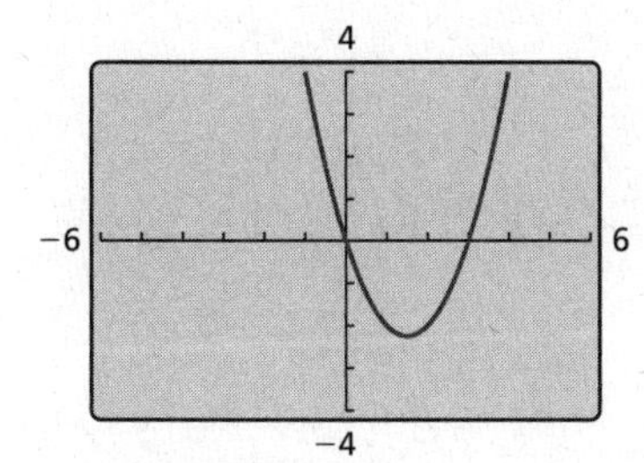

d.

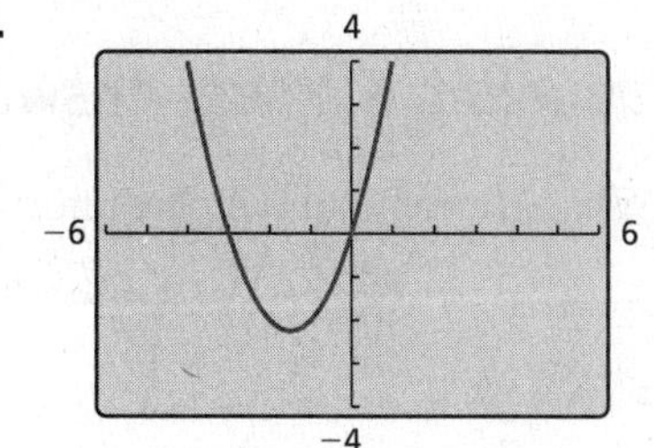

e.

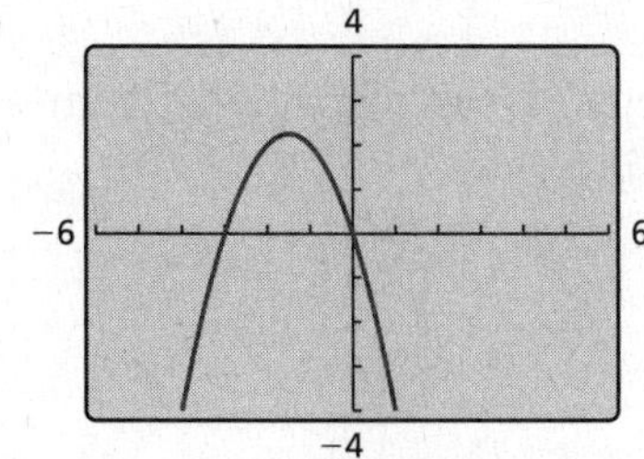

f.

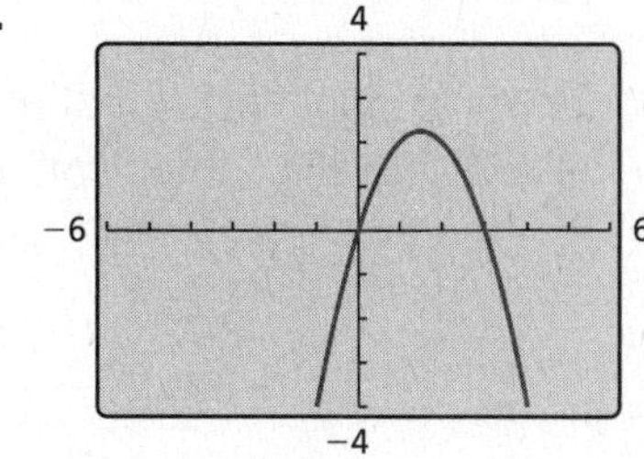

1 EXPLORATION: Using Zeros to Write Functions (continued)

g.

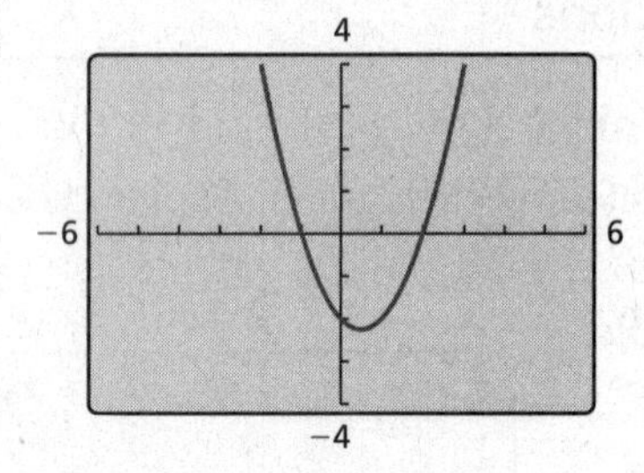

h.

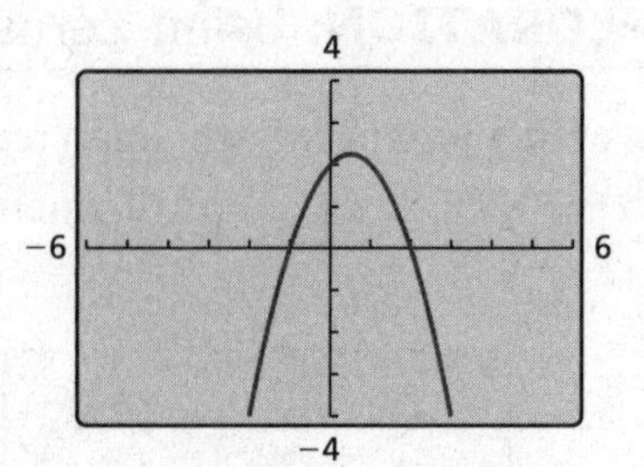

Communicate Your Answer

2. What are some of the characteristics of the graph of $f(x) = a(x - p)(x - q)$?

3. Consider the graph of $f(x) = a(x - p)(x - q)$.

a. Does changing the sign of a change the x-intercepts? Does changing the sign of a change the y-intercept? Explain your reasoning.

b. Does changing the value of p change the x-intercepts? Does changing the value of p change the y-intercept? Explain your reasoning.

Name______________________________ Date__________

8.5 Notetaking with Vocabulary

For use after Lesson 8.5

In your own words, write the meaning of each vocabulary term.

intercept form

Core Concepts

Graphing $f(x) = a(x - p)(x - q)$

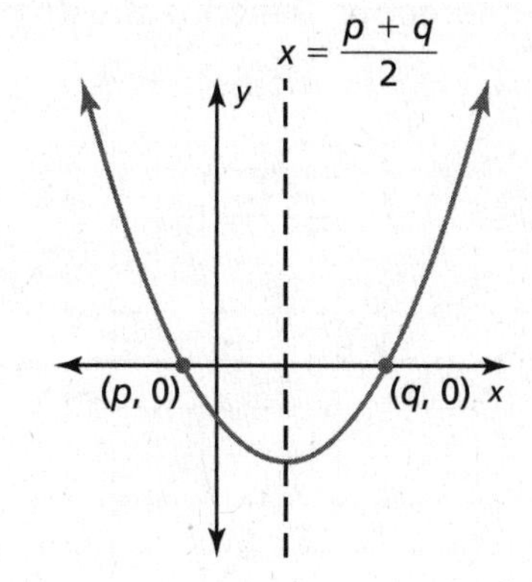

- The x-intercepts are p and q.
- The axis of symmetry is halfway between $(p, 0)$ and $(q, 0)$. So, the axis of symmetry is $x = \frac{p + q}{2}$.
- The graph opens up when $a > 0$, and the graph opens down when $a < 0$.

Notes:

Factors and Zeros

For any factor $x - n$ of a polynomial, n is a zero of the function defined by the polynomial.

Notes:

Extra Practice

In Exercises 1 and 2, find the *x*-intercepts and axis of symmetry of the graph of the function.

1. $y = (x + 2)(x - 4)$

2. $y = -3(x - 2)(x - 3)$

In Exercises 3–6, graph the quadratic function. Label the vertex, axis of symmetry, and *x*-intercepts. Describe the domain and range of the function.

3. $m(x) = (x + 5)(x + 1)$

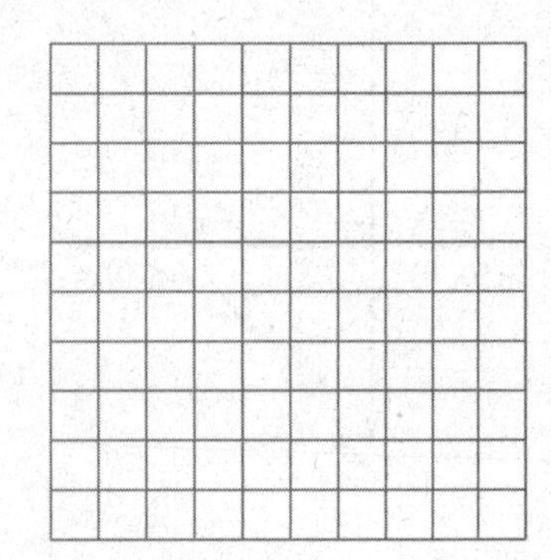

4. $y = -4(x - 3)(x - 1)$

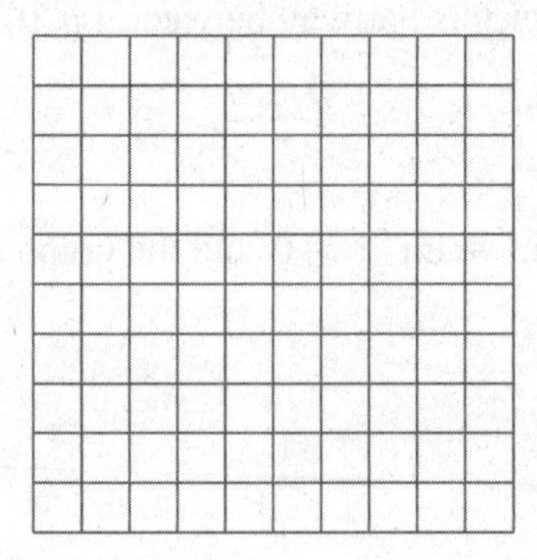

5. $y = x^2 - 4$

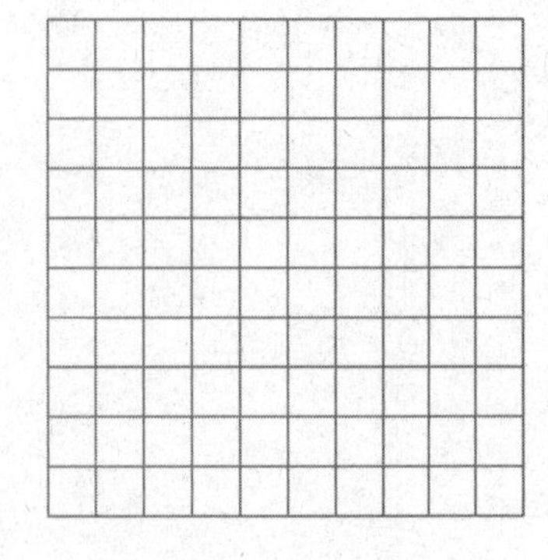

6. $f(x) = x^2 + 2x - 15$

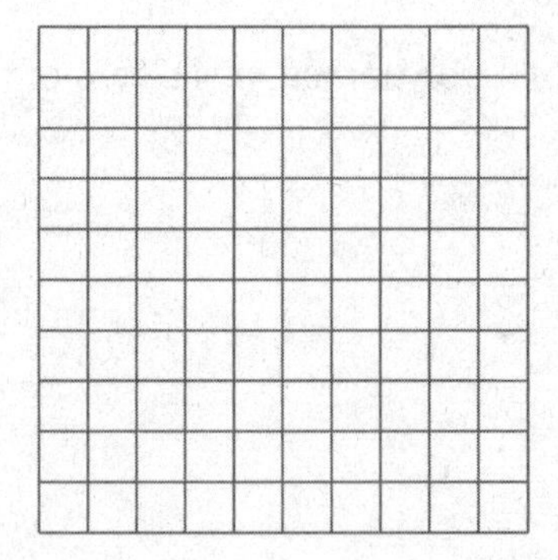

Name____ Date______________

8.5 Notetaking with Vocabulary (continued)

In Exercises 7 and 8, find the zero(s) of the function.

7. $y = 6x^2 - 6$

8. $y = x^2 + 9x + 20$

In Exercises 9–12, use zeros to graph the function.

9. $f(x) = x^2 - 3x - 10$

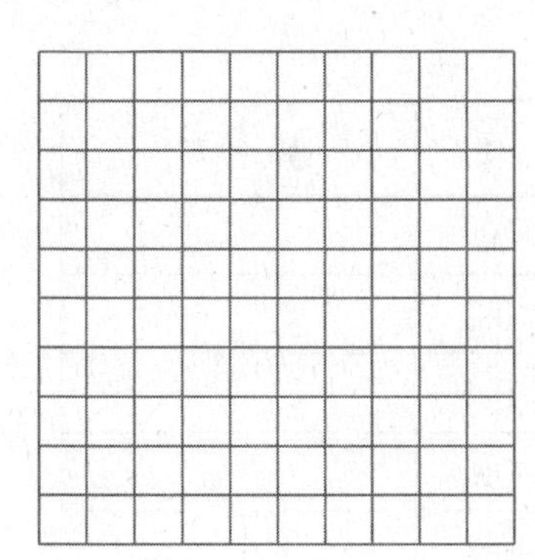

10. $f(x) = -2(x + 3)(x - 1)$

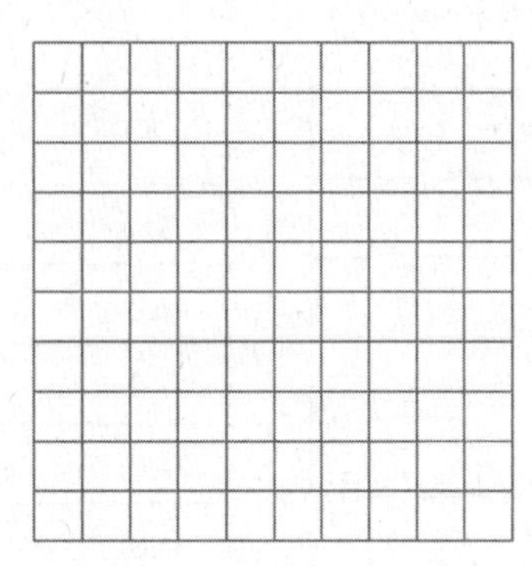

11. $f(x) = x^3 - 9x$

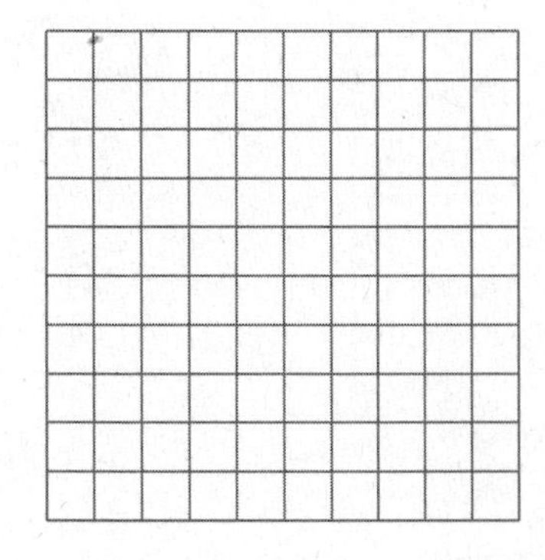

12. $f(x) = 2x^3 - 12x^2 + 10x$

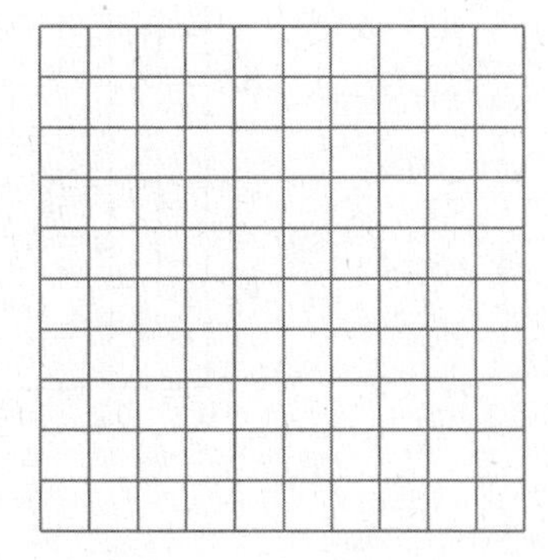

Name ____ Date __________

8.6 Comparing Linear, Exponential, and Quadratic Functions

For use with Exploration 8.6

Essential Question How can you compare the growth rates of linear, exponential, and quadratic functions?

1 EXPLORATION: Comparing Speeds

Go to *BigIdeasMath.com* for an interactive tool to investigate this exploration.

Work with a partner. Three cars start traveling at the same time. The distance traveled in t minutes is y miles. Complete each table and sketch all three graphs in the same coordinate plane. Compare the speeds of the three cars. Which car has a constant speed? Which car is accelerating the most? Explain your reasoning.

t	$y = t$
0	
0.2	
0.4	
0.6	
0.8	
1.0	

t	$y = 2^t - 1$
0	
0.2	
0.4	
0.6	
0.8	
1.0	

t	$y = t^2$
0	
0.2	
0.4	
0.6	
0.8	
1.0	

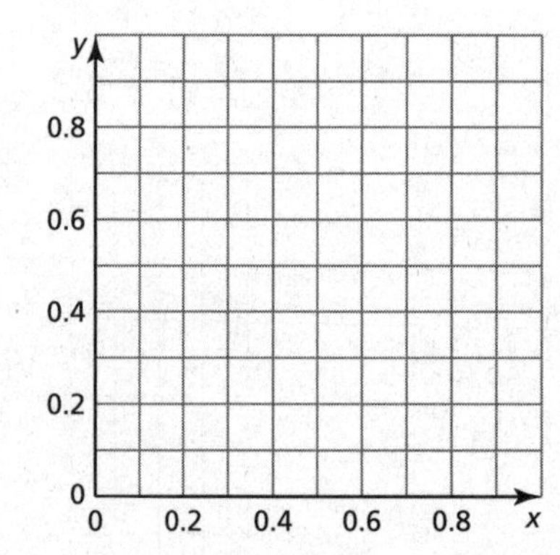

8.6 Comparing Linear, Exponential, and Quadratic Functions (continued)

2 EXPLORATION: Comparing Speeds

Work with a partner. Analyze the speeds of the three cars over the given time periods. The distance traveled in t minutes is y miles. Which car eventually overtakes the others?

t	$y = t$
1	
2	
3	
4	
5	
6	
7	
8	
9	

t	$y = 2^t - 1$
1	
2	
3	
4	
5	
6	
7	
8	
9	

t	$y = t^2$
1	
2	
3	
4	
5	
6	
7	
8	
9	

Communicate Your Answer

3. How can you compare the growth rates of linear, exponential, and quadratic functions?

4. Which function has a growth rate that is eventually much greater than the growth rates of the other two functions? Explain your reasoning.

Name ______________________________ Date __________

8.6 Notetaking with Vocabulary
For use after Lesson 8.6

In your own words, write the meaning of each vocabulary term.

average rate of change

Core Concepts

Linear, Exponential, and Quadratic Functions

Linear Function

$y = mx + b$

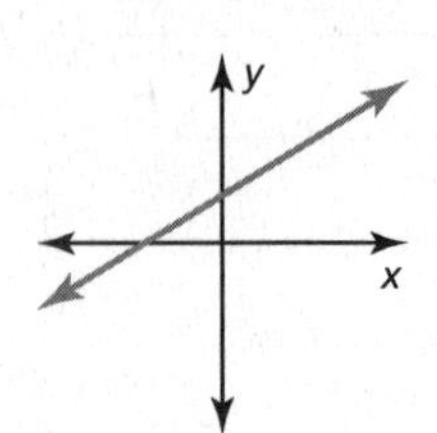

Exponential Function

$y = ab^x$

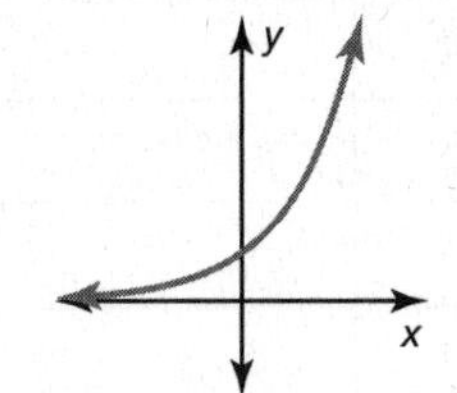

Quadratic Function

$y = ax^2 + bx + c$

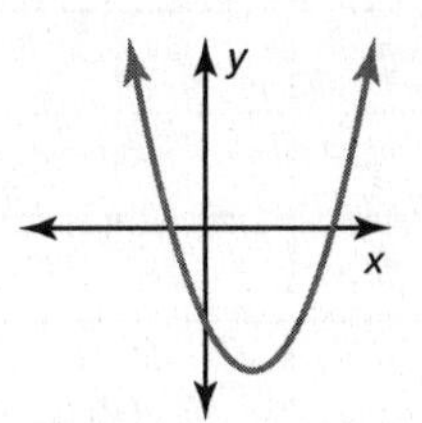

Notes:

Differences and Ratios of Functions

You can use patterns between consecutive data pairs to determine which type of function models the data. The differences of consecutive y-values are called *first differences.* The differences of consecutive first differences are called *second differences.*

- **Linear Function** The first differences are constant.
- **Exponential Function** Consecutive y-values have a common *ratio.*
- **Quadratic Function** The second differences are constant.

In all cases, the differences of consecutive x-values need to be constant.

Notes:

Comparing Functions Using Average Rates of Change

- Over the same interval, the average rate of change of a function increasing quadratically eventually exceeds the average rate of change of a function increasing linearly. So, the value of the quadratic function eventually exceeds the value of the linear function.
- Over the same interval, the average rate of change of a function increasing exponentially eventually exceeds the average rate of change of a function increasing linearly or quadratically. So, the value of the exponential function eventually exceeds the value of the linear or quadratic function.

Notes:

Extra Practice

In Exercises 1–4, plot the points. Tell whether the points appear to represent a *linear*, an *exponential*, or a *quadratic* function.

1. $(-3, 2), (-2, 4), (-4, 4), (-1, 8), (-5, 8)$

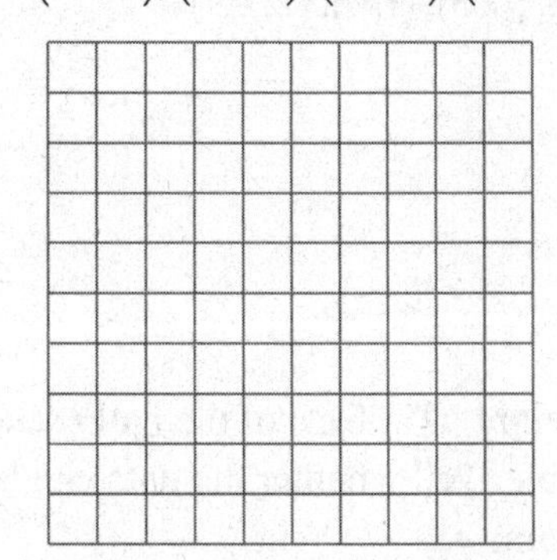

2. $(-3, 1), (-2, 2), (-1, 4), (0, 8), (2, 14)$

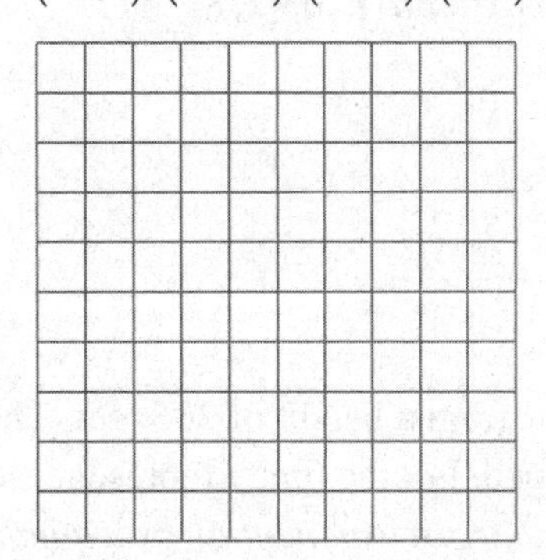

3. $(4, 0), (2, 1), (0, 3), (-1, 6), (-2, 10)$

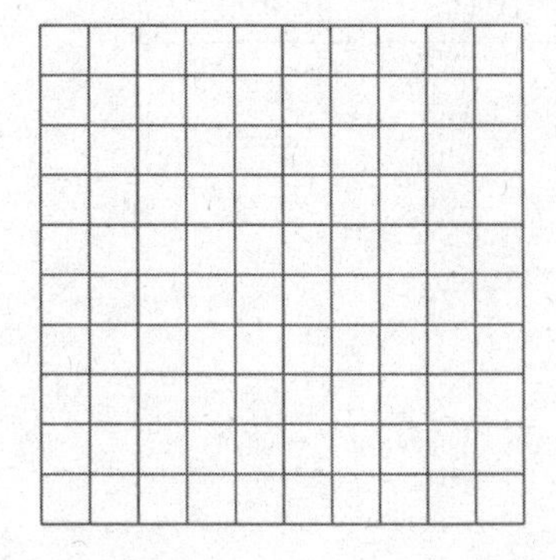

4. $(2, -4), (0, -2), (-2, 0), (-4, 2), (-6, 4)$

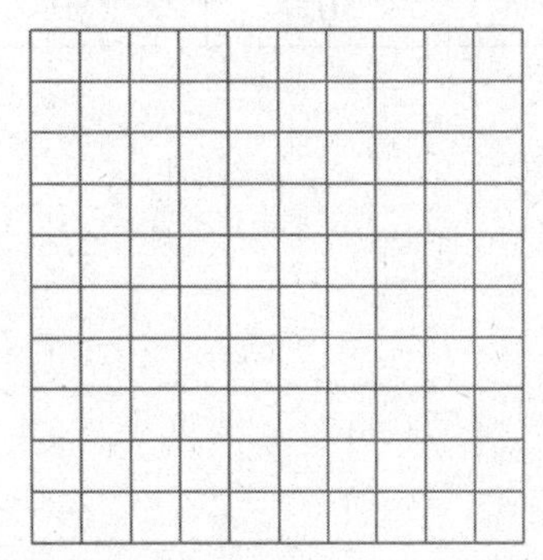

8.6 Notetaking with Vocabulary (continued)

In Exercises 5 and 6, tell whether the table of values represents a *linear*, an *exponential*, or a *quadratic* function.

5.

x	-2	-1	0	1	2
y	7	4	1	-2	-5

6.

x	-2	-1	0	1	2
y	6	2	0	2	6

In Exercises 7 and 8, tell whether the data represent a *linear*, an *exponential*, or a *quadratic* function. Then write the function.

7. $(-2, -4), (-1, -1), (0, 2), (1, 5), (2, 8)$

8. $(-2, -9), (-1, 0), (0, 3), (1, 0), (2, -9)$

9. A ball is dropped from a height of 305 feet. The table shows the height h (in feet) of the ball t seconds after being dropped. Let the time t represent the independent variable. Tell whether the data can be modeled by a *linear*, an *exponential*, or a *quadratic* function. Explain.

Time, t	0	1	2	3	4
Height, h	305	289	241	161	49

Chapter 9 Maintaining Mathematical Proficiency

Factor the trinomial.

1. $x^2 - 6x + 9$

2. $x^2 + 4x + 4$

3. $x^2 - 14x + 49$

4. $x^2 + 22x + 121$

5. $x^2 - 24x + 144$

6. $x^2 + 26x + 169$

Solve the system of linear equations by graphing.

7. $y = 2x - 1$
$y = -3x + 9$

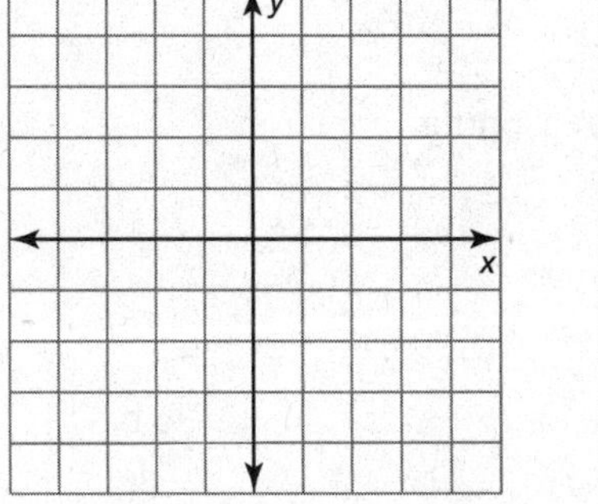

8. $y = -\frac{1}{2}x - 1$
$y = \frac{1}{4}x - 4$

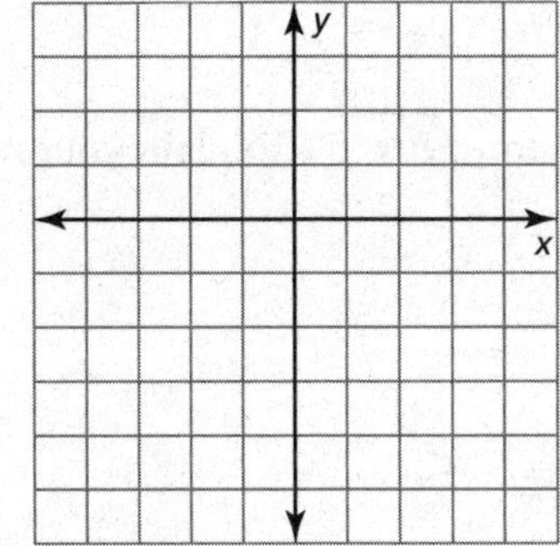

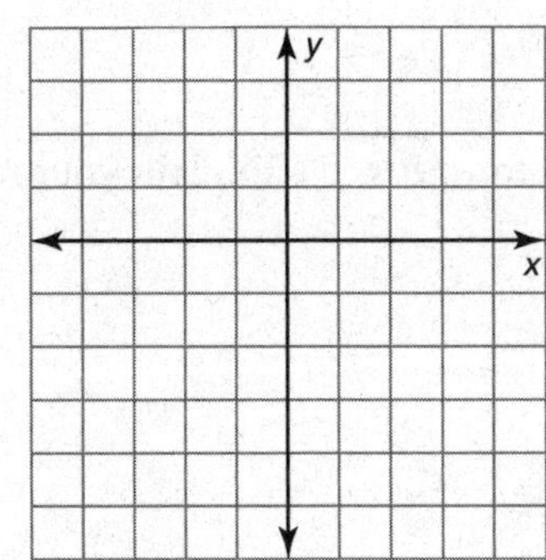

9. $y = 2x + 3$
$y = -3x - 2$

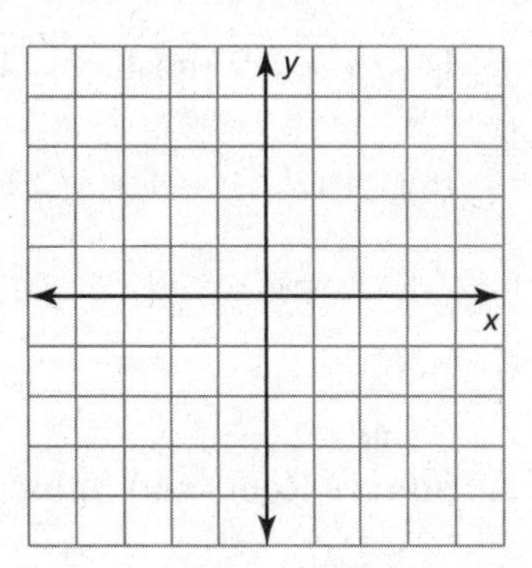

10. $y = x + 3$
$y = -\frac{1}{3}x - 1$

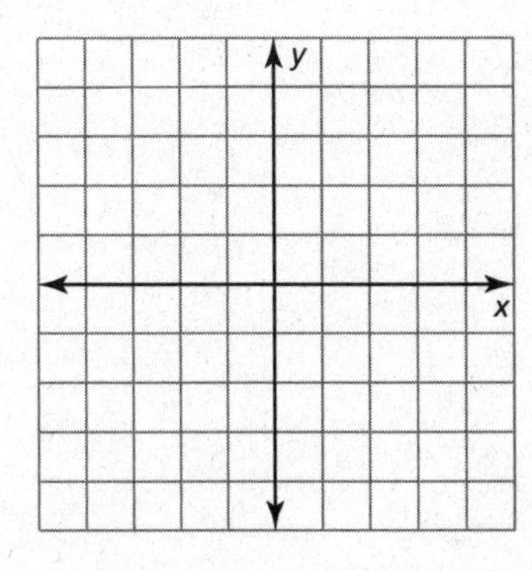

11. $y = x + 1$
$y = 3x - 1$

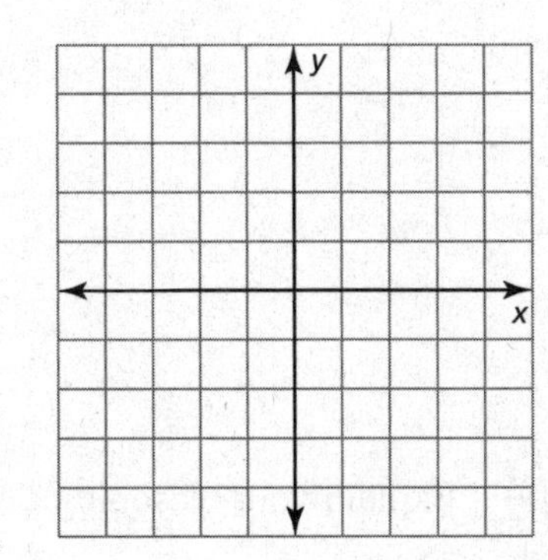

12. $y = 2x - 3$
$y = x + 1$

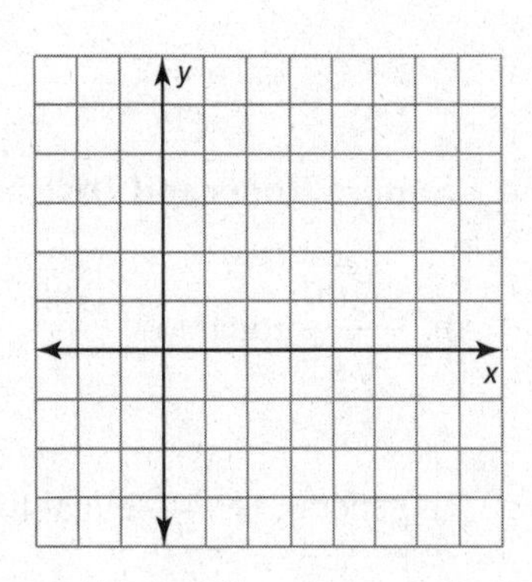

Name ______________________________ Date __________

9.1 Properties of Radicals
For use with Exploration 9.1

Essential Question How can you multiply and divide square roots?

1 EXPLORATION: Operations with Square Roots

Work with a partner. For each operation with square roots, compare the results obtained using the two indicated orders of operations. What can you conclude?

a. Square Roots and Addition

Is $\sqrt{36} + \sqrt{64}$ equal to $\sqrt{36 + 64}$?

In general, is $\sqrt{a} + \sqrt{b}$ equal to $\sqrt{a + b}$? Explain your reasoning.

b. Square Roots and Multiplication

Is $\sqrt{4} \bullet \sqrt{9}$ equal to $\sqrt{4 \bullet 9}$?

In general, is $\sqrt{a} \bullet \sqrt{b}$ equal to $\sqrt{a \bullet b}$? Explain your reasoning.

c. Square Roots and Subtraction

Is $\sqrt{64} - \sqrt{36}$ equal to $\sqrt{64 - 36}$?

In general, is $\sqrt{a} - \sqrt{b}$ equal to $\sqrt{a - b}$? Explain your reasoning.

d. Square Roots and Division

Is $\frac{\sqrt{100}}{\sqrt{4}}$ equal to $\sqrt{\frac{100}{4}}$?

In general, is $\frac{\sqrt{a}}{\sqrt{b}}$ equal to $\sqrt{\frac{a}{b}}$? Explain your reasoning.

Name____ Date__________

2 EXPLORATION: Writing Counterexamples

Work with a partner. A **counterexample** is an example that proves that a general statement is *not* true. For each general statement in Exploration 1 that is not true, write a counterexample different from the example given.

Communicate Your Answer

3. How can you multiply and divide square roots?

4. Give an example of multiplying square roots and an example of dividing square roots that are different from the examples in Exploration 1.

5. Write an algebraic rule for each operation.

a. the product of square roots

b. the quotient of square roots

Name ______________________________ Date __________

9.1 Notetaking with Vocabulary
For use after Lesson 9.1

In your own words, write the meaning of each vocabulary term.

counterexample

radical expression

simplest form

rationalizing the denominator

conjugates

like radicals

Core Concepts

Product Property of Square Roots

Words The square root of a product equals the product of the square roots of the factors.

Numbers $\sqrt{9 \cdot 5} = \sqrt{9} \cdot \sqrt{5} = 3\sqrt{5}$

Algebra $\sqrt{ab} = \sqrt{a} \cdot \sqrt{b}$, where $a, b \geq 0$

Notes:

Name________________________________ Date__________

9.1 Notetaking with Vocabulary (continued)

Quotient Property of Square Roots

Words The square root of a quotient equals the quotient of the square roots of the numerator and denominator.

Numbers $\sqrt{\frac{3}{4}} = \frac{\sqrt{3}}{\sqrt{4}} = \frac{\sqrt{3}}{2}$

Algebra $\sqrt{\frac{a}{b}} = \frac{\sqrt{a}}{\sqrt{b}}$, where $a \geq 0$ and $b > 0$

Notes:

Extra Practice

In Exercises 1–12, simplify the expression.

1. $\sqrt{24}$

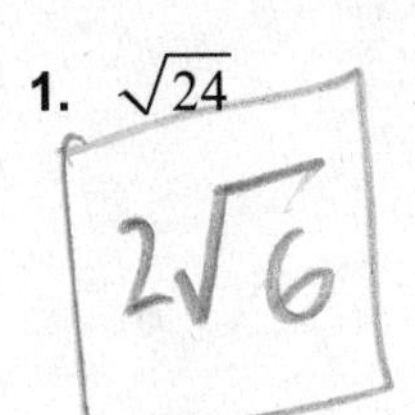

2. $-\sqrt{48}$

3. $\sqrt{162g^6}$

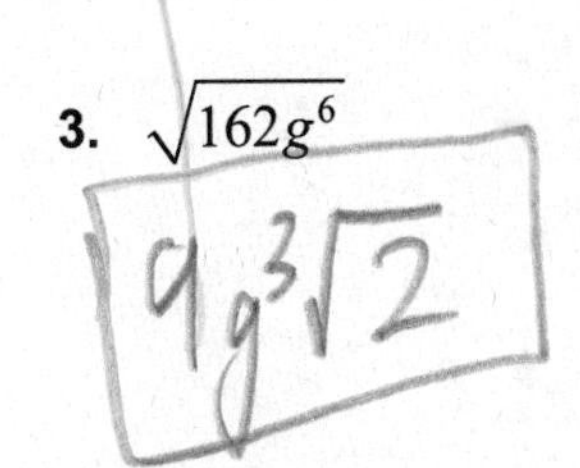

4. $-\sqrt{512h^7}$

5. $\sqrt{\frac{25}{64}}$

6. $-\sqrt{\frac{6}{49}}$

7. $-\sqrt{\frac{196}{r^4}}$

$-\frac{14}{r^2}$

8. $\sqrt{\frac{49x^3}{64y^2}}$

9. $\sqrt[3]{-135}$

$-3\sqrt[3]{5}$

10. $\sqrt[3]{729}$

11. $-\sqrt[3]{-192x^5}$

12. $\sqrt[3]{\frac{12a^6}{512b^4}}$

Name ______________________________ Date ________

9.1 Notetaking with Vocabulary (continued)

In Exercises 13–20, simplify the expression.

13. $\dfrac{\sqrt{15}}{\sqrt{500}}$ **14.** $\sqrt{\dfrac{8}{100}}$ **15.** $\dfrac{\sqrt{3x^2y^3}}{\sqrt{80xy^3}}$ **16.** $\dfrac{8}{\sqrt[3]{16}}$

17. $\dfrac{5}{-3 - 3\sqrt{3}}$ **18.** $\dfrac{3}{4 + 4\sqrt{5}}$ **19.** $\dfrac{4}{\sqrt{2} - 5\sqrt{3}}$ **20.** $\dfrac{\sqrt{5}}{\sqrt{3} + \sqrt{5}}$

21. The ratio of the length to the width of a *golden rectangle* is $\left(1 + \sqrt{5}\right) : 2$. The length of a golden rectangle is 62 meters. What is the width? Round your answer to the nearest meter.

In Exercises 22–27, simplify the expression.

22. $3\sqrt{8} + 3\sqrt{2}$ **23.** $2\sqrt{18} - 2\sqrt{20} - 2\sqrt{5}$ **24.** $3\sqrt{12} + 3\sqrt{18} + 2\sqrt{27}$

25. $2\sqrt{5}\left(\sqrt{6} + 2\right)$ **26.** $\left(\sqrt{7} - \sqrt{3}\right)\left(\sqrt{7} + \sqrt{3}\right)$ **27.** $\sqrt[3]{2}\left(\sqrt[3]{108} - \sqrt[3]{135}\right)$

Name____ Date__________

9.2 Solving Quadratic Equations by Graphing

For use with Exploration 9.2

Essential Question How can you use a graph to solve a quadratic equation in one variable?

1 EXPLORATION: Solving a Quadratic Equation by Graphing

Go to *BigIdeasMath.com* for an interactive tool to investigate this exploration.

Work with a partner.

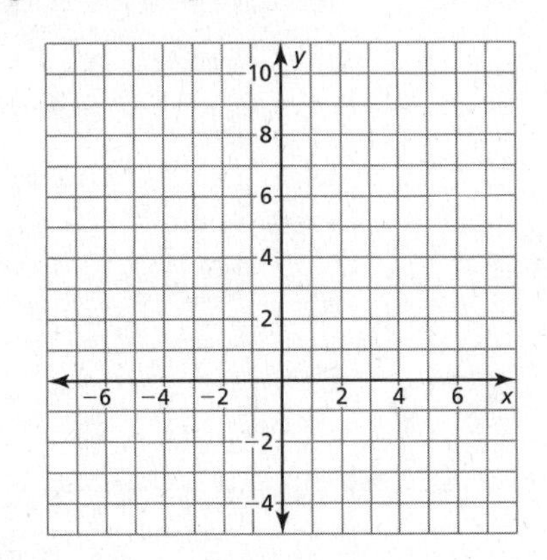

a. Sketch the graph of $y = x^2 - 2x$.

b. What is the definition of an x-intercept of a graph? How many x-intercepts does this graph have? What are they?

c. What is the definition of a solution of an equation in x? How many solutions does the equation $x^2 - 2x = 0$ have? What are they?

d. Explain how you can verify the solutions you found in part (c).

2 EXPLORATION: Solving Quadratic Equations by Graphing

Go to *BigIdeasMath.com* for an interactive tool to investigate this exploration.

Work with a partner. Solve each equation by graphing.

a. $x^2 - 4 = 0$

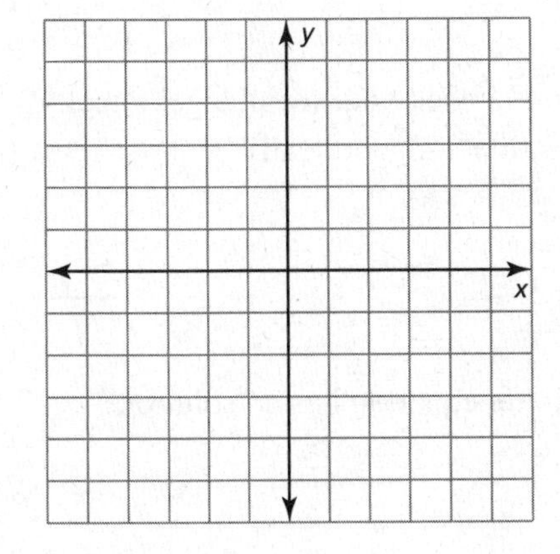

b. $x^2 + 3x = 0$

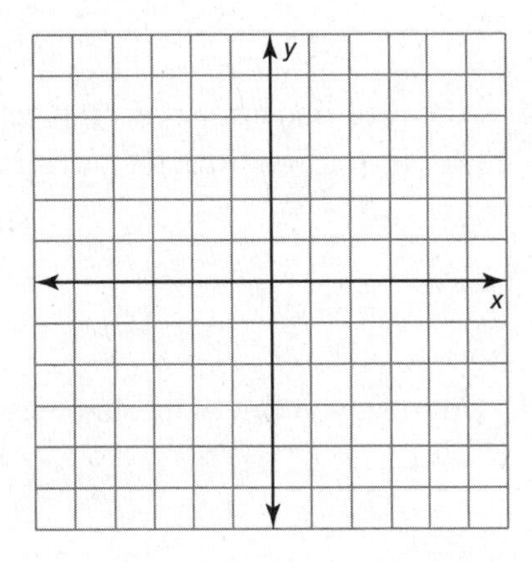

Name ______________________________ Date __________

2 EXPLORATION: Solving Quadratic Equations by Graphing (continued)

c. $-x^2 + 2x = 0$

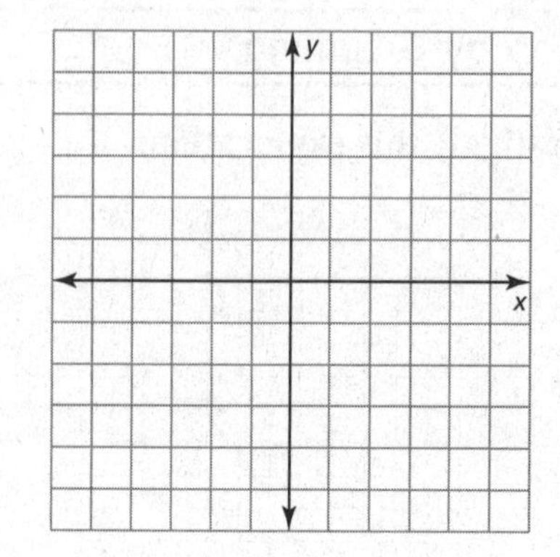

d. $x^2 - 2x + 1 = 0$

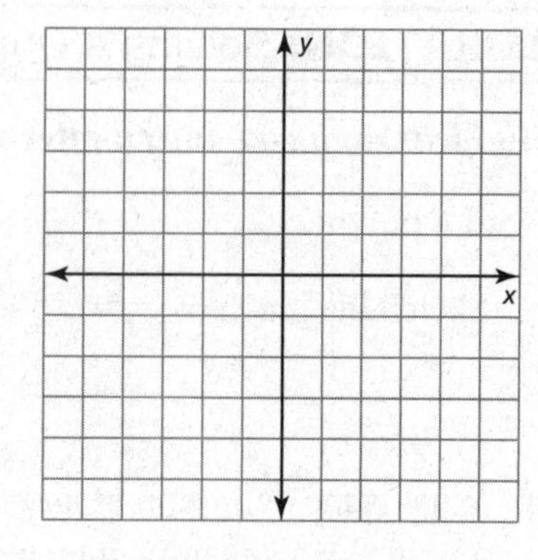

e. $x^2 - 3x + 5 = 0$

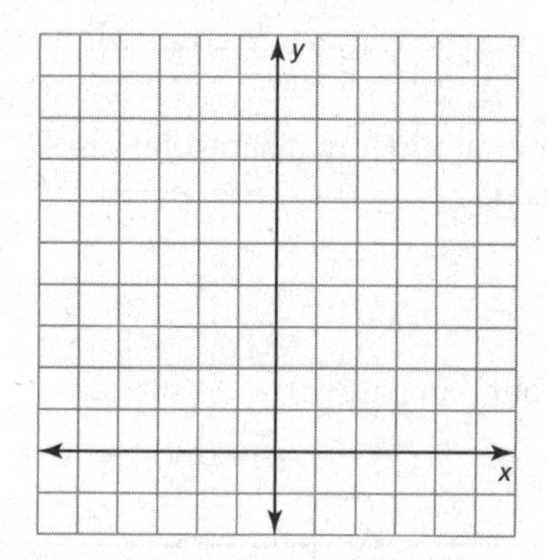

f. $-x^2 + 3x - 6 = 0$

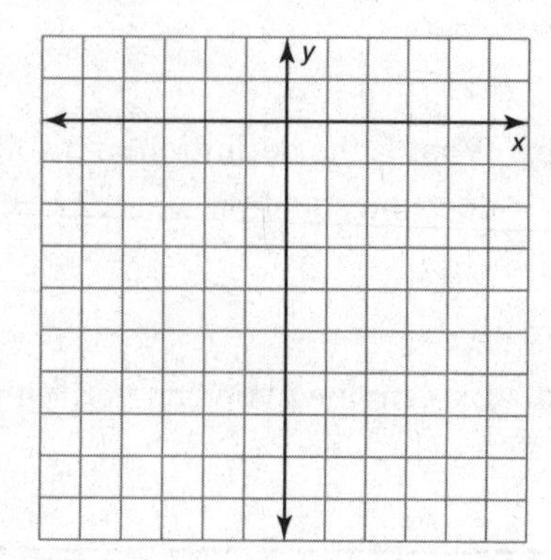

Communicate Your Answer

3. How can you use a graph to solve a quadratic equation in one variable?

4. After you find a solution graphically, how can you check your result algebraically? Check your solutions for parts (a)–(d) in Exploration 2 algebraically.

5. How can you determine graphically that a quadratic equation has no solution?

Name____ Date__________

9.2 Notetaking with Vocabulary

For use after Lesson 9.2

In your own words, write the meaning of each vocabulary term.

quadratic equation

Core Concepts

Solving Quadratic Equations by Graphing

Step 1 Write the equation in standard form, $ax^2 + bx + c = 0$.

Step 2 Graph the related function $y = ax^2 + bx + c$.

Step 3 Find the x-intercepts, if any.

The solutions, or *roots*, of $ax^2 + bx + c = 0$ are the x-intercepts of the graph.

Notes:

Number of Solutions of a Quadratic Equation

A quadratic equation has:

- two real solutions when the graph of its related function has two x-intercepts.
- one real solution when the graph of its related function has one x-intercept.
- no real solutions when the graph of its related function has no x-intercepts.

Notes:

Extra Practice

In Exercises 1–9, solve the equation by graphing.

1. $x^2 + 4x = 0$

2. $-x^2 = -2x + 1$

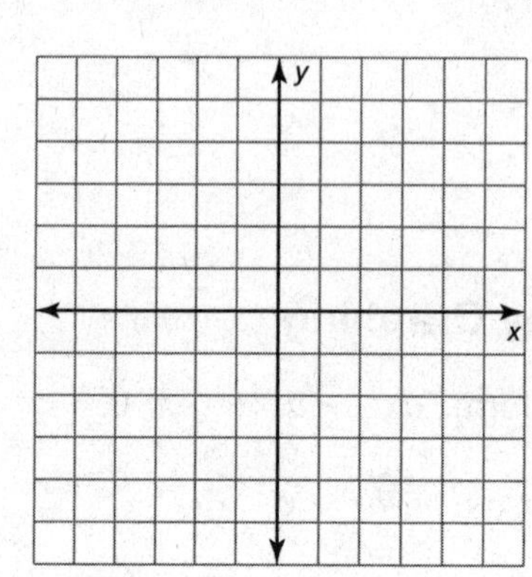

3. $x^2 + 2x + 4 = 0$

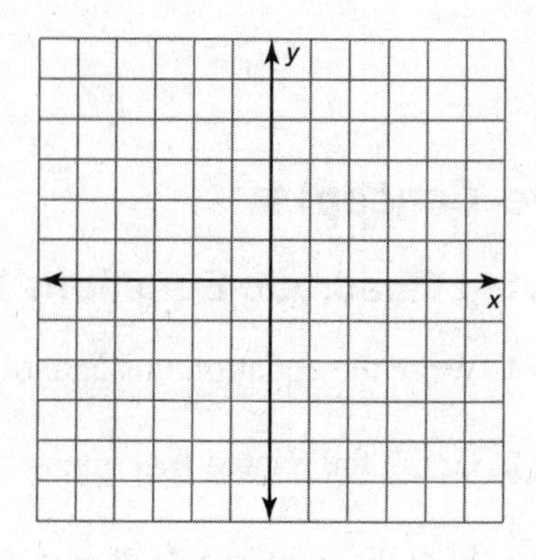

4. $x^2 - 5x + 4 = 0$

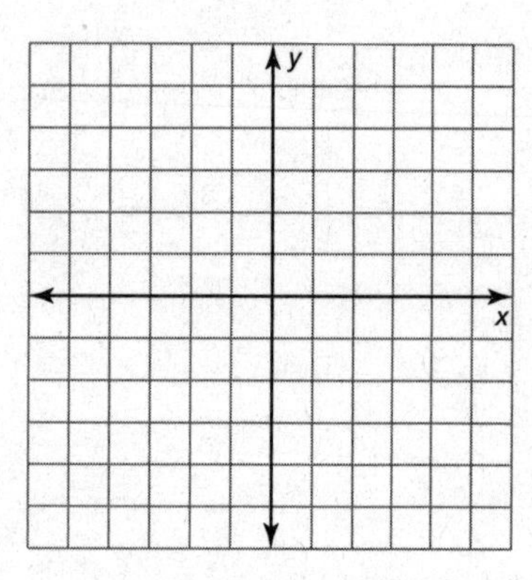

5. $x^2 + 6x + 9 = 0$

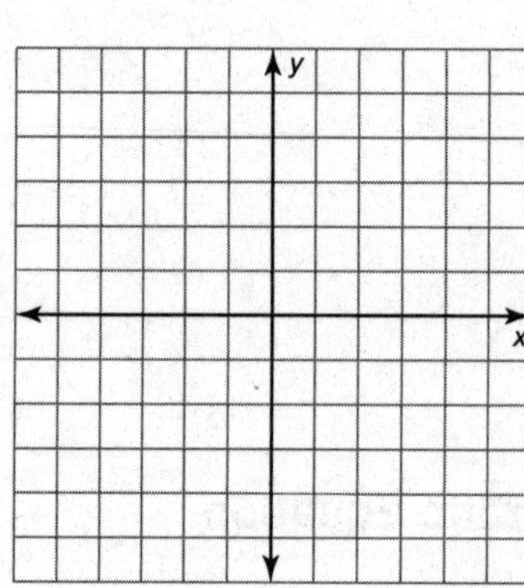

6. $x^2 = 2x - 6$

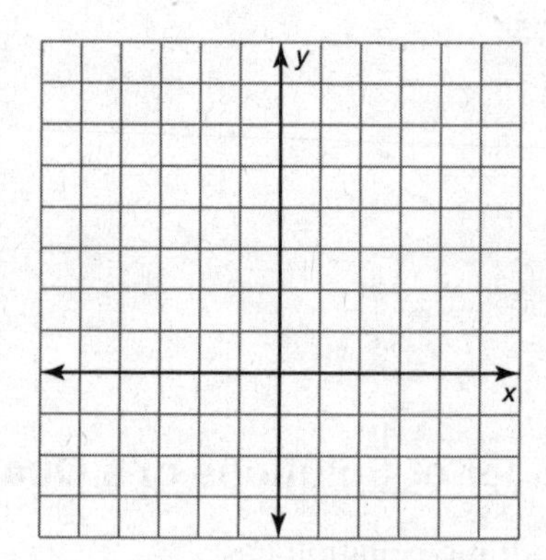

7. $x^2 - x - 12 = 0$

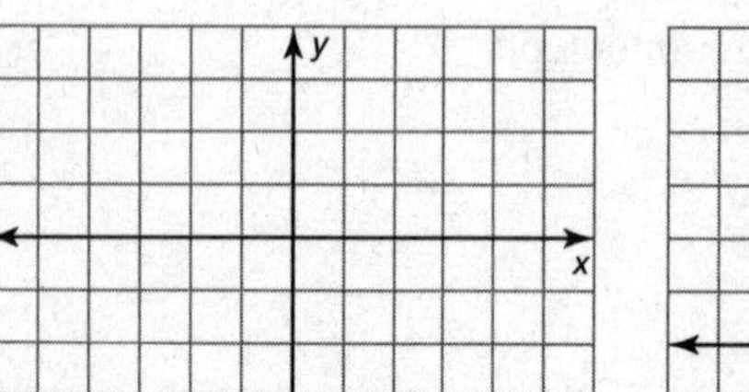

8. $x^2 - 10x + 25 = 0$

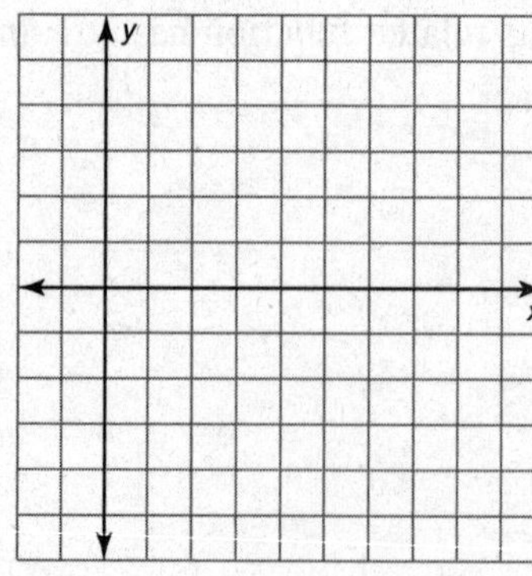

9. $x^2 + 4 = 0$

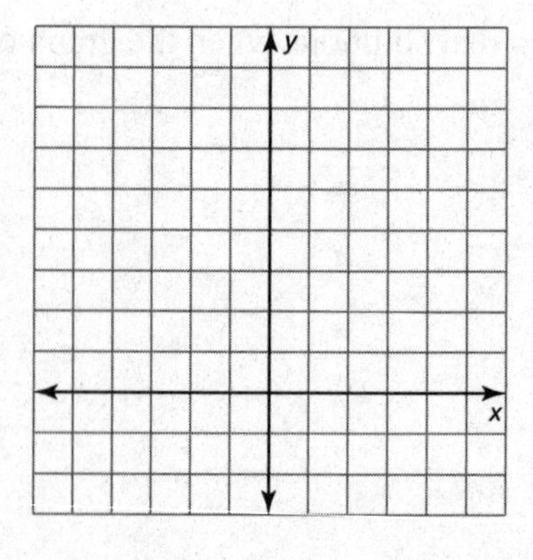

Name____ Date__________

9.2 Notetaking with Vocabulary (continued)

In Exercises 10–15, find the zero(s) of *f*.

10.

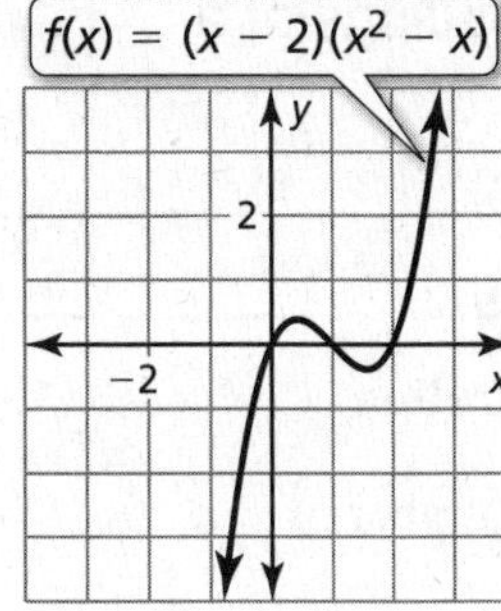

11.

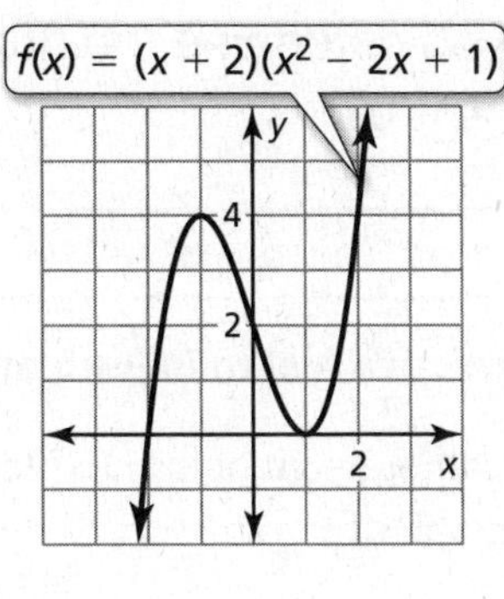

12.

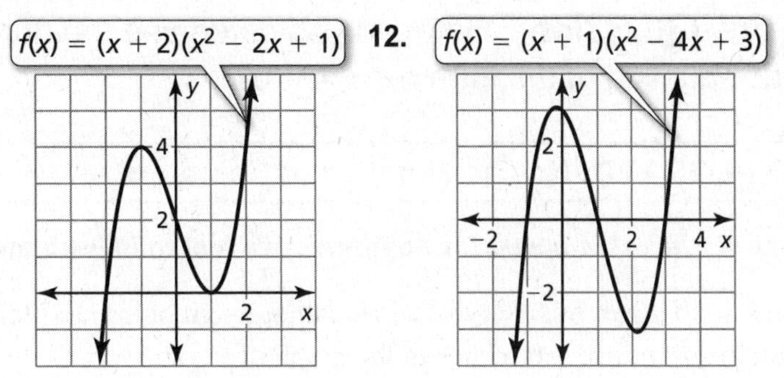

13.

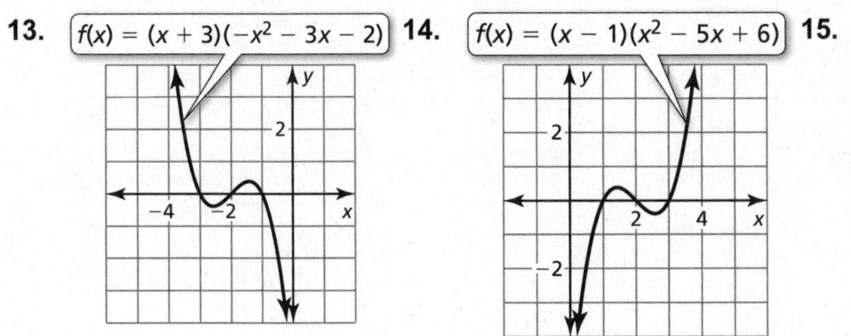

14.

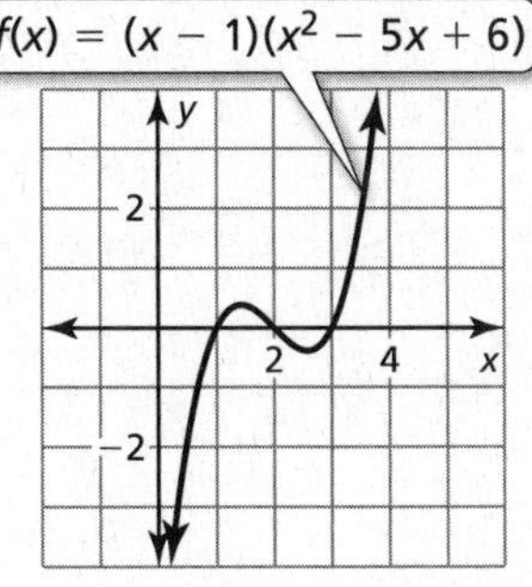

15.

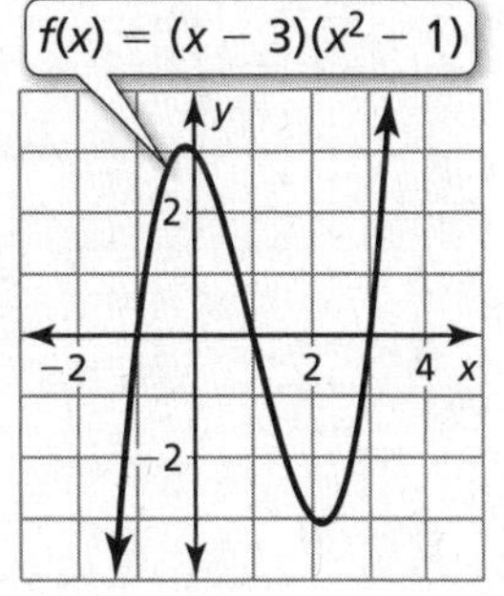

In Exercises 16–18, approximate the zeros of *f* to the nearest tenth.

16.

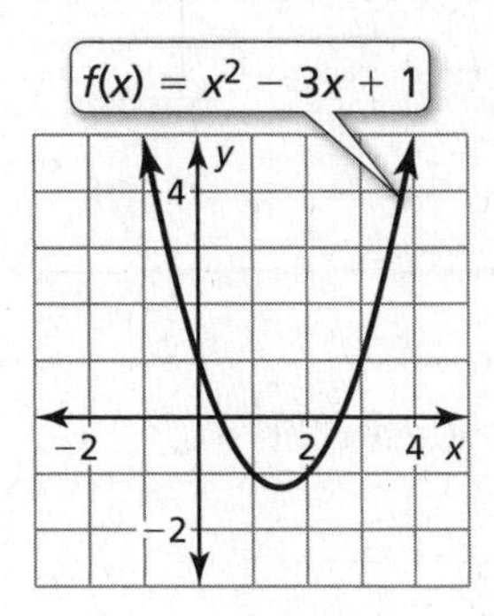

17.

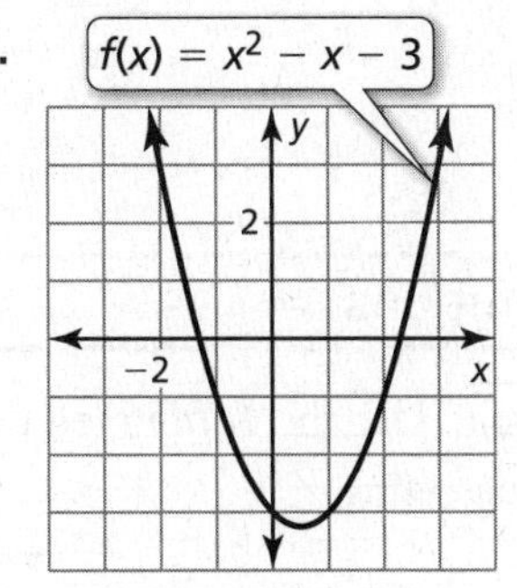

18.

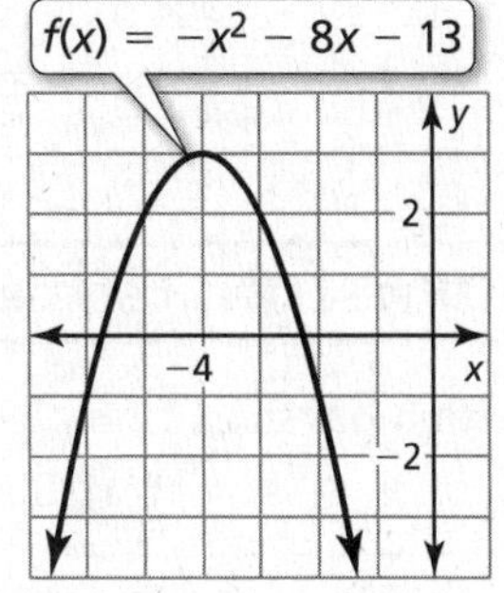

Name ______________________________ Date __________

9.3 Solving Quadratic Equations Using Square Roots
For use with Exploration 9.3

Essential Question How can you determine the number of solutions of a quadratic equation of the form $ax^2 + c = 0$?

1 EXPLORATION: The Number of Solutions of $ax^2 + c = 0$

Go to *BigIdeasMath.com* for an interactive tool to investigate this exploration.

Work with a partner. Solve each equation by graphing. Explain how the number of solutions of $ax^2 + c = 0$ relates to the graph of $y = ax^2 + c$.

a. $x^2 - 4 = 0$

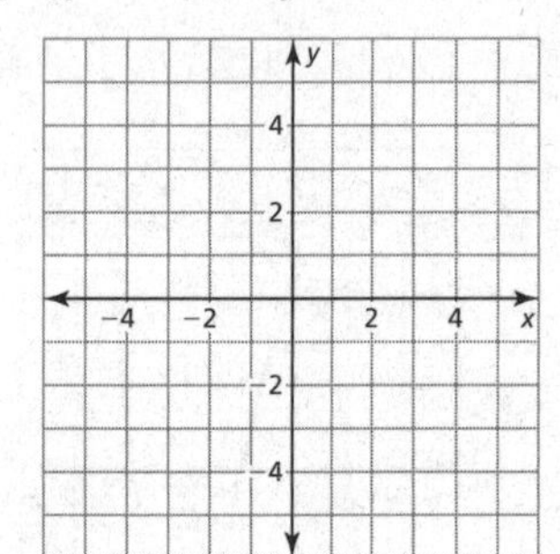

b. $2x^2 + 5 = 0$

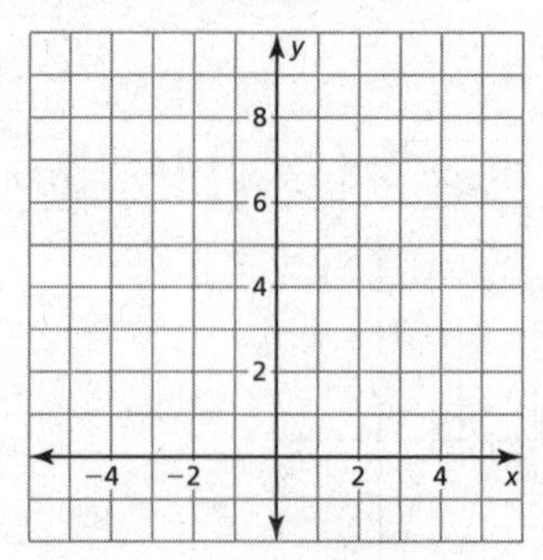

c. $x^2 = 0$

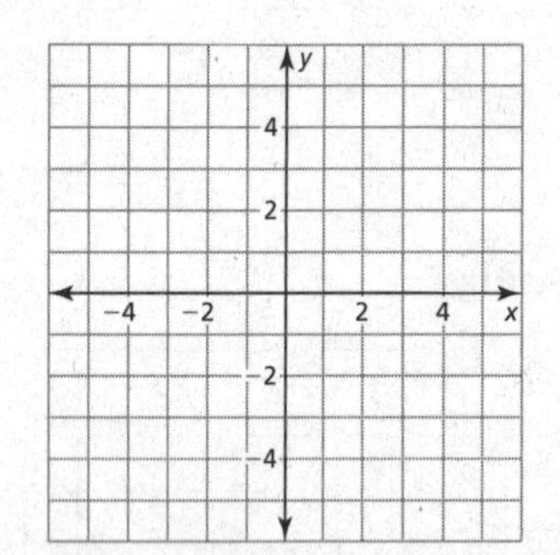

d. $x^2 - 5 = 0$

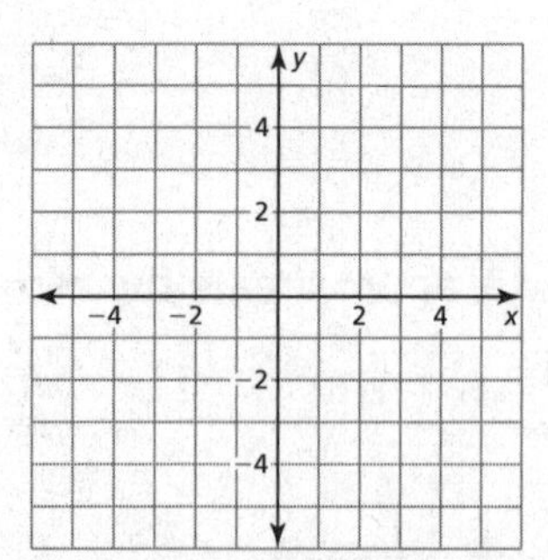

2 EXPLORATION: Estimating Solutions

Work with a partner. Complete each table. Use the completed tables to estimate the solutions of $x^2 - 5 = 0$. Explain your reasoning.

a.

x	$x^2 - 5$
2.21	
2.22	
2.23	
2.24	
2.25	
2.26	

b.

x	$x^2 - 5$
−2.21	
−2.22	
−2.23	
−2.24	
−2.25	
−2.26	

3 EXPLORATION: Using Technology to Estimate Solutions

Work with a partner. Two equations are equivalent when they have the same solutions.

a. Are the equations $x^2 - 5 = 0$ and $x^2 = 5$ equivalent? Explain your reasoning.

b. Use the square root key on a calculator to estimate the solutions of $x^2 - 5 = 0$. Describe the accuracy of your estimates in Exploration 2.

c. Write the exact solutions of $x^2 - 5 = 0$.

Communicate Your Answer

4. How can you determine the number of solutions of a quadratic equation of the form $ax^2 + c = 0$?

5. Write the exact solutions of each equation. Then use a calculator to estimate the solutions.

a. $x^2 - 2 = 0$

b. $3x^2 - 18 = 0$

c. $x^2 = 8$

Name ____ Date __________

9.3 Notetaking with Vocabulary

For use after Lesson 9.3

In your own words, write the meaning of each vocabulary term.

square root

zero of a function

Core Concepts

Solutions of $x^2 = d$

- When $d > 0$, $x^2 = d$ has two real solutions, $x = \pm\sqrt{d}$.
- When $d = 0$, $x^2 = d$ has one real solution, $x = 0$.
- When $d < 0$, $x^2 = d$ has no real solutions.

Notes:

Extra Practice

In Exercises 1–18, solve the equation using square roots.

1. $x^2 + 49 = 0$ **2.** $x^2 - 25 = 0$ **3.** $x^2 + 6 = 6$

4. $2x^2 + 84 = 0$ **5.** $2x^2 - 72 = 0$ **6.** $-x^2 - 12 = -12$

7. $8x^2 - 49 = 151$ **8.** $-3x^2 + 16 = -11$ **9.** $81x^2 - 49 = -24$

10. $16x^2 - 1 = 0$ **11.** $25x^2 + 9 = 0$ **12.** $16 - 2x^2 = 16$

13. $(x - 4)^2 = 0$ **14.** $(x + 2)^2 = 196$ **15.** $(2x + 7)^2 = 49$

9.3 Notetaking with Vocabulary (continued)

16. $16(x - 3)^2 = 25$

17. $81(3x + 1)^2 = 49$

18. $(4x - 3)^2 = 64$

In Exercises 19–24, solve the equation using square roots. Round your solutions to the nearest hundredth.

19. $x^2 + 6 = 8$

20. $x^2 - 12 = 3$

21. $x^2 + 25 = 49$

22. $3x^2 - 4 = 14$

23. $6x^2 + 5 = 20$

24. $20 - 4x^2 = 18$

25. A ball is dropped from a window at a height of 81 feet. The function $h = -16x^2 + 81$ represents the height (in feet) of the ball after x seconds. How long does it take for the ball to hit the ground?

26. The volume of a cone with height h and radius r is given by the formula $V = \frac{1}{3}\pi r^2 h$. Solve the formula for r. Then find the radius of a cone with volume 27π cubic inches and height 4 inches.

Name ______________________________ Date __________

9.4 Solving Quadratic Equations by Completing the Square

For use with Exploration 9.4

Essential Question How can you use "completing the square" to solve a quadratic equation?

1 EXPLORATION: Solving by Completing the Square

Go to *BigIdeasMath.com* for an interactive tool to investigate this exploration.

Work with a partner.

a. Write the equation modeled by the algebra tiles. This is the equation to be solved.

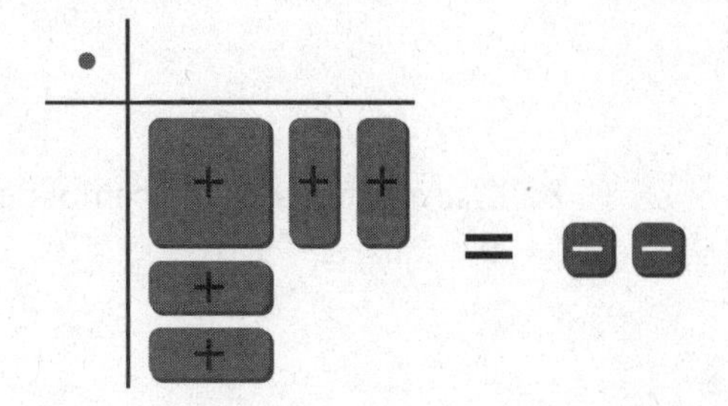

b. Four algebra tiles are added to the left side to "complete the square." Why are four algebra tiles also added to the right side?

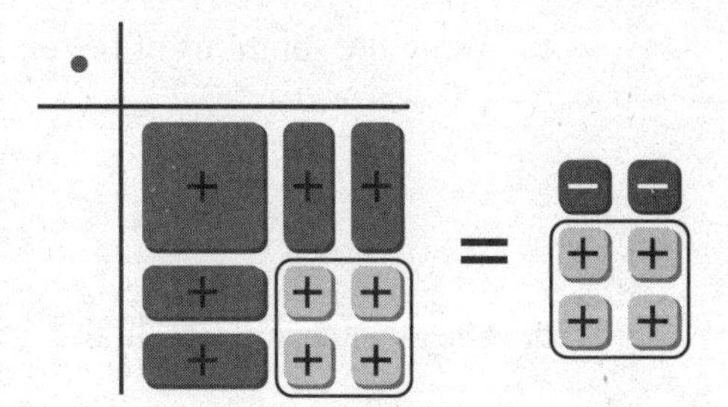

c. Use algebra tiles to label the dimensions of the square on the left side and simplify on the right side.

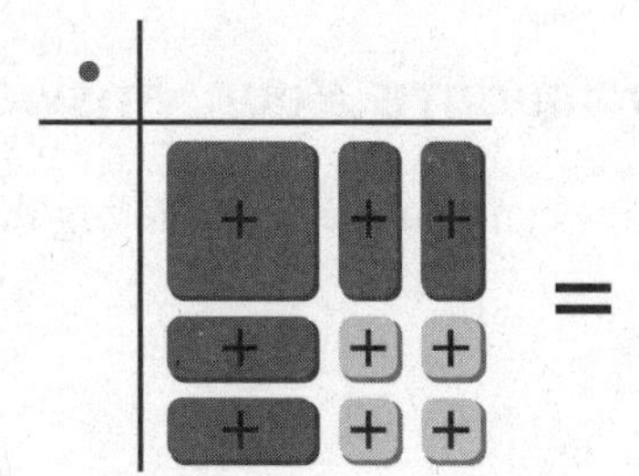

d. Write the equation modeled by the algebra tiles so that the left side is the square of a binomial. Solve the equation using square roots.

Name ______________________________ Date __________

2 EXPLORATION: Solving by Completing the Square

Go to *BigIdeasMath.com* for an interactive tool to investigate this exploration.

Work with a partner.

a. Write the equation modeled by the algebra tiles.

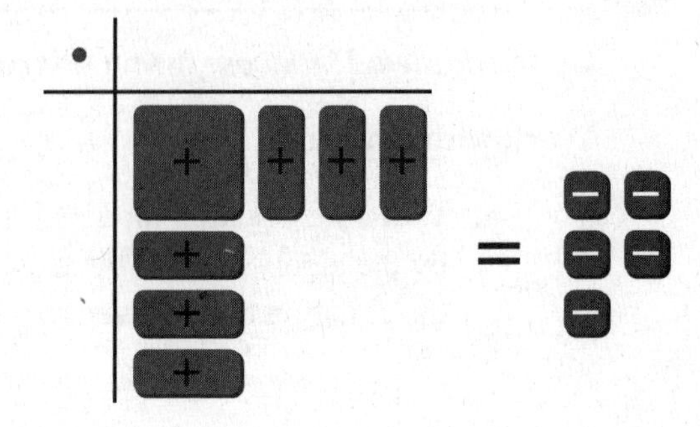

b. Use algebra tiles to "complete the square."

c. Write the solutions of the equation.

d. Check each solution in the original equation.

Communicate Your Answer

3. How can you use "completing the square" to solve a quadratic equation?

4. Solve each quadratic equation by completing the square.

a. $x^2 - 2x = 1$ **b.** $x^2 - 4x = -1$ **c.** $x^2 + 4x = -3$

Name______________________________ Date__________

9.4 Notetaking with Vocabulary

For use after Lesson 9.4

In your own words, write the meaning of each vocabulary term.

completing the square

Core Concepts

Completing the Square

Words To complete the square for an expression of the form $x^2 + bx$, follow these steps.

Step 1 Find one-half of b, the coefficient of x.

Step 2 Square the result from Step 1.

Step 3 Add the result from Step 2 to $x^2 + bx$.

Factor the resulting expression as the square of a binomial.

Algebra $x^2 + bx + \left(\frac{b}{2}\right)^2 = \left(x + \frac{b}{2}\right)^2$

Notes:

Name ______________________________ Date __________

Extra Practice

In Exercises 1–6, complete the square for the expression. Then factor the trinomial.

1. $x^2 + 12x$

2. $x^2 - 14x$

3. $x^2 + 4x$

4. $x^2 + 18x$

5. $x^2 - 7x$

6. $x^2 + 11x$

In Exercises 7–18, solve the equation by completing the square. Round your solutions to the nearest hundredth, if necessary.

7. $x^2 - 8x = -15$

8. $x^2 + 2x = 3$

9. $x^2 + 7x = 30$

10. $x^2 - 26x = -9$

11. $x^2 - 12x = 10$

12. $x^2 - 15x = 18$

13. $x^2 - 12x + 9 = 0$

14. $x^2 + 14x - 10 = 0$

15. $x^2 + 2x - 99 = 0$

16. $10x^2 - 13x - 9 = 0$

17. $3x^2 + 6x - 1 = 0$

18. $12x^2 - 8x - 2 = 0$

In Exercises 19–24, determine whether the quadratic function has a maximum or minimum value. Then find the value.

19. $y = -x^2 + 4x + 3$

20. $y = x^2 + 6x + 10$

21. $y = -x^2 + 8x - 2$

22. $y = x^2 - 10x + 8$

23. $y = 3x^2 + 3x - 1$

24. $y = -4x^2 + 8x + 12$

25. A diver jumps off a diving board. The function $h = -16x^2 + 6x + 5$ represents the height (in feet) of the diver after x seconds. What is the maximum height above the water of the diver? How many seconds did it take for the diver to reach the maximum height? Round your answers to the nearest hundredth.

Name ______________________________ Date __________

9.5 Solving Quadratic Equations Using the Quadratic Formula

For use with Exploration 9.5

Essential Question How can you derive a formula that can be used to write the solutions of any quadratic equation in standard form?

1 EXPLORATION: Deriving the Quadratic Formula

Work with a partner. The following steps show a method of solving $ax^2 + bx + c = 0$. Explain what was done in each step.

$ax^2 + bx + c = 0$ 1. Write the equation.

$4a^2x^2 + 4abx + 4ac = 0$ 2. ______________________

$4a^2x^2 + 4abx + 4ac + b^2 = b^2$ 3. ______________________

$4a^2x^2 + 4abx + b^2 = b^2 - 4ac$ 4. ______________________

$(2ax + b)^2 = b^2 - 4ac$ 5. ______________________

$2ax + b = \pm\sqrt{b^2 - 4ac}$ 6. ______________________

$2ax = -b \pm \sqrt{b^2 - 4ac}$ 7. ______________________

Quadratic Formula: $x = \dfrac{-b \pm \sqrt{b^2 - 4ac}}{2a}$ 8. ______________________

Name____ Date__________

2 EXPLORATION: Deriving the Quadratic Formula by Completing the Square

Work with a partner.

a. Solve $ax^2 + bx + c = 0$ by completing the square. (*Hint:* Subtract c from each side, divide each side by a, and then proceed by completing the square.)

b. Compare this method with the method in Exploration 1. Explain why you think $4a$ and b^2 were chosen in Steps 2 and 3 of Exploration 1.

Communicate Your Answer

3. How can you derive a formula that can be used to write the solutions of any quadratic equation in standard form?

4. Use the Quadratic Formula to solve each quadratic equation.

a. $x^2 + 2x - 3 = 0$ **b.** $x^2 - 4x + 4 = 0$ **c.** $x^2 + 4x + 5 = 0$

5. Use the Internet to research *imaginary numbers*. How are they related to quadratic equations?

Name ______________________________ Date __________

9.5 Notetaking with Vocabulary
For use after Lesson 9.5

In your own words, write the meaning of each vocabulary term.

Quadratic Formula

discriminant

Core Concepts

Quadratic Formula

The real solutions of the quadratic equation $ax^2 + bx + c = 0$ are

$$x = \frac{-b \pm \sqrt{b^2 - 4ac}}{2a} \quad \text{Quadratic Formula}$$

where $a \neq 0$ and $b^2 - 4ac \geq 0$.

Notes:

Interpreting the Discriminant

$b^2 - 4ac > 0$

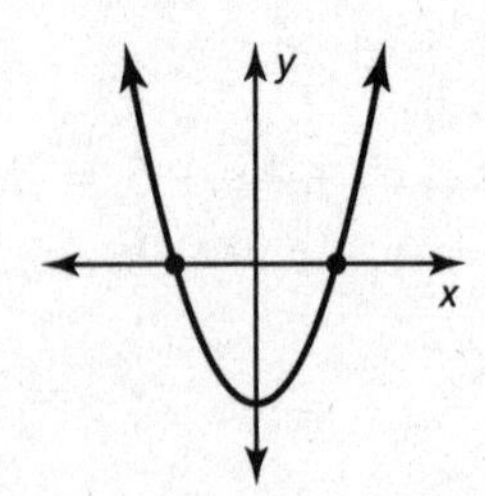

- two real solutions
- two x-intercepts

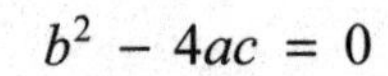

$b^2 - 4ac = 0$

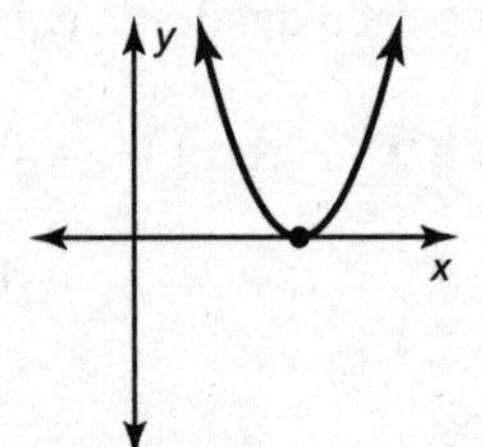

- one real solution
- one x-intercept

$b^2 - 4ac < 0$

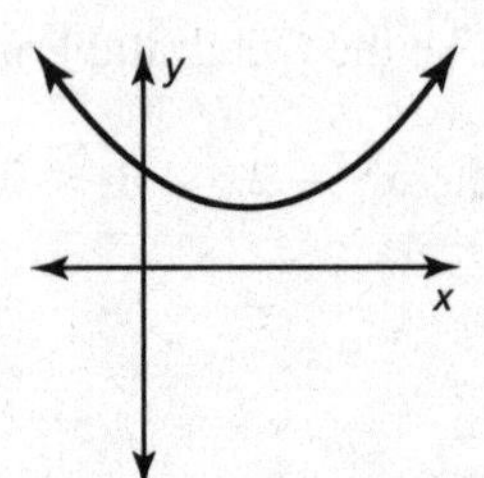

- no real solutions
- no x-intercepts

Notes:

Name________________________________ Date__________

9.5 Notetaking with Vocabulary (continued)

Methods for Solving Quadratic Equations

Method	Advantages	Disadvantages
Factoring (*Lessons 7.5–7.8*)	• Straightforward when the equation can be factored easily	• Some equations are not factorable.
Graphing (*Lesson 9.2*)	• Can easily see the number of solutions • Use when approximate solutions are sufficient. • Can use a graphing calculator	• May not give exact solutions
Using Square Roots (*Lesson 9.3*)	• Used to solve equations of the form $x^2 = d$.	• Can only be used for certain equations
Completing the Square (*Lesson 9.4*)	• Best used when $a = 1$ and b is even	• May involve difficult calculations
Quadratic Formula (*Lesson 9.5*)	• Can be used for any quadratic equation • Gives exact solutions	• Takes time to do calculations

Notes:

Extra Practice

In Exercises 1–6, solve the equation using the Quadratic Formula. Round your solutions to the nearest tenth, if necessary.

1. $x^2 - 10x + 16 = 0$

2. $x^2 + 2x - 8 = 0$

3. $3x^2 - x - 2 = 0$

4. $x^2 + 6x = -13$

5. $-3x^2 + 5x - 1 = -7$

6. $-4x^2 + 8x + 12 = 6$

7. A square pool has a side length of x feet. A uniform border around the pool is 1 foot wide. The total area of the pool and the border is 361 square feet. What is the area of the pool?

In Exercises 8–10, determine the number of real solutions of the equation.

8. $-x^2 + 6x + 3 = 0$ **9.** $x^2 + 6x + 9 = 0$ **10.** $x^2 + 3x + 8 = 0$

In Exercises 11–13 find the number of *x*-intercepts of the graph of the function.

11. $y = -x^2 + 4x + 3$ **12.** $y = x^2 + 14x + 49$ **13.** $y = -x^2 - 8x - 18$

In Exercises 14–16, solve the equation using any method. Explain your choice of method.

14. $x^2 - 4x + 4 = 16$ **15.** $x^2 - 8x + 7 = 0$ **16.** $3x^2 + x - 5 = 0$

Name____ Date__________

9.6 Solving Nonlinear Systems of Equations

For use with Exploration 9.6

Essential Question How can you solve a system of two equations when one is linear and the other is quadratic?

1 EXPLORATION: Solving a System of Equations

Go to *BigIdeasMath.com* for an interactive tool to investigate this exploration.

Work with a partner. Solve the system of equations by graphing each equation and finding the points of intersection.

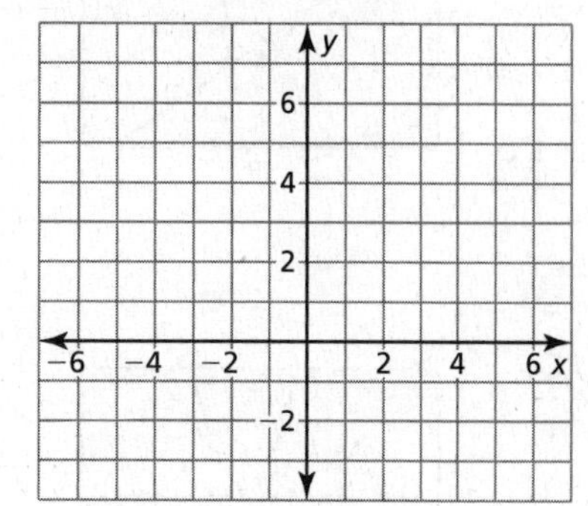

System of Equations

$y = x + 2$ Linear

$y = x^2 + 2x$ Quadratic

2 EXPLORATION: Analyzing Systems of Equations

Work with a partner. Match each system of equations with its graph (shown on the next page). Then solve the system of equations.

a. $y = x^2 - 4$
$y = -x - 2$

b. $y = x^2 - 2x + 2$
$y = 2x - 2$

c. $y = x^2 + 1$
$y = x - 1$

d. $y = x^2 - x - 6$
$y = 2x - 2$

2 EXPLORATION: Analyzing Systems of Equations (continued)

A.

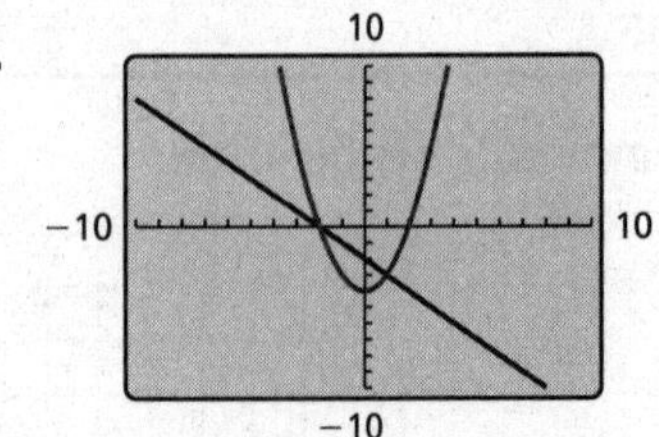

B.

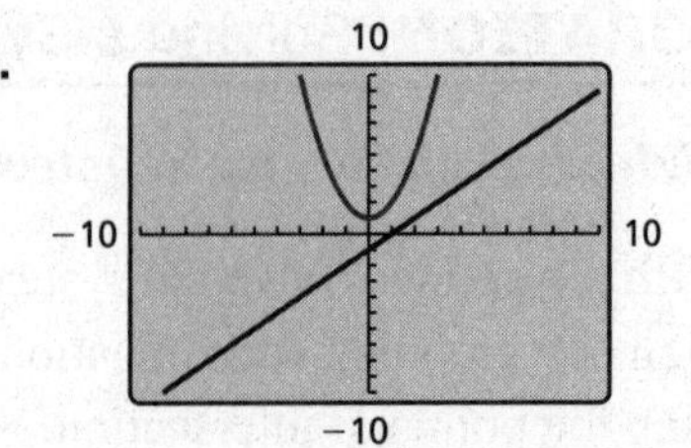

C.

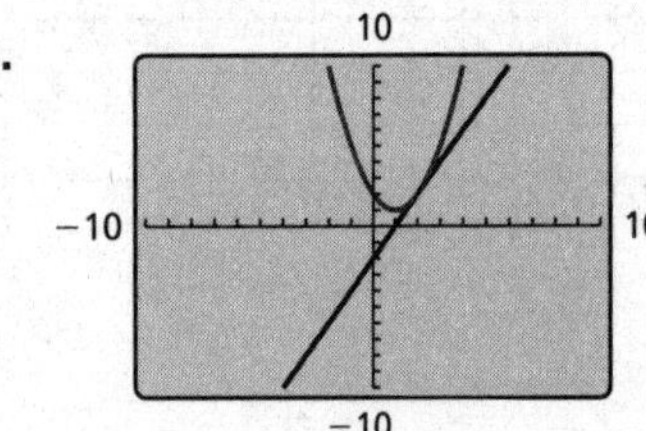

D.

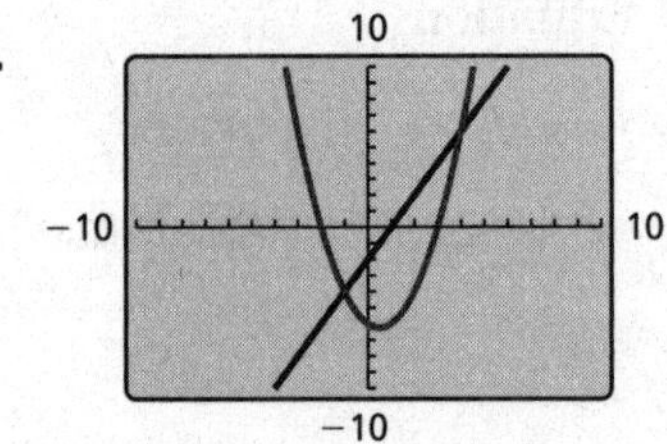

Communicate Your Answer

3. How can you solve a system of two equations when one is linear and the other is quadratic?

4. Write a system of equations (one linear and one quadratic) that has (a) no solutions, (b) one solution, and (c) two solutions. Your systems should be different from those in Explorations 1 and 2.

Name____ Date__________

9.6 Notetaking with Vocabulary
For use after Lesson 9.6

In your own words, write the meaning of each vocabulary term.

system of nonlinear equations

Notes:

Name ______________________________ Date __________

Extra Practice

In Exercises 1–6, solve the system by graphing.

1. $y = x^2 + 5x + 6$

$y = -x + 1$

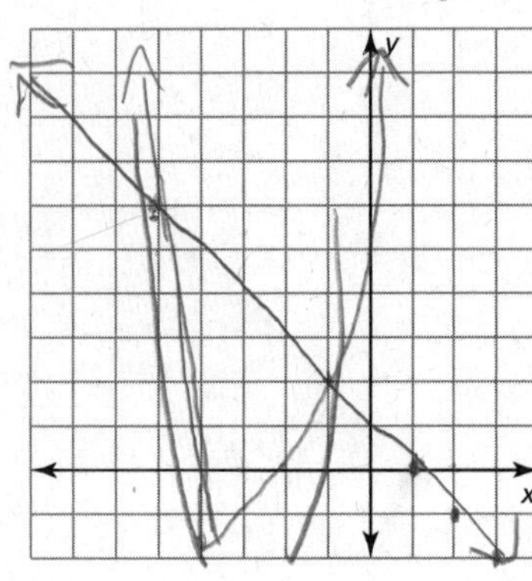

2. $y = x^2 + x - 3$

$y = x + 1$

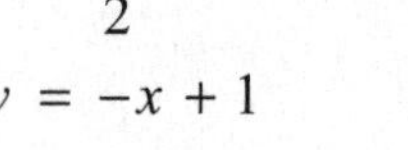

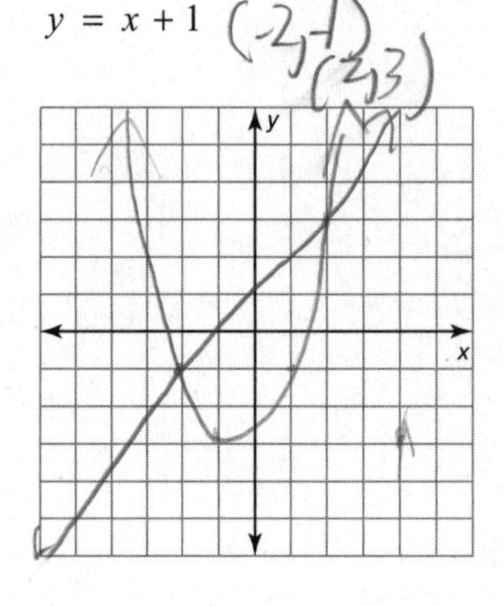

3. $y = \frac{1}{2}x^2 - 2x + 1$

$y = -x + 1$

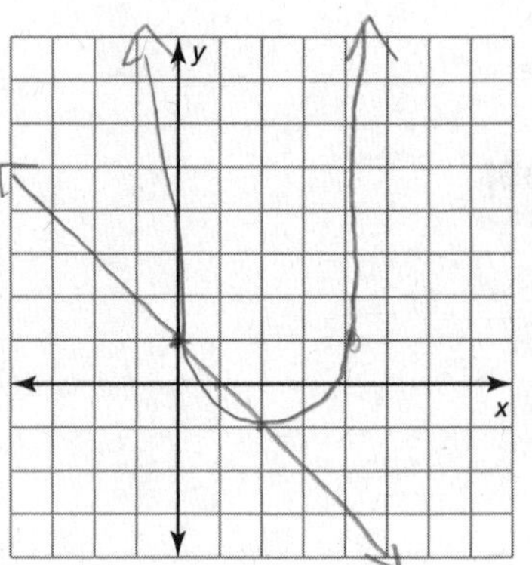

4. $y = -3x^2 - 3x + 2$

$y = 2x$

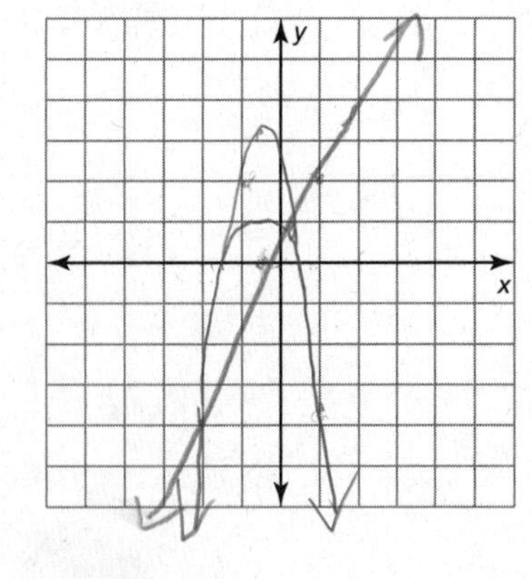

5. $y = -\frac{1}{3}x^2 + x - 2$

$y = -2$

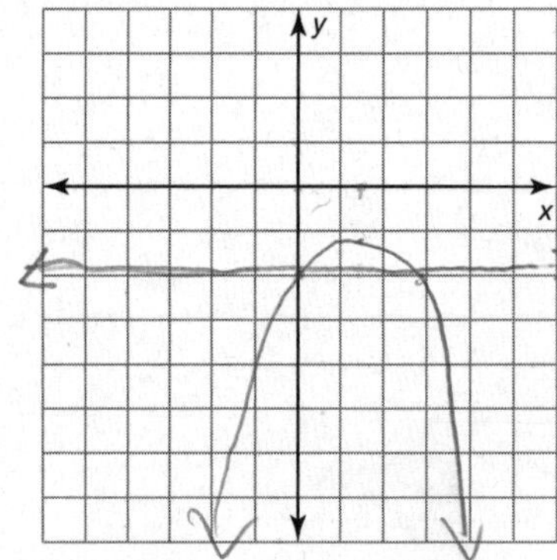

6. $y = 6x^2 + 3x - 5$

$y = -3x - 5$

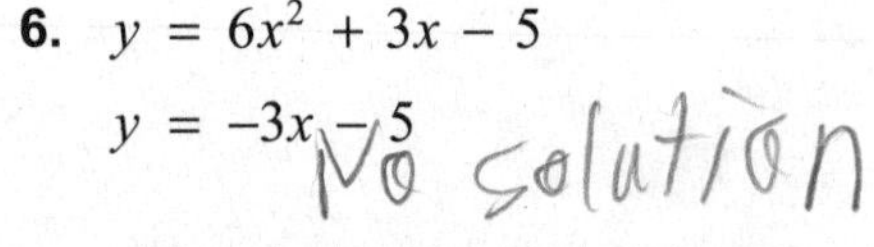

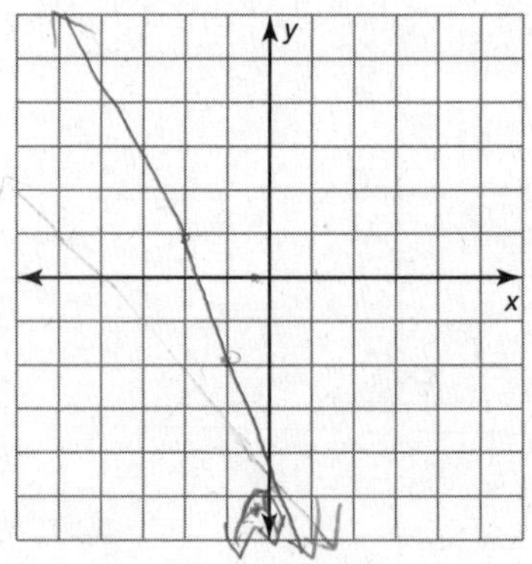

In Exercises 7–9, solve the equation by substitution.

7. $y - 2 = x^2$

$y = 6$

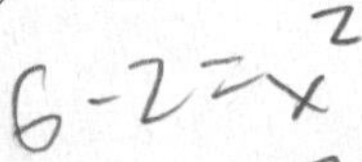

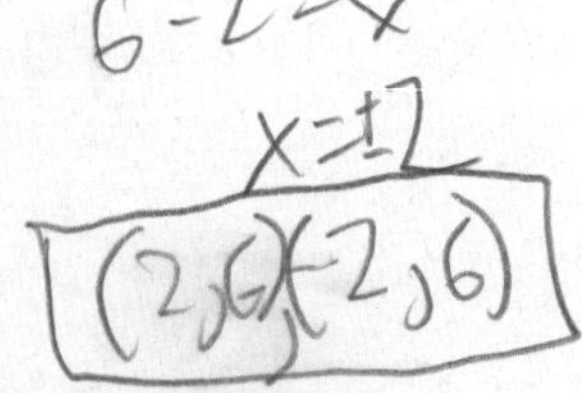

8. $y = -2x^2$

$y = 3x + 2$

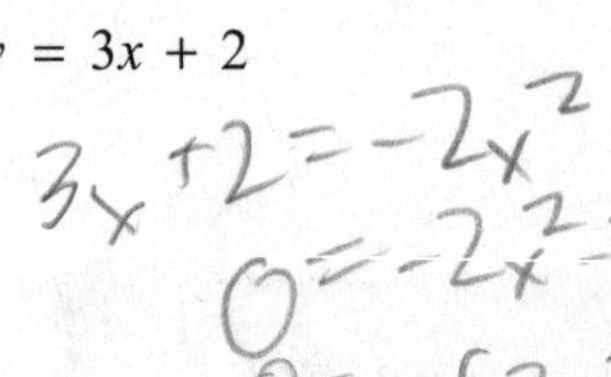

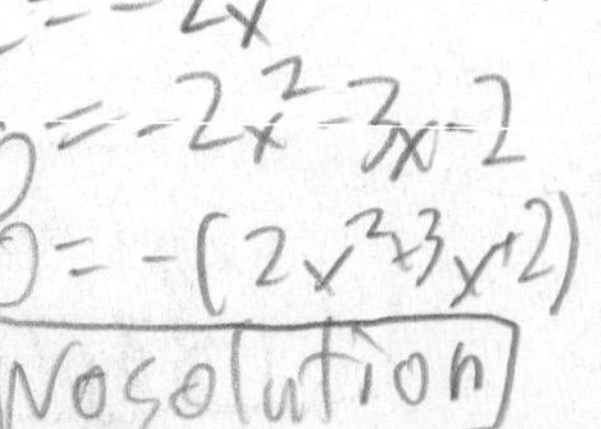

9. $y = x - 4$

$y = x^2 + 3x - 4$

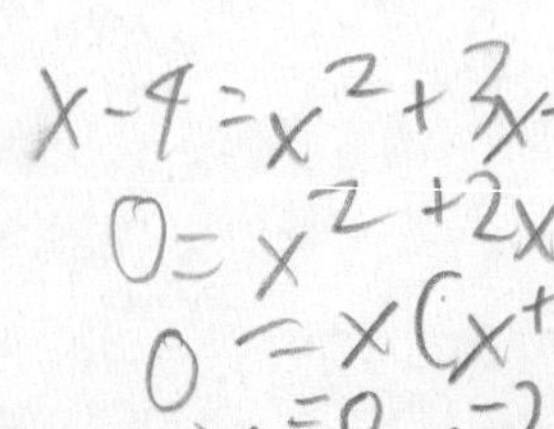

Name______________________________ Date__________

9.6 Notetaking with Vocabulary (continued)

In Exercises 10–12, solve the equation by elimination.

10. $y = x^2$
$y = x - 3$

11. $y = x^2 + 3x - 5$
$y = 3x - 1$

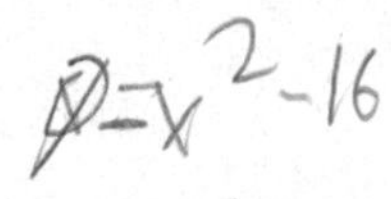

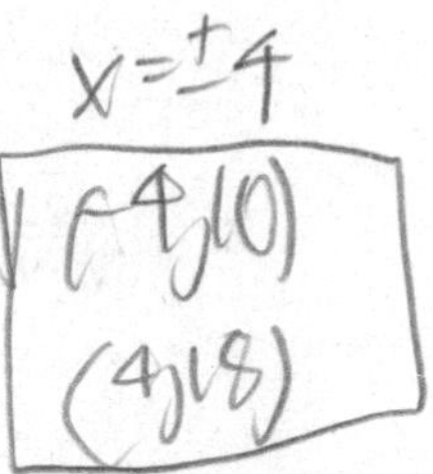

12. $y = x^2 + x - 2$
$y = x + 14$

In Exercises 13–18, solve the equation. Round your solution(s) to the nearest hundredth, if necessary.

13. $-6x + 14 = x^2 - 9x + 16$

14. $-x^2 + 4x = -2x + 8$

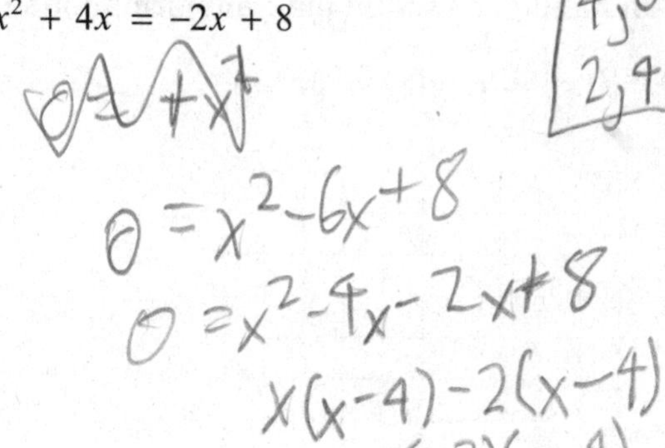

15. $4x^2 - 9 = 4x - 1$

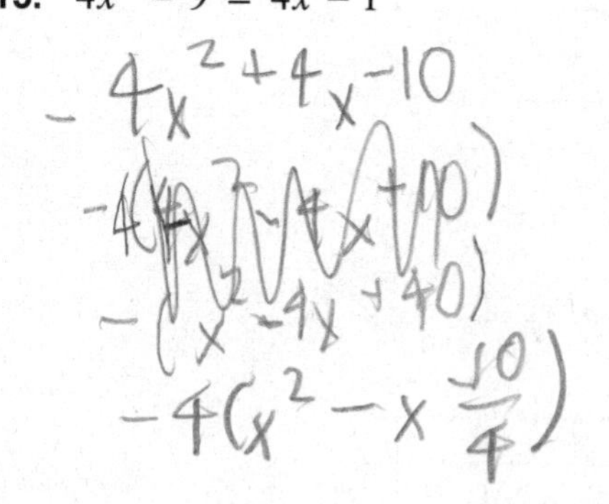

16. $-\frac{1}{2}x + 1 = -x^2 + 4x$

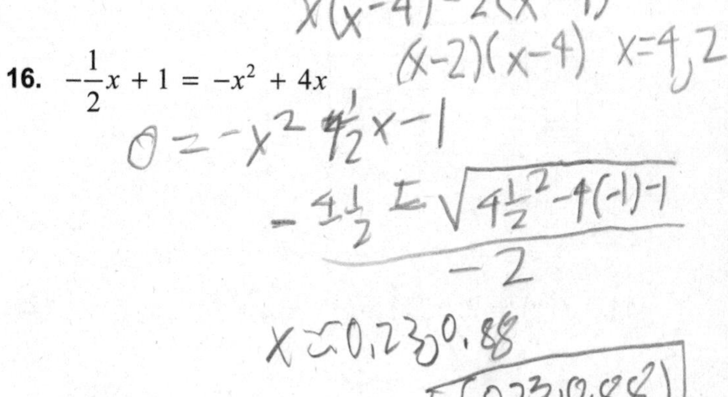

17. $2x^2 - 4 = -x^2 + 6$

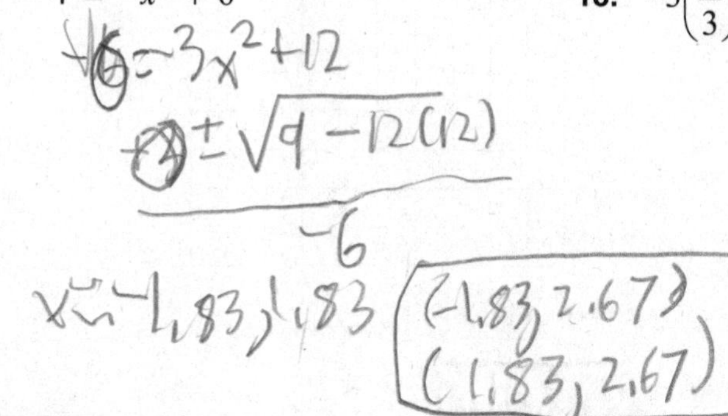

18. $-3\left(\frac{2}{3}\right)^x + 2 = x^2 - 2$

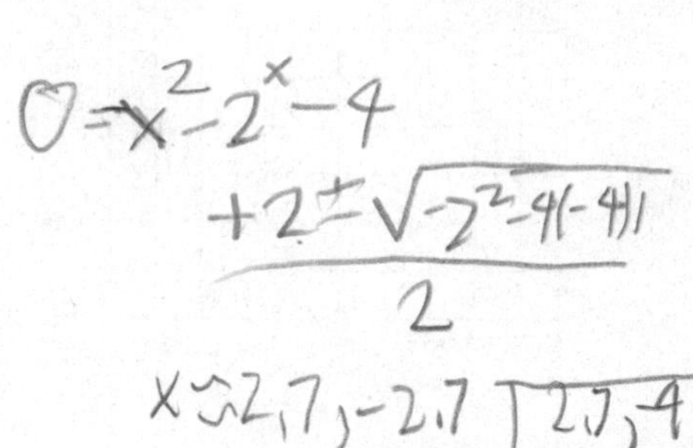

Name ______________________________ Date __________

Chapter 10 Maintaining Mathematical Proficiency

Evaluate the expression.

1. $6\sqrt{36} + 4$

2. $-7 - \sqrt{\frac{49}{4}}$

3. $4\left(\frac{\sqrt{25}}{4} - 6\right)$

4. $-3\left(4\sqrt{16} + 24\right)$

5. $9 - 4\sqrt{81}$

6. $-\sqrt{\frac{225}{9}} + 35$

7. $4\left(3 - \frac{\sqrt{36}}{6}\right)$

8. $2\left(2\sqrt{100} + 12\right)$

Graph *f* and *g*. Describe the transformations from the graph of *f* to the graph of *g*.

9. $f(x) = x;\ g(x) = 3x - 1$

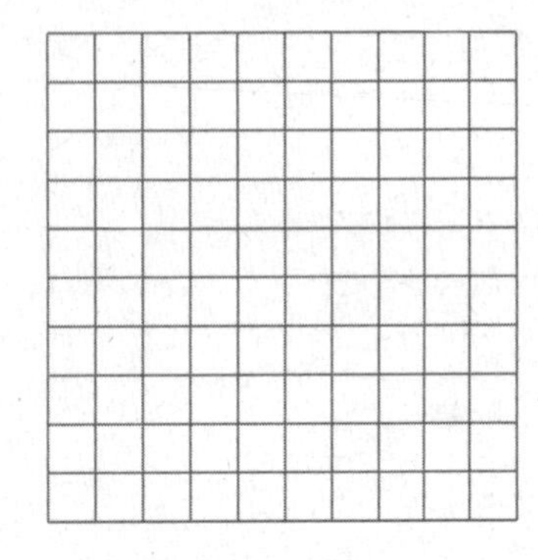

10. $f(x) = x;\ g(x) = \frac{1}{2}x + 3$

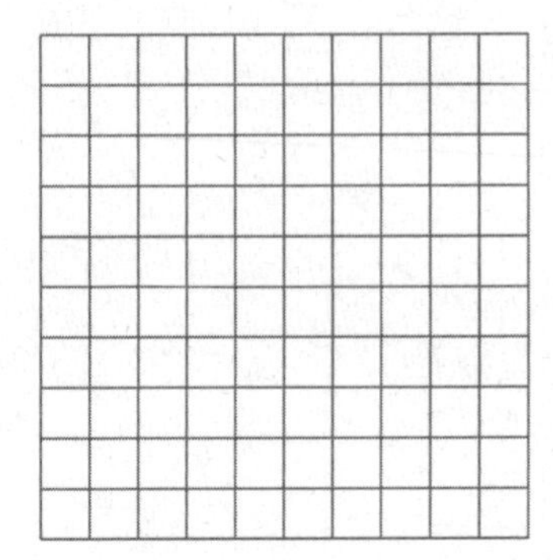

11. $f(x) = x;\ g(x) = -x - 2$

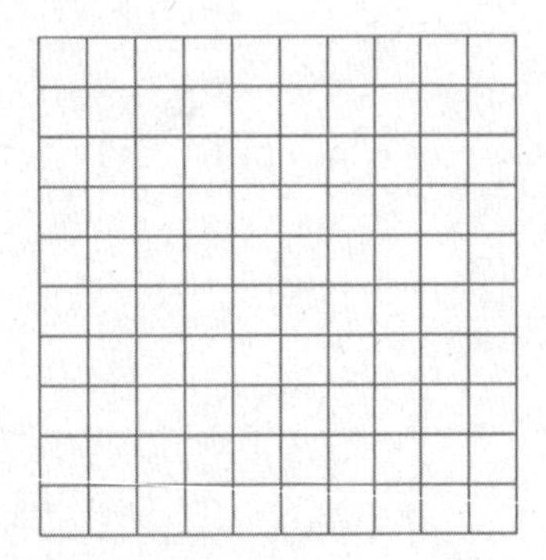

12. $f(x) = x;\ g(x) = -\frac{1}{4}x + 3$

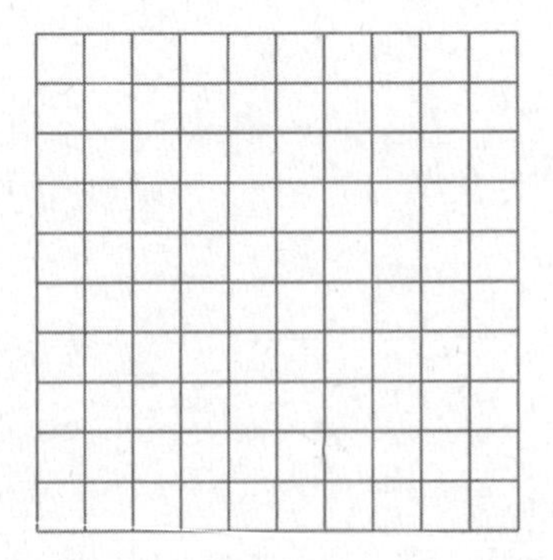

Name______________________________ Date__________

10.1 Graphing Square Root Functions

For use with Exploration 10.1

Essential Question What are some of the characteristics of the graph of a square root function?

1 EXPLORATION: Graphing Square Root Functions

Work with a partner.

- Make a table of values for each function.
- Use the table to sketch the graph of each function.
- Describe the domain of each function.
- Describe the range of each function.

a. $y = \sqrt{x}$

x						
y						

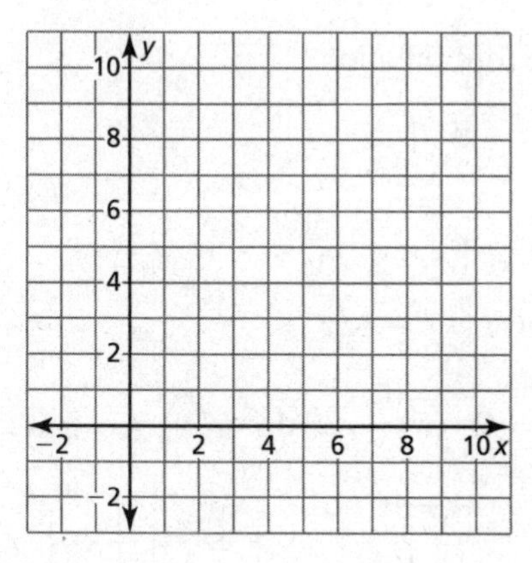

b. $y = \sqrt{x + 2}$

x						
y						

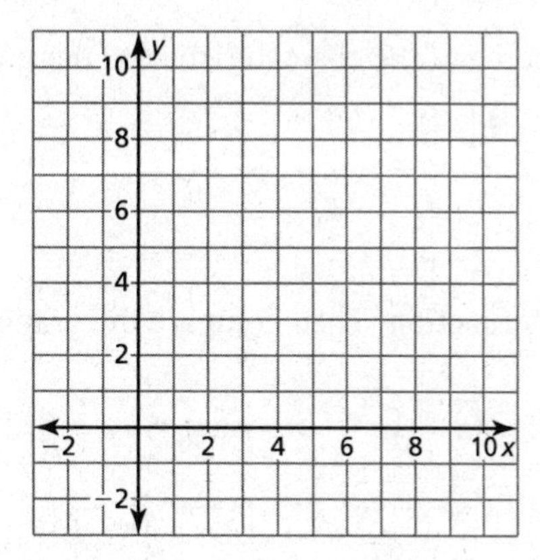

2 EXPLORATION: Writing Square Root Functions

Work with a partner. Write a square root function, $y = f(x)$, that has the given values. Then use the function to complete the table.

a.

x	$f(x)$
−4	0
−3	
−2	
−1	$\sqrt{3}$
0	2
1	

b.

x	$f(x)$
−4	0
−3	
−2	
−1	$1 + \sqrt{3}$
0	3
1	

Communicate Your Answer

3. What are some of the characteristics of the graph of a square root function?

4. Graph each function. Then compare the graph to the graph of $f(x) = \sqrt{x}$.

a. $g(x) = \sqrt{x - 1}$ **b.** $g(x) = \sqrt{x} - 1$ **c.** $g(x) = 2\sqrt{x}$ **d.** $g(x) = -2\sqrt{x}$

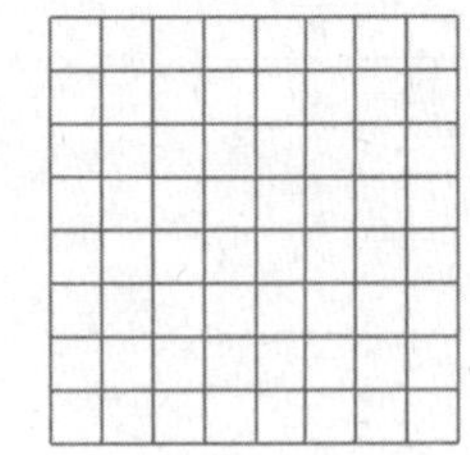

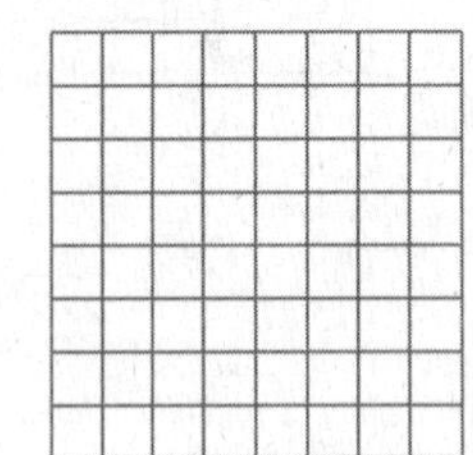

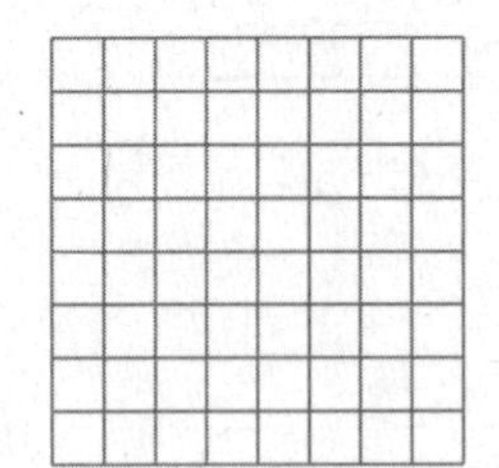

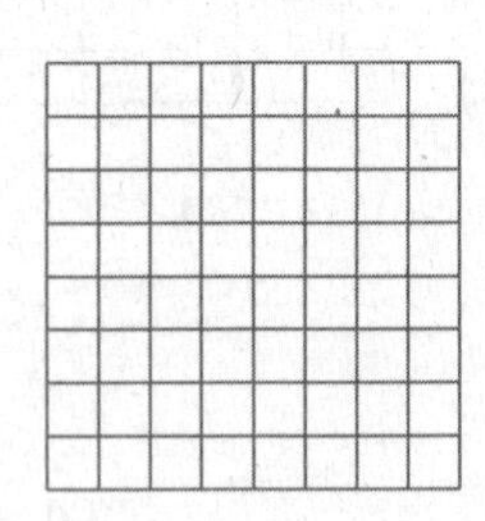

Name____ Date__________

10.1 Notetaking with Vocabulary

For use after Lesson 10.1

In your own words, write the meaning of each vocabulary term.

square root function

radical function

Core Concepts

Square Root Functions

A **square root function** is a function that contains a square root with the independent variable in the radicand. The parent function for the family of square root functions is $f(x) = \sqrt{x}$. The domain of f is $x \geq 0$, and the range of f is $y \geq 0$.

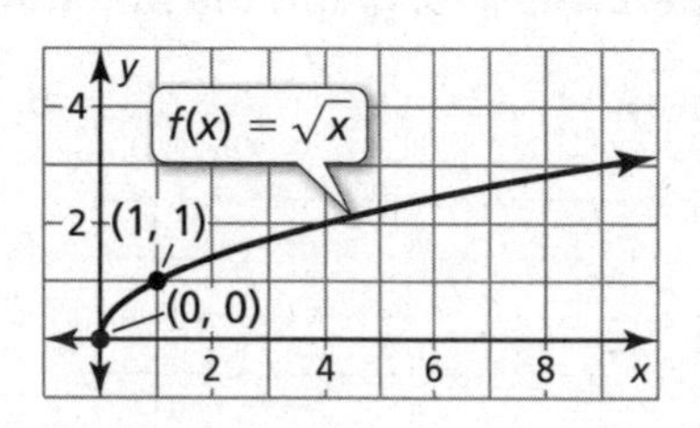

Notes:

Transformation	f(x) Notation	Examples	
Horizontal Translation Graph shifts left or right.	$f(x - h)$	$g(x) = \sqrt{x-2}$	**2 units right**
		$g(x) = \sqrt{x+3}$	**3 units left**
Vertical Translation Graph shifts up or down.	$f(x) + k$	$g(x) = \sqrt{x} + 7$	**7 units up**
		$g(x) = \sqrt{x} - 1$	**1 unit down**
Reflection Graph flips over x- or y-axis.	$f(-x)$	$g(x) = \sqrt{-x}$	**in the y-axis**
	$-f(x)$	$g(x) = -\sqrt{x}$	**in the x-axis**
Horizontal Stretch or Shrink Graph stretches away from or shrinks toward y-axis.	$f(ax)$	$g(x) = \sqrt{3x}$	**shrink by a factor of $\frac{1}{3}$**
		$g(x) = \sqrt{\frac{1}{2}x}$	**stretch by a factor of 2**
Vertical Stretch or Shrink Graph stretches away from or shrinks toward x-axis.	$a \bullet f(x)$	$g(x) = 4\sqrt{x}$	**stretch by a factor of 4**
		$g(x) = \frac{1}{5}\sqrt{x}$	**shrink by a factor of $\frac{1}{5}$**

Notes:

Name ____________________ Date ________

Extra Practice

In Exercises 1–3, describe the domain of the function.

1. $y = 4\sqrt{-x}$

2. $y = \sqrt{x - 3}$

3. $f(x) = \sqrt{\frac{1}{3}x + 4}$

In Exercises 4–6, graph the function. Describe the range.

4. $y = \sqrt{3x}$

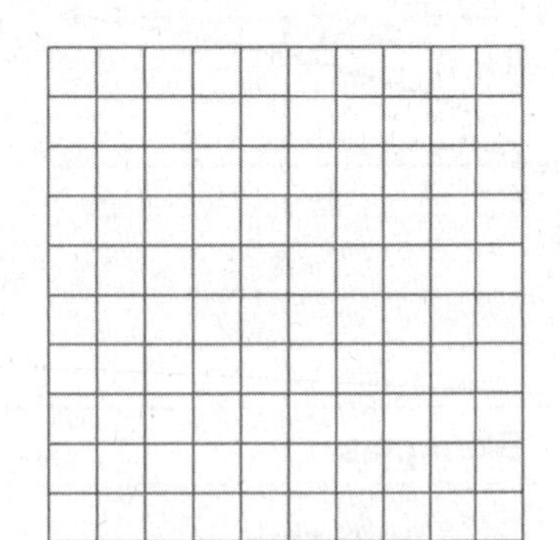

5. $y = 2\sqrt{-x}$

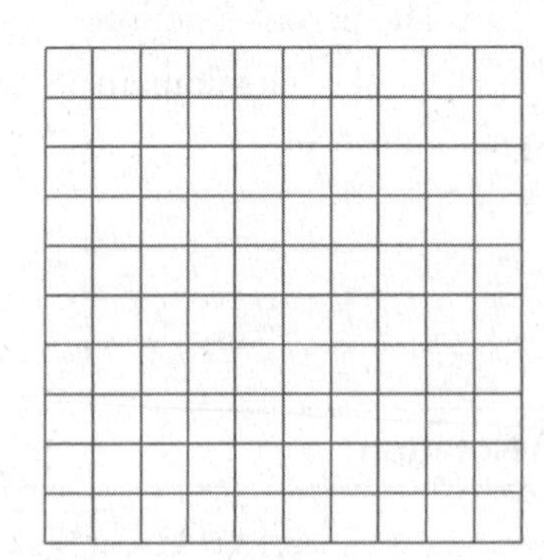

6. $g(x) = \sqrt{x + 3} - 1$

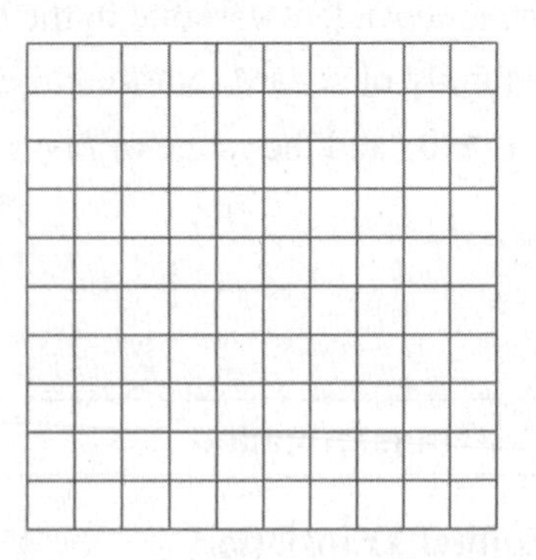

In Exercises 7–9, graph the function. Compare the graph to the graph of $f(x) = \sqrt{x}$.

7. $r(x) = \sqrt{-\frac{1}{2}x}$

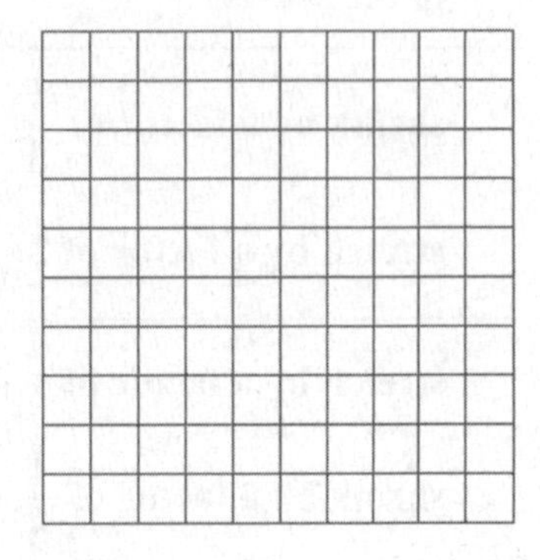

8. $s(x) = -\sqrt{x} - 2$

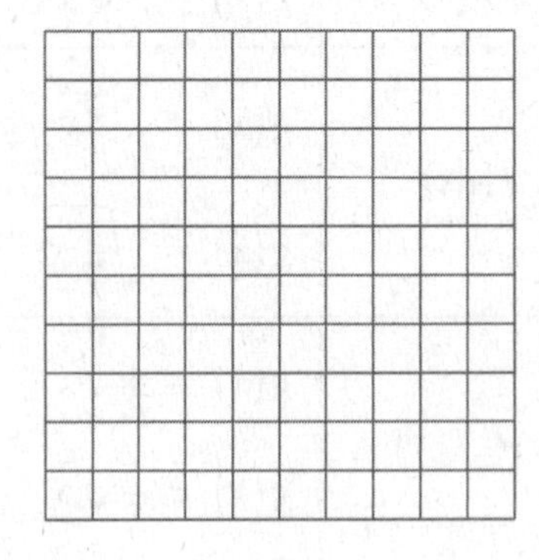

9. $t(x) = \sqrt{x + 4}$

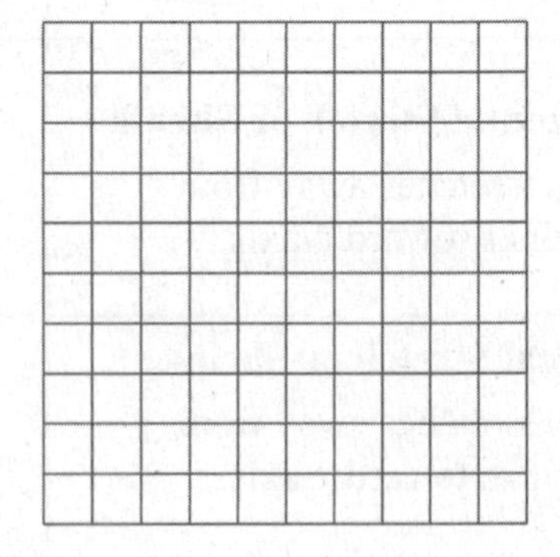

10.1 Notetaking with Vocabulary (continued)

In Exercises 10–12, describe the transformations from the graph of $f(x) = \sqrt{x}$ to the graph the of *h*. Then graph *h*.

10. $h(x) = \frac{1}{2}\sqrt{x + 2} - 2$ **11.** $h(x) = 2\sqrt{x - 3} + 1$ **12.** $h(x) = -\sqrt{x + 4} - 4$

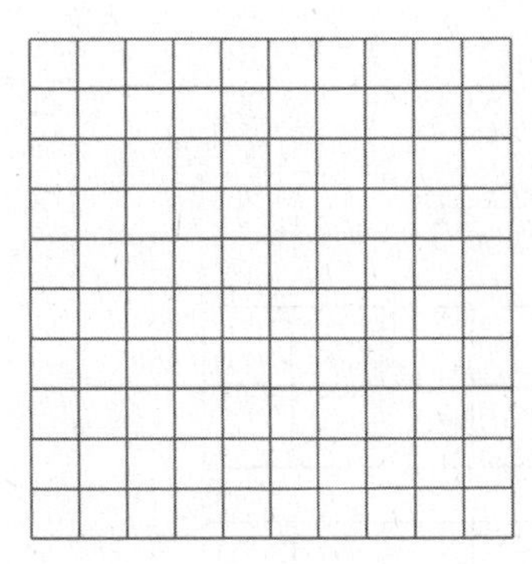

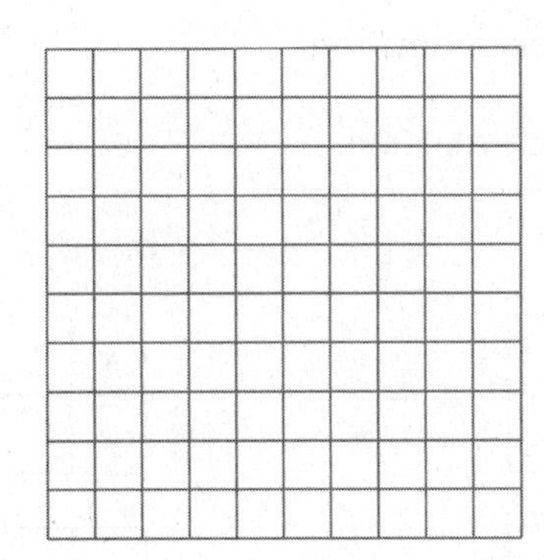

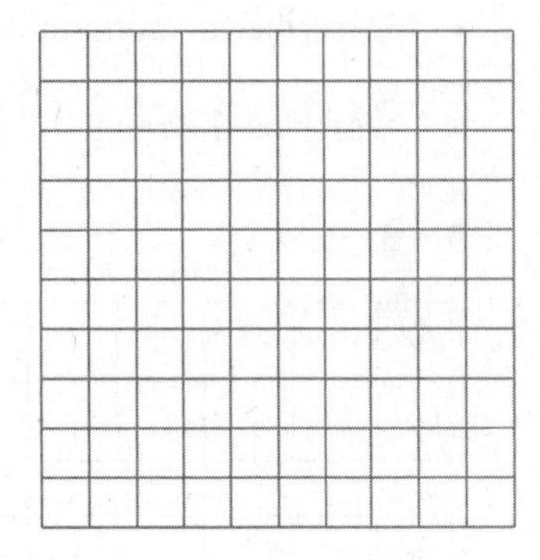

13. The model $S(d) = \sqrt{30df}$ represents the speed S (in miles per hour) of a car before it skids to a stop, where f is the drag factor of the road surface and d is the length (in feet) of the skid marks. The drag factor of Road Surface C is 0.8. The graph shows the speed of the car on Road Surface D. Compare the speeds by finding and interpreting their average rates of change over the interval $d = 0$ to $d = 20$.

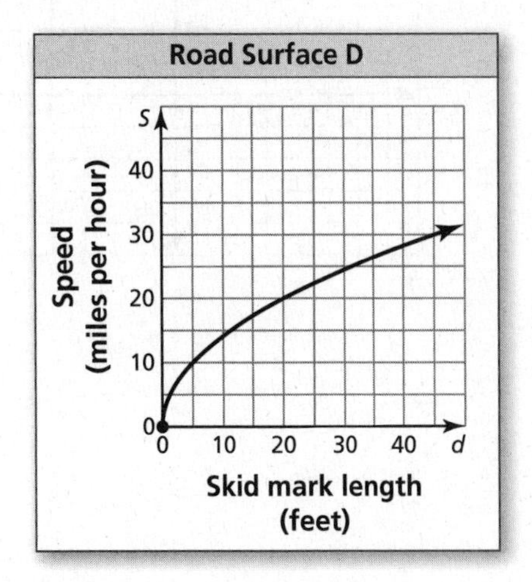

Name ______________________________ Date __________

10.2 Graphing Cube Root Functions

For use with Exploration 10.2

Essential Question What are some of the characteristics of the graph of a cube root function?

1 EXPLORATION: Graphing Cube Root Functions

Work with a partner.

- Make a table of values for each function. Use positive and negative values of x.
- Use the table to sketch the graph of each function.
- Describe the domain of each function.
- Describe the range of each function.

a. $y = \sqrt[3]{x}$

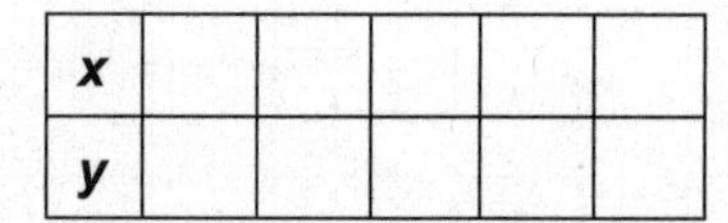

x					
y					

x					
y					

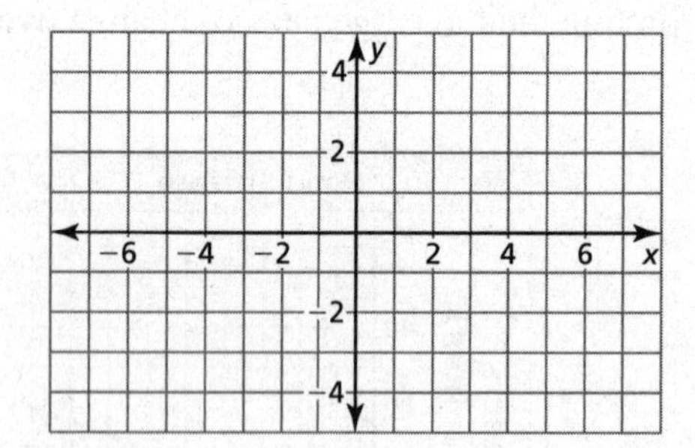

b. $y = \sqrt[3]{x + 3}$

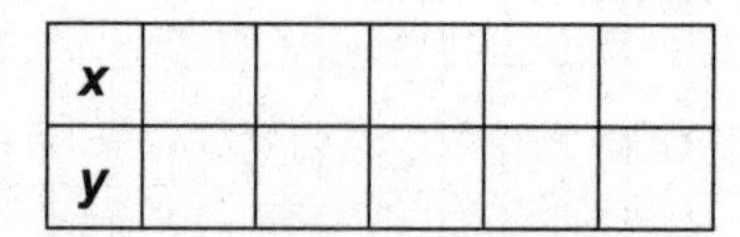

x					
y					

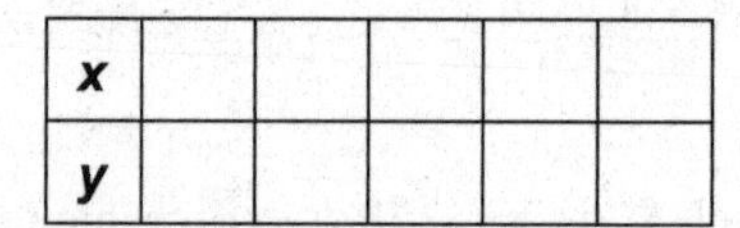

x					
y					

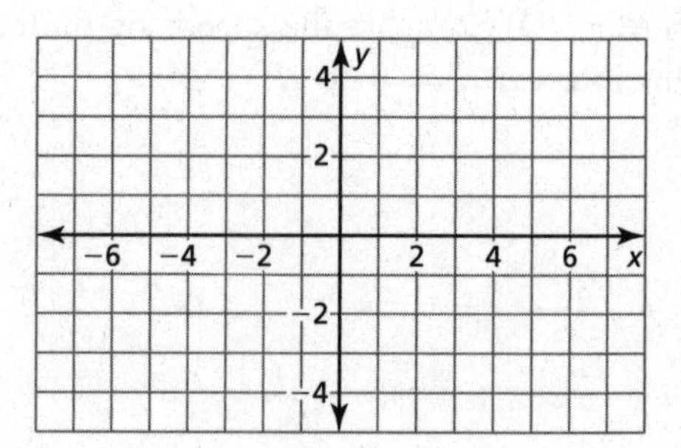

Name_______________________________________ Date__________

10.2 Graphing Cube Root Functions (continued)

2 EXPLORATION: Writing Cube Root Functions

Work with a partner. Write a cube root function, $y = f(x)$, that has the given values. Then use the function to complete the table.

a.

x	$f(x)$
−4	0
−3	
−2	
−1	$\sqrt[3]{3}$
0	

x	$f(x)$
1	
2	
3	
4	2
5	

b.

x	$f(x)$
−4	1
−3	
−2	
−1	$1 + \sqrt[3]{3}$
0	

x	$f(x)$
1	
2	
3	
4	3
5	

Communicate Your Answer

3. What are some of the characteristics of the graph of a cube root function?

4. Graph each function. Then compare the graph to the graph of $f(x) = \sqrt[3]{x}$.

a. $g(x) = \sqrt[3]{x - 1}$ **b.** $g(x) = \sqrt[3]{x} - 1$ **c.** $g(x) = 2\sqrt[3]{x}$ **d.** $g(x) = -2\sqrt[3]{x}$

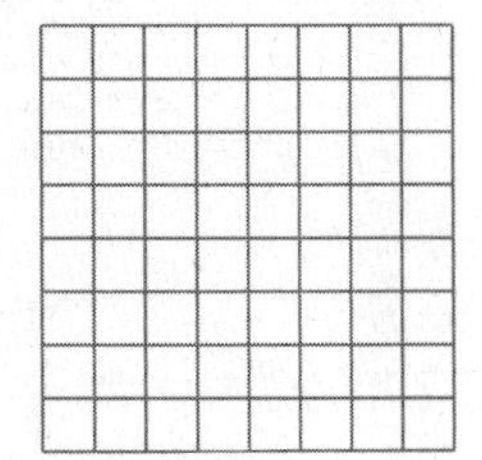
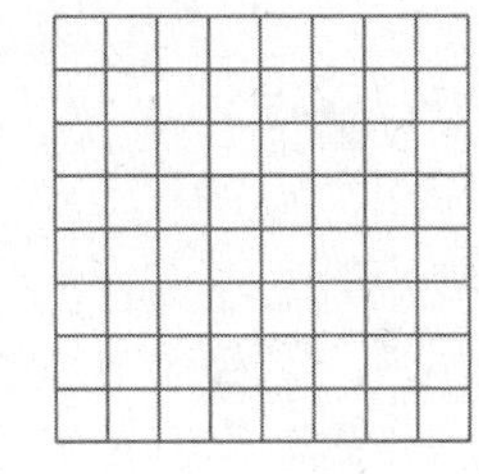
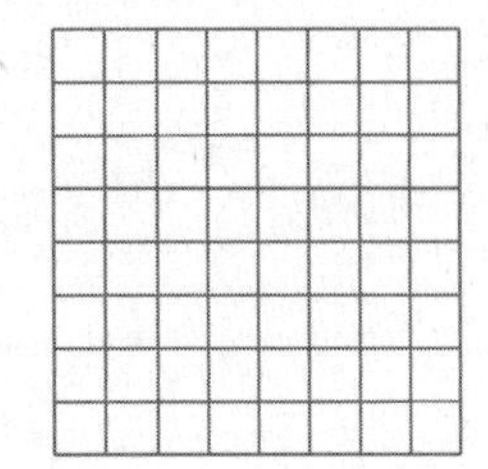
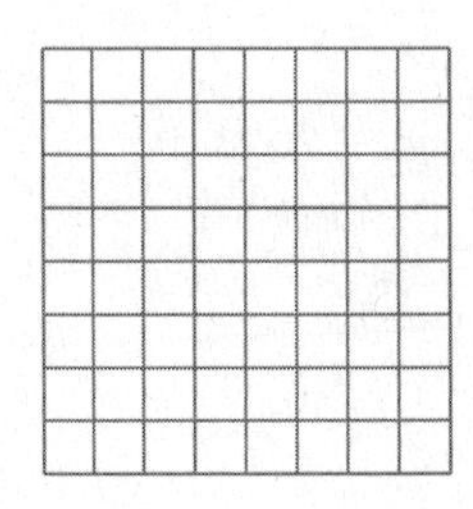

Name ______________________________ Date __________

10.2 Notetaking with Vocabulary

For use after Lesson 10.2

In your own words, write the meaning of each vocabulary term.

cube root function

Core Concepts

Cube Root Functions

A **cube root function** is a radical function with an index of 3. The parent function for the family of cube root functions is $f(x) = \sqrt[3]{x}$. The domain and range of f are all real numbers.

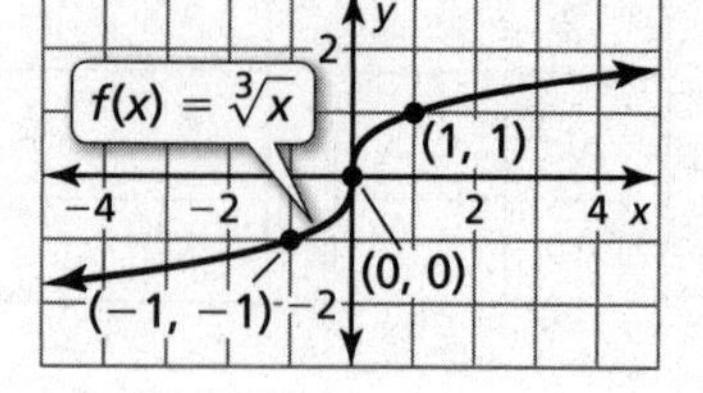

Notes:

Extra Practice

In Exercises 1–6, graph the function. Compare the graph to the graph of $f(x) = \sqrt[3]{x}$.

1. $h(x) = \sqrt[3]{x - 3}$

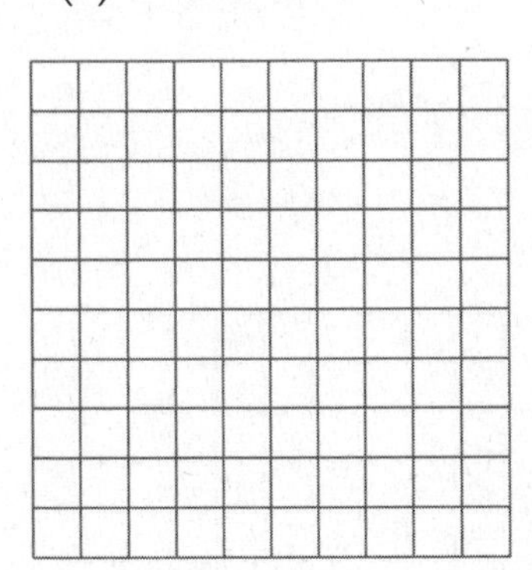

2. $g(x) = \sqrt[3]{x} + 2$

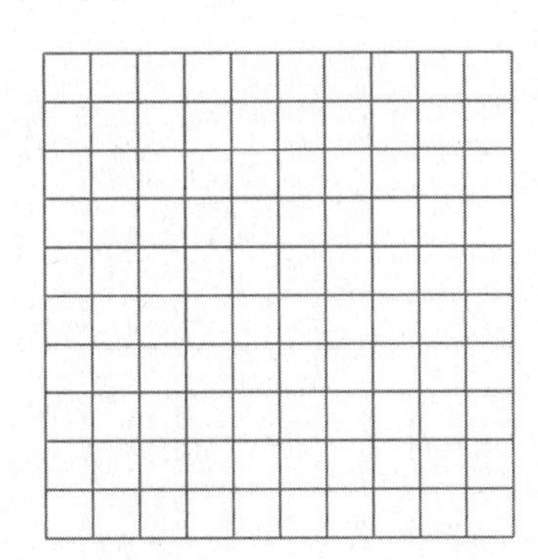

3. $j(x) = 4\sqrt[3]{x}$

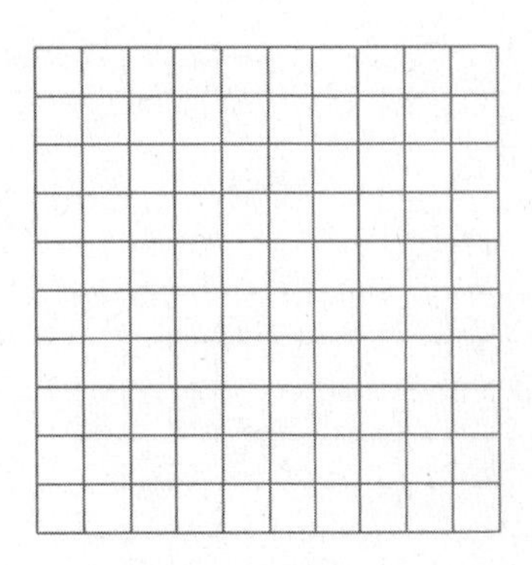

4. $r(x) = -\sqrt[3]{x - 3}$

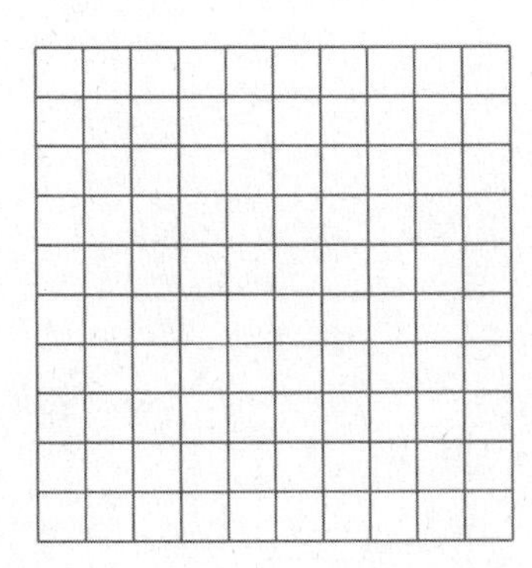

5. $s(x) = 2\sqrt[3]{x} - 1$

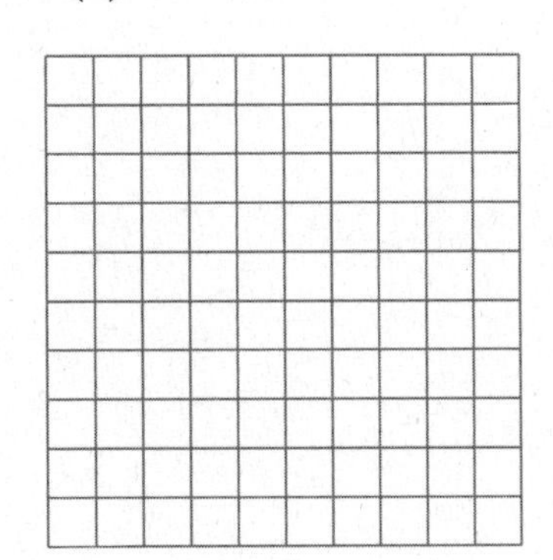

6. $t(x) = \sqrt[3]{-6x} - 2$

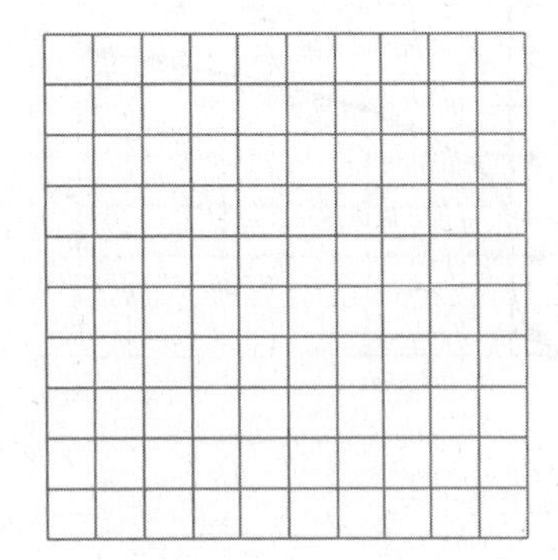

Name ____ Date __________

10.2 Notetaking with Vocabulary (continued)

In Exercises 7–9, describe the transformations from the graph of $f(x) = \sqrt[3]{x}$ to the graph of the given function. Then graph the given function.

7. $p(x) = \sqrt[3]{x - 1} + 1$ **8.** $q(x) = -4\sqrt[3]{x + 2} + 3$ **9.** $r(x) = \frac{1}{2}\sqrt[3]{x + 1} + 4$

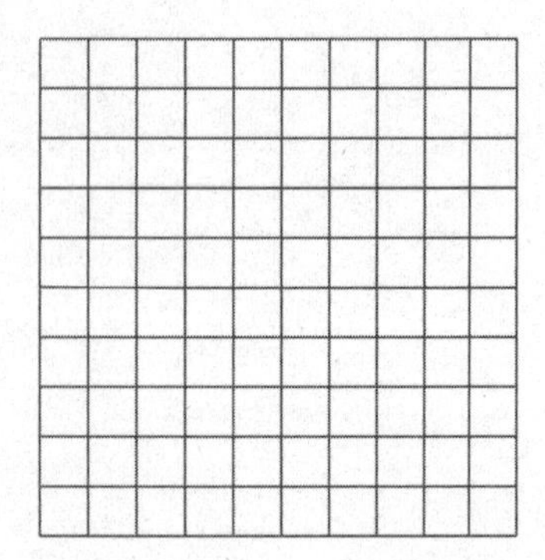

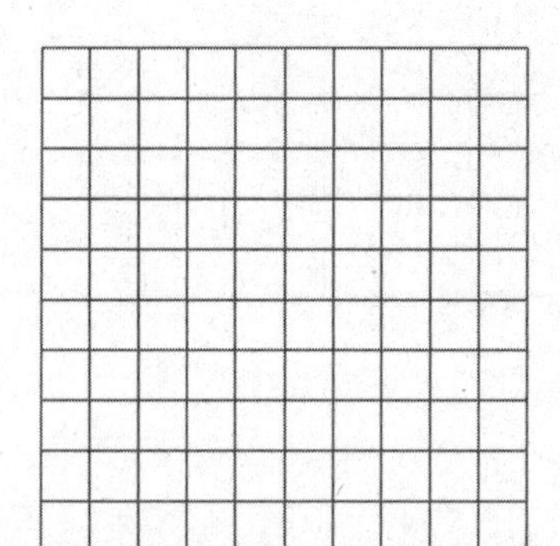

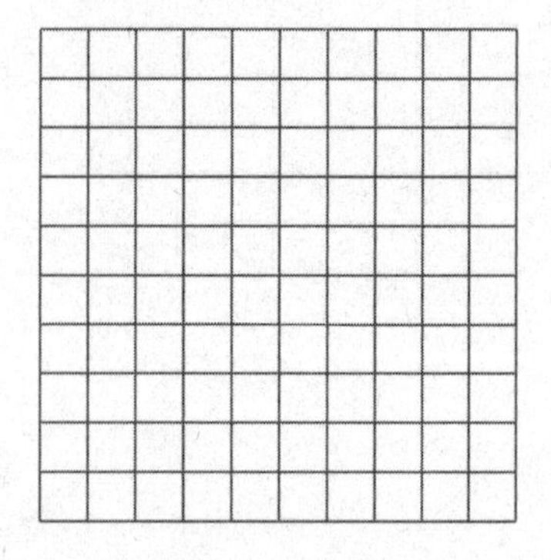

10. The graph of cube root function g is shown. Compare the average rate of change of g to the average rate of change of $h(x) = 2\sqrt[3]{x}$ over the interval $x = 0$ to $x = 8$.

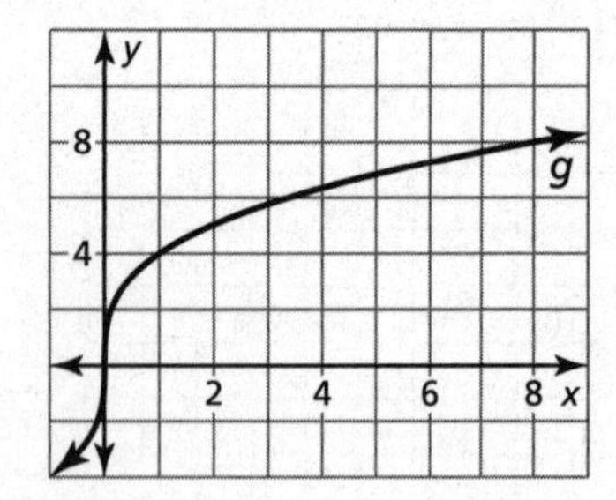

11. The edge length s of a regular tetrahedron is approximately given by $s = \sqrt[3]{8.49V}$, where V is the volume of the tetrahedron. Use a graphing calculator to graph the function. Estimate the volume of a regular tetrahedron with an edge length of 24 inches.

Name____ Date__________

10.3 Solving Radical Equations

For use with Exploration 10.3

Essential Question How can you solve an equation that contains square roots?

1 EXPLORATION: Analyzing a Free-Falling Object

Go to *BigIdeasMath.com* for an interactive tool to investigate this exploration.

Work with a partner. The table shows the time t (in seconds) that it takes a free-falling object (with no air resistance) to fall d feet.

d (feet)	t (seconds)
0	0.00
32	1.41
64	2.00
96	2.45
128	2.83
160	3.16
192	3.46
224	3.74
256	4.00
288	4.24
320	4.47

a. Use the data in the table to sketch the graph of t as a function of d. Use the coordinate plane below.

b. Use your graph to estimate the time it takes the object to fall 240 feet.

c. The relationship between d and t is given by the function

$$r = \sqrt{\frac{d}{16}}.$$

Use this function to check you estimate in part (b).

d. It takes 5 seconds for the object to hit the ground. How far did it fall? Explain your reasoning.

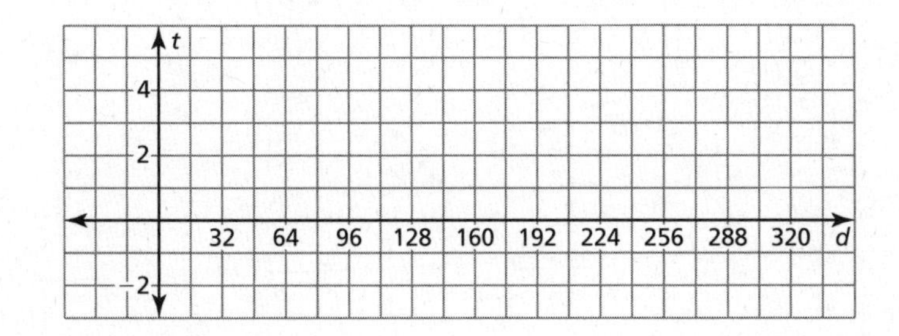

Name ______________________________ Date __________

2 EXPLORATION: Solving a Square Root Equation

Work with a partner. The speed s (in feet per second) of the free-falling object in Exploration 1 is given by the function

$$s = \sqrt{64d}.$$

Find the distance the object has fallen when it reaches each speed.

a. $s = 8$ ft/sec

b. $s = 16$ ft/sec

c. $s = 24$ ft/sec

Communicate Your Answer

3. How can you solve an equation that contains square roots?

4. Use your answer to Question 3 to solve each equation.

a. $5 = \sqrt{x + 20}$

b. $4 = \sqrt{x - 18}$

c. $\sqrt{x} + 2 = 3$

d. $-3 = -2\sqrt{x}$

Name____ Date__________

10.3 Notetaking with Vocabulary
For use after Lesson 10.3

In your own words, write the meaning of each vocabulary term.

radical equation

Core Concepts

Squaring Each Side of an Equation

Words If two expressions are equal, then their squares are also equal.

Algebra If $a = b$, then $a^2 = b^2$.

Notes:

Name ______________________________ Date __________

10.3 Notetaking with Vocabulary (continued)

Extra Practice

In Exercises 1–21, solve the equation. Check your solution(s).

1. $\sqrt{x} = 4$

2. $8 = \sqrt{n} - 3$

3. $3\sqrt{a} - 15 = -6$

4. $\sqrt{s - 3} + 7 = 11$

5. $6\sqrt{t - 2} = 12$

6. $3\sqrt{3x - 6} + 2 = 20$

7. $\sqrt{d} = \sqrt{5d - 8}$

8. $\sqrt{3c - 2} = \sqrt{4c - 6}$

9. $\sqrt{4b - 4} = \sqrt{2b + 4}$

10. $\sqrt{z - 12} = \sqrt{\frac{z}{3} - 3}$

11. $\sqrt{\frac{2v}{3} + 10} = \sqrt{4v - 10}$

12. $\sqrt{3w + 1} - \sqrt{6w} = 0$

10.3 Notetaking with Vocabulary (continued)

13. $5 = \sqrt[3]{x}$

14. $-3 = \sqrt[3]{x + 2}$

15. $\sqrt[3]{7m - 3} = \sqrt[3]{m + 9}$

16. $k + 6 = \sqrt{2k + 15}$

17. $\sqrt{-1 - 2b} = b$

18. $\sqrt{3p + 19} = p - 3$

19. $r - 1 = \sqrt{r + 5}$

20. $\sqrt{2x - 1} + 6 = 3$

21. $k - 1 = \sqrt{5k - 9}$

22. The period P (in seconds) of a pendulum is given by the function $P = 2\pi\sqrt{\frac{L}{32}}$, where L is the pendulum length (in feet). A pendulum has a period of 16 seconds. Is this pendulum 16 times as long as a pendulum with a period of 4 seconds? Explain your reasoning.

Name ______________________________ Date __________

10.4 Inverse of a Function

For use with Exploration 10.4

Essential Question How are a function and its inverse related?

1 EXPLORATION: Exploring Inverse Functions

Work with a partner. The functions f and g are *inverses* of each other. Compare the tables of values of the two functions. How are the functions related?

x	0	0.5	1	1.5	2	2.5	3	3.5
$f(x)$	0	0.25	1	2.25	4	6.25	9	12.25

x	0	0.25	1	2.25	4	6.25	9	12.25
$g(x)$	0	0.5	1	1.5	2	2.5	3	3.5

2 EXPLORATION: Exploring Inverse Functions

Go to *BigIdeasMath.com* for an interactive tool to investigate this exploration.

Work with a partner.

a. Plot the two sets of points represented by the tables in Exploration 1. Use the coordinate plane below.

b. Connect each set of points with a smooth curve.

c. Describe the relationship between the two graphs.

d. Write an equation for each function.

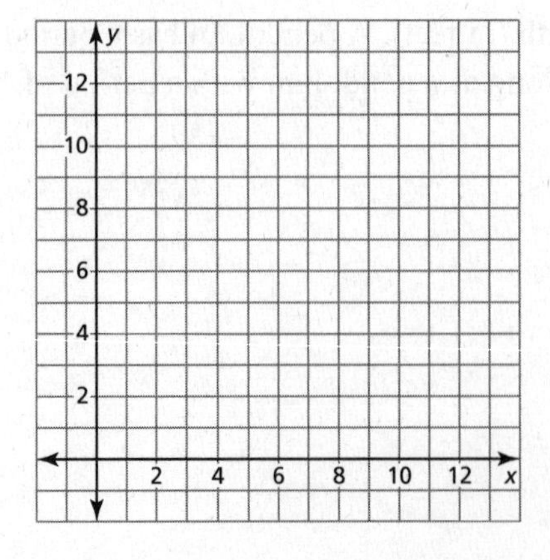

Name______________________________ Date__________

Communicate Your Answer

3. How are a function and its inverse related?

4. A table of values for a function f is given. Create a table of values for a function g, the inverse of f.

x	0	1	2	3	4	5	6	7
$f(x)$	1	2	3	4	5	6	7	8

x								
$g(x)$								

5. Sketch the graphs of $f(x) = x + 4$ and its inverse in the same coordinate plane. Then write an equation of the inverse of f. Explain your reasoning.

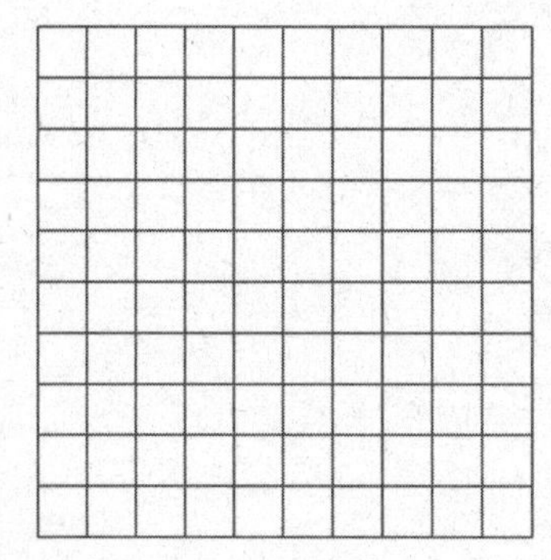

Name ______________________________ Date ________

10.4 Notetaking with Vocabulary
For use after Lesson 10.4

In your own words, write the meaning of each vocabulary term.

inverse relation

inverse function

Core Concepts

Inverse Relation

When a relation contains (a, b), the inverse relation contains (b, a).

Notes:

Finding Inverses of Functions Algebraically

Step 1 Set y equal to $f(x)$.

Step 2 Switch x and y in the equation.

Step 3 Solve the equation for y.

Notes:

Horizontal Line Test

The inverse of a function f is also a function if and only if no horizontal line intersects the graph of f more than once.

Notes:

Name____ Date__________

Extra Practice

In Exercises 1 and 2, find the inverse of the relation.

1. $(1,-1),(2,5),(4,-2),(6,8),(8,9)$

2.

Input	−3	−1	0	1	3
Output	4	2	2	5	3

Input					
Output					

In Exercises 3–5, solve $y = f(x)$ for x. Then find the input when the output is 3.

3. $f(x) = x + 3$

4. $f(x) = 3x - 2$

5. $f(x) = 4x^2$

In Exercises 6–11, find the inverse of the function. Then graph the function and its inverse.

6. $f(x) = 3x - 1$

7. $f(x) = -3x + 2$

8. $f(x) = \frac{1}{2}x + 2$

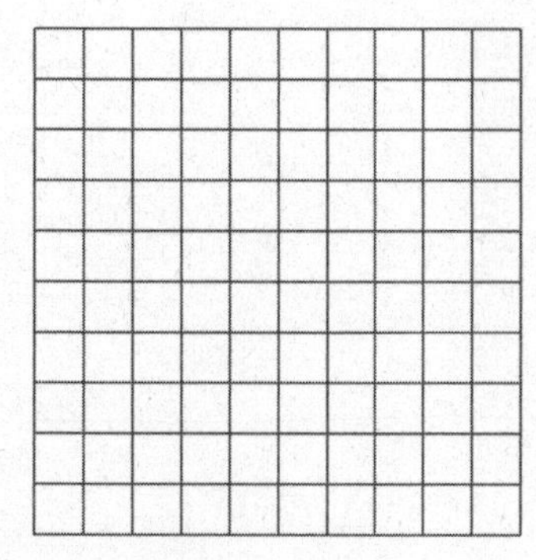

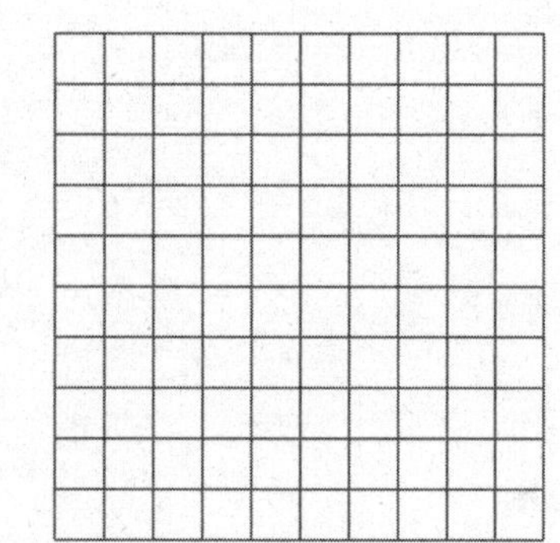

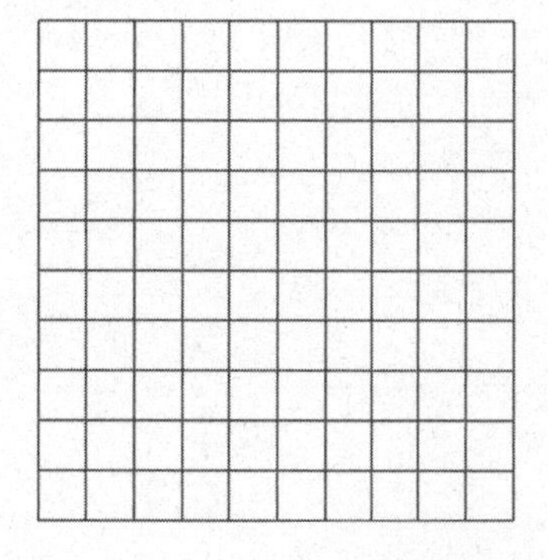

9. $f(x) = 2x^2, x \geq 0$

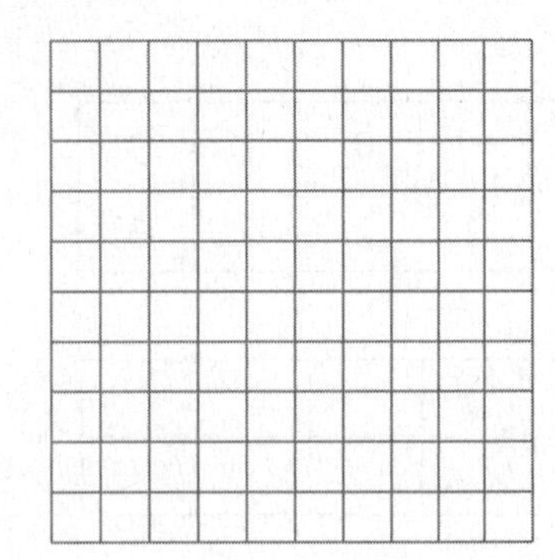

10. $f(x) = -x^2 + 5, x \leq 0$

11. $f(x) = 16x^2 + 3, x \geq 0$

In Exercises 12–17, determine whether the inverse of *f* is a function. Then find the inverse.

12. $f(x) = \sqrt{x + 4}$

13. $f(x) = \sqrt{3x - 9}$

14. $f(x) = 2\sqrt{x - 4}$

15. $f(x) = 3x^2$

16. $f(x) = 5x^2 - 1$

17. $f(x) = -\sqrt{2x + 3} - 5$

Name______________________________ Date__________

Chapter 11 Maintaining Mathematical Proficiency

The table shows the results of a survey. Display the data in a histogram.

1.

Movies attended last month	Frequency
0–1	16
2–3	12
4–5	8

2.

Hours of homework	Frequency
0–1	8
2–3	15
4–5	4
6–7	1

The table shows the results of a survey. Display the data in a circle graph.

3.

Favorite ice cream flavor	Vanilla	Chocolate	Strawberry	Butter Pecan
Students	5	6	4	3

4.

Favorite Sport	Baseball	Tennis	Basketball	Soccer	Golf
Students	10	4	8	7	2

Name ______________________________ Date __________

11.1 Measures of Center and Variation

For use with Exploration 11.1

Essential Question How can you describe the variation of a data set?

1 EXPLORATION: Describing the Variation of Data

Work with a partner. The graphs show the weights of the players on a professional football team and a professional baseball team.

Weights of Players on a Football Team

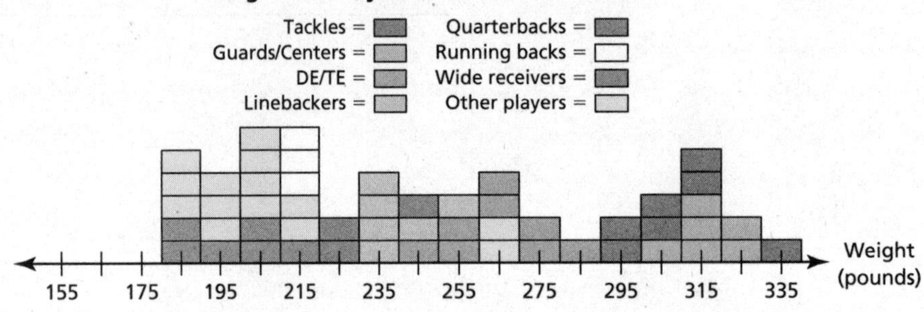

Weights of Players on a Baseball Team

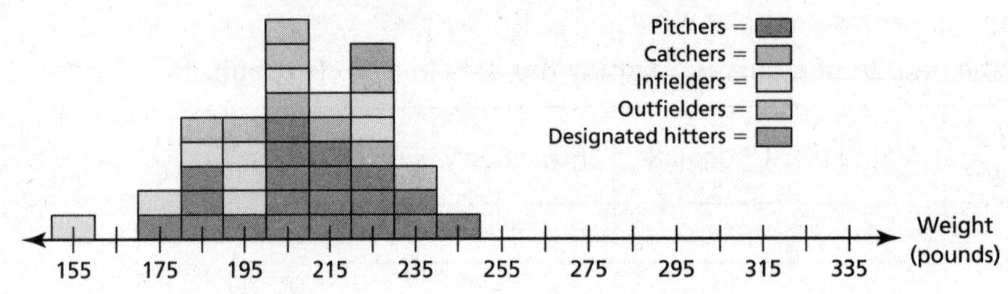

a. Describe the data in each graph in terms of how much the weights vary from the mean. Explain your reasoning.

b. Compare how much the weights of the players on the football team vary from the mean to how much the weights of the players on the baseball team vary from the mean.

c. Does there appear to be a correlation between the body weights and the positions of players in professional football? in professional baseball? Explain.

Name____ Date__________

2 EXPLORATION: Describing the Variation of Data

Go to *BigIdeasMath.com* for an interactive tool to investigate this exploration.

Work with a partner. The weights (in pounds) of the players on a professional basketball team by position are as follows.

Power forwards: 235, 255, 295, 245; small forwards: 235, 235; centers: 255, 245, 325; point guards: 205, 185, 205; shooting guards: 205, 215, 185.

Make a graph that represents the weights and positions of the players. Does there appear to be a correlation between the body weights and the positions of players in professional basketball? Explain your reasoning

Communicate Your Answer

3. How can you describe the variation of a data set?

Name ______________________________ Date __________

11.1 Notetaking with Vocabulary
For use after Lesson 11.1

In your own words, write the meaning of each vocabulary term.

measure of center

mean

median

mode

outlier

measure of variation

range

standard deviation

data transformation

Notes:

Core Concepts

Mean

The **mean** of a numerical data set is the sum of the data divided by the number of data values. The symbol $\overline{x}$ represents the mean. It is read as "*x*-bar."

Median

The **median** of a numerical data set is the middle number when the values are written in numerical order. When a data set has an even number of values, the median is the mean of the two middle values.

Mode

The **mode** of a data set is the value or values that occur most often. There may be one mode, no mode, or more than one mode.

Notes:

Standard Deviation

The **standard deviation** of a numerical data set is a measure of how much a typical value in the data set differs from the mean. The symbol σ represents the standard deviation. It is read as "sigma." It is given by

$$\sigma = \sqrt{\frac{(x_1 - \overline{x})^2 + (x_2 - \overline{x})^2 + \cdots + (x_n - \overline{x})^2}{n}}$$

where n is the number of values in the data set. The deviation of a data value x is the difference of the data value and the mean of the data set, $x - \overline{x}$.

Step 1 Find the mean, $\overline{x}$.

Step 2 Find the deviation of each data value, $x - \overline{x}$.

Step 3 Square each deviation, $(x - \overline{x})^2$.

Step 4 Find the mean of the squared deviations. This is called the *variance*.

Step 5 Take the square root of the variance.

Notes:

Name ______________________________ Date __________

11.1 Notetaking with Vocabulary (continued)

Data Transformations Using Addition

When a real number k is added to each value in a numerical data set

- the measures of center of the new data set can be found by adding k to the original measures of center.
- the measures of variation of the new data set are the *same* as the original measures of variation.

Data Transformations Using Multiplication

When each value in a numerical data set is multiplied by a real number k, where $k > 0$, the measures of center and variation can be found by multiplying the original measures by k.

Notes:

Extra Practice

1. Consider the data set: 2, 5, 16, 2, 2, 7, 3, 4, 4.

a. Find the mean, median, and mode of the data set.

b. Determine which measure of center best represents the data. Explain.

2. The table shows the masses of eight gorillas.

Masses (kilograms)							
160	157	162	158	44	160	159	161

a. Identify the outlier. How does the outlier affect the mean, median, and mode?

b. Describe one possible explanation for the outlier.

11.1 Notetaking with Vocabulary (continued)

3. The heights of the members of two girls' basketball teams are shown. Find the range of the heights for each team. Compare your results.

Team A Heights (inches)									
58	75	60	48	56	78	60	57	54	59

Team B Heights (inches)									
49	50	70	56	58	66	64	57	62	63

4. Consider the data in Exercise 3.

a. Find the standard deviation of the heights of Team A. Interpret your result.

b. Find the standard deviation of the heights of Team B. Interpret your result.

c. Compare the standard deviations for Team A and Team B. What can you conclude?

5. Find the values of the measures shown when each value in the data set increases by 8.

Mean: 42 Median: 40 Mode: 38
Range: 15 Standard deviation: 4.9

Name ________________________________ Date __________

11.2 Box-and-Whisker Plots

For use with Exploration 11.2

Essential Question How can you use a box-and-whisker plot to describe a data set?

1 EXPLORATION: Drawing a Box-and-Whisker Plot

Go to *BigIdeasMath.com* for an interactive tool to investigate this exploration.

Work with a partner. The numbers of first cousins of the students in a ninth-grade class are shown. A *box-and-whisker plot* is one way to represent the data visually.

Numbers of First Cousins			
3	10	18	8
9	3	0	32
23	19	13	8
6	3	3	10
12	45	1	5
13	24	16	14

a. Order the data on a strip of grid paper with 24 equally spaced boxes.

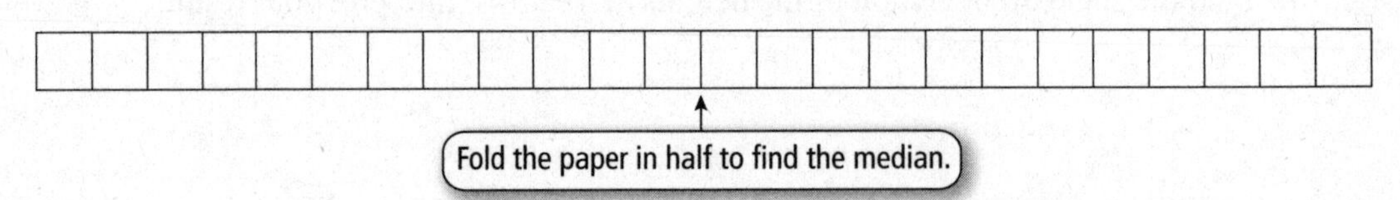

b. Fold the paper in half again to divide the data into four groups. Because there are 24 numbers in the data set, each group should have 6 numbers. Find the least value, the greatest value, the first quartile, and the third quartile.

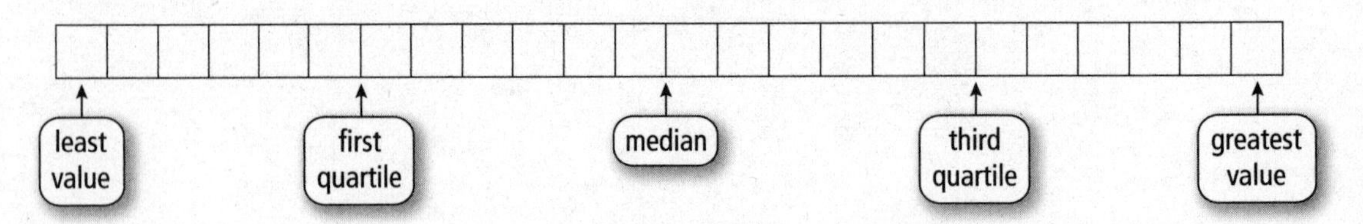

1 EXPLORATION: Drawing a Box-and-Whisker Plot (continued)

c. Explain how the box-and-whisker plot shown represents the data set.

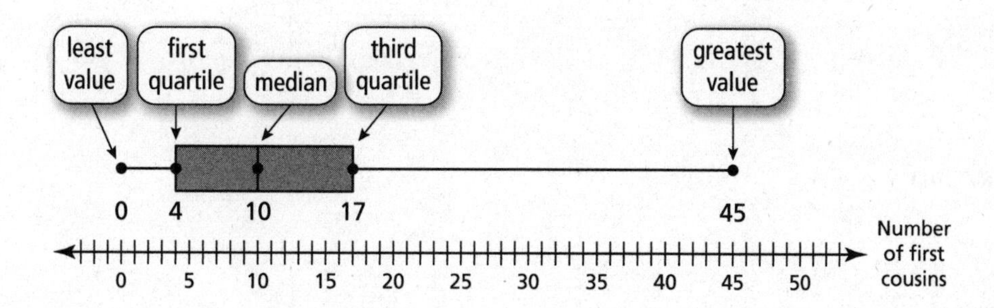

Communicate Your Answer

2. How can you use a box-and-whisker plot to describe a data set?

3. Interpret each box-and-whisker plot.

a. body mass indices (BMI) of students in a ninth-grade class

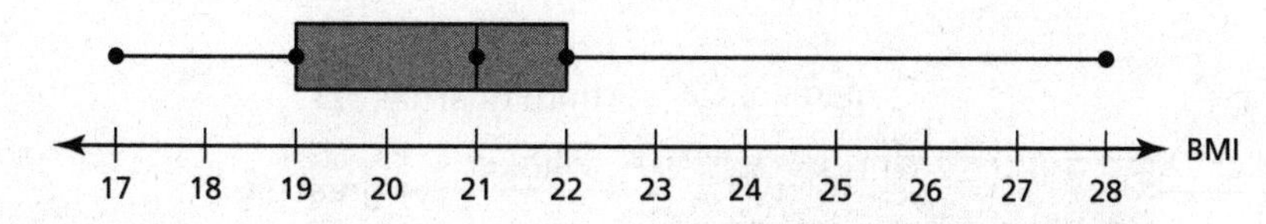

b. heights of roller coasters at an amusement park

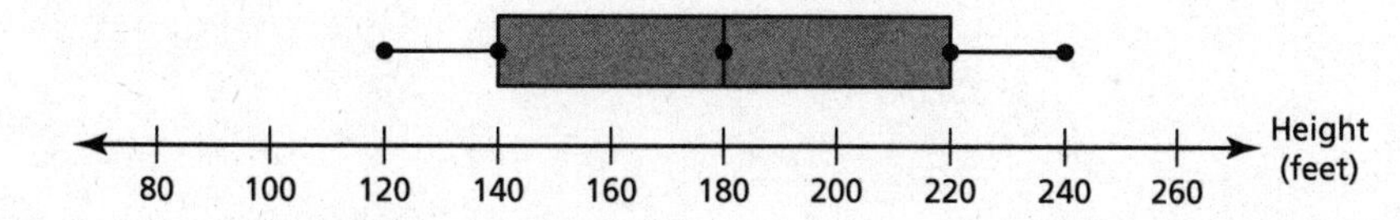

Name ____ Date __________

11.2 Notetaking with Vocabulary

For use after Lesson 11.2

In your own words, write the meaning of each vocabulary term.

box-and-whisker plot

quartile

five-number summary

interquartile range

Core Concepts

Box-and-Whisker Plot

A **box-and-whisker plot** shows the variability of a data set along a number line using the least value, the greatest value, and the *quartiles* of the data. **Quartiles** divide the data set into four equal parts. The median (second quartile, Q2) divides the data set into two halves. The median of the lower half is the first quartile, Q1. The median of the upper half is the third quartile, Q3.

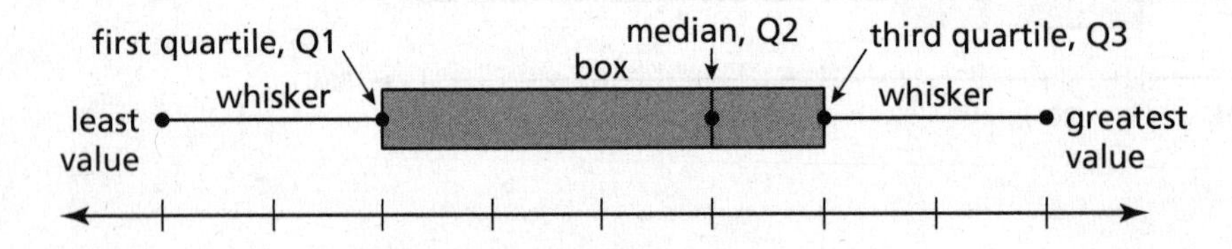

The five numbers that make up a box-and-whisker plot are called the **five-number summary** of the data set.

Notes:

11.2 Notetaking with Vocabulary (continued)

Shapes of Box-and-Whisker Plots

Skewed left

- The left whisker is longer than the right whisker.
- Most of the data are on the right side of the plot.

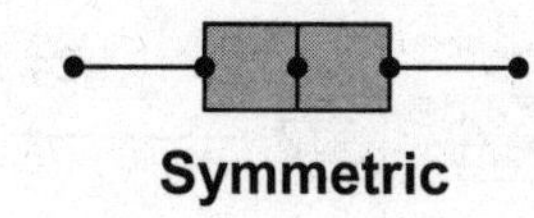

Symmetric

- The whiskers are about the same length.
- The median is in the middle of the plot.

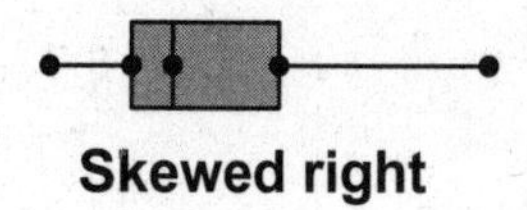

Skewed right

- The right whisker is longer than the left whisker.
- Most of the data are on the left side of the plot.

Notes:

Extra Practice

In Exercises 1 and 2, make a box-and-whisker plot that represents the data.

1. Hours of sleep: 7, 9, 8, 8, 8, 6, 6, 5, 4

2. Algebra test scores: 71, 92, 84, 76, 88, 96, 84, 63, 82

Name ______________________________ Date __________

3. The box-and-whisker plot represents the prices (in dollars) of soccer balls at different sporting goods stores.

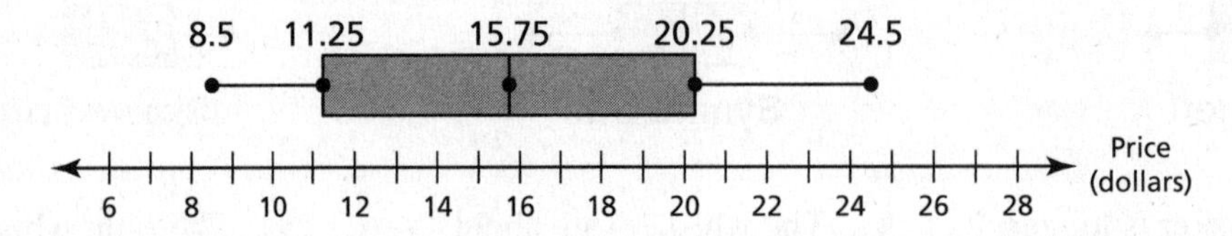

a. Find and interpret the range of the data.

b. Describe the distribution of the data.

c. Find and interpret the interquartile range of the data.

d. Are the data more spread out below Q1 or above Q3? Explain.

4. The double box-and-whisker plot represents the number of tornados per month for a year for two states.

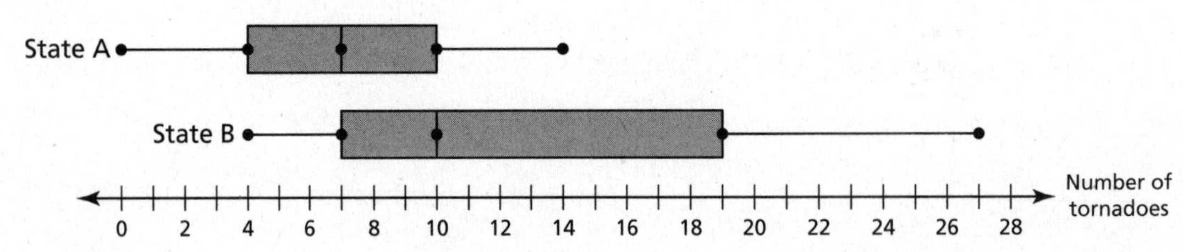

a. Identify the shape of each distribution.

b. Which state's tornadoes are more spread out? Explain.

c. Which state had the single least number of tornadoes in a month during the year? Explain.

Name ____________________ Date __________

11.3 Shapes of Distributions

For use with Exploration 11.3

Essential Question How can you use a histogram to characterize the basic shape of a distribution?

1 EXPLORATION: Analyzing a Famous Symmetric Distribution

Work with a partner. A famous data set was collected in Scotland in the mid-1800s. It contains the chest sizes, measured in inches, of 5738 men in the Scottish Militia. Estimate the percent of the chest sizes that lie within (a) 1 standard deviation of the mean, (b) 2 standard deviations of the mean, and (c) 3 standard deviations of the mean. Explain your reasoning.

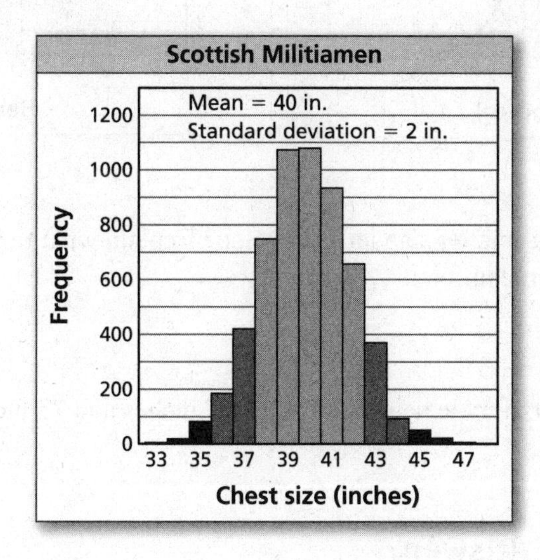

Name ______________________________ Date __________

2 EXPLORATION: Comparing Two Symmetric Distributions

Work with a partner. The graphs show the distributions of the heights of 250 adult American males and 250 adult American females.

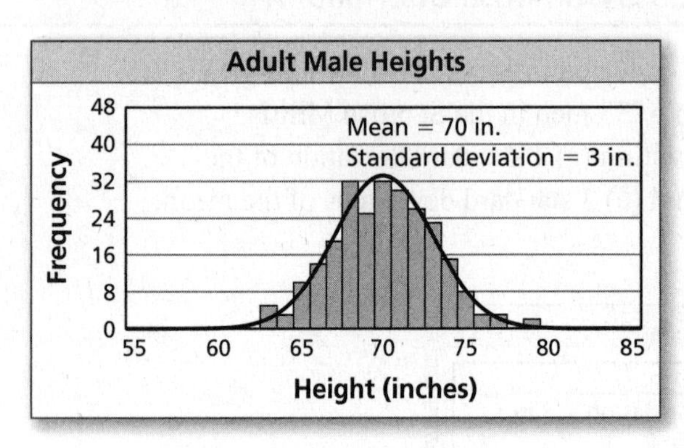

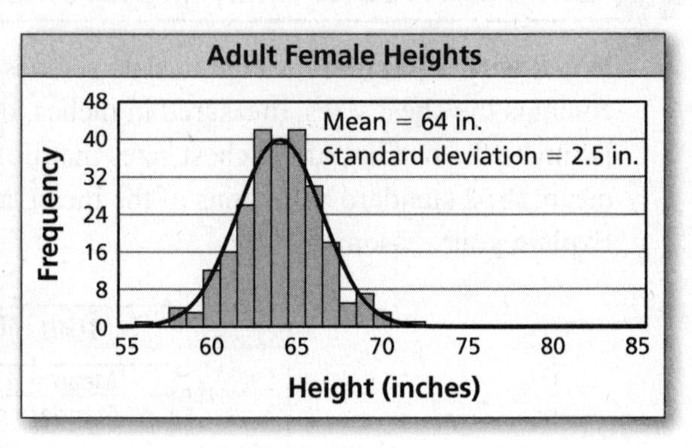

a. Which data set has a smaller standard deviation? Explain what this means in the context of the problem.

b. Estimate the percent of male heights between 67 inches and 73 inches.

Communicate Your Answer

3. How can you use a histogram to characterize the basic shape of a distribution?

4. All three distributions in Explorations 1 and 2 are roughly symmetric. The histograms are called "bell-shaped."

a. What are the characteristics of a symmetric distribution?

b. Why is a symmetric distribution called "bell-shaped?"

c. Give two other real-life examples of symmetric distributions.

Name____ Date__________

11.3 Notetaking with Vocabulary

For use after Lesson 11.3

In your own words, write the meaning of each vocabulary term.

histogram

frequency table

Core Concepts

Symmetric and Skewed Distributions

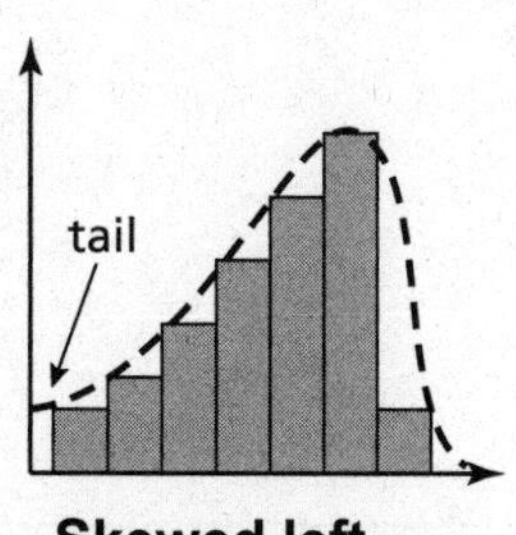

Skewed left

- The "tail" of the graph extends to the left.
- Most of the data are on the right.

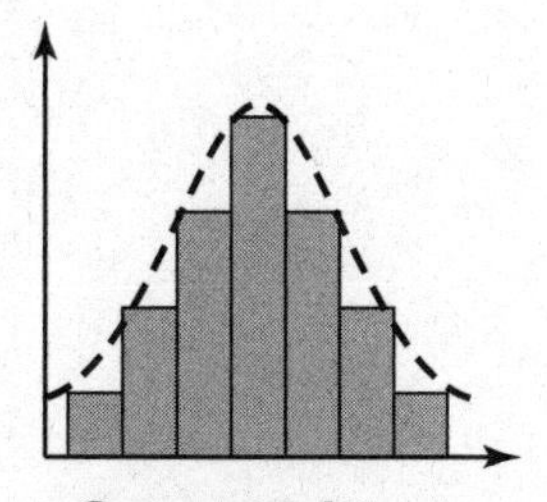

Symmetric

- The data on the right of the distribution are approximately a mirror image of the data on the left of the distribution.

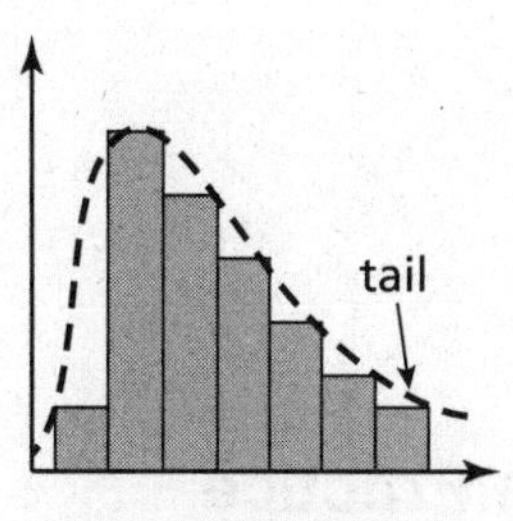

Skewed right

- The "tail" of the graph extends to the right.
- Most of the data are on the left.

Notes:

11.3 Notetaking with Vocabulary (continued)

Choosing Appropriate Measures

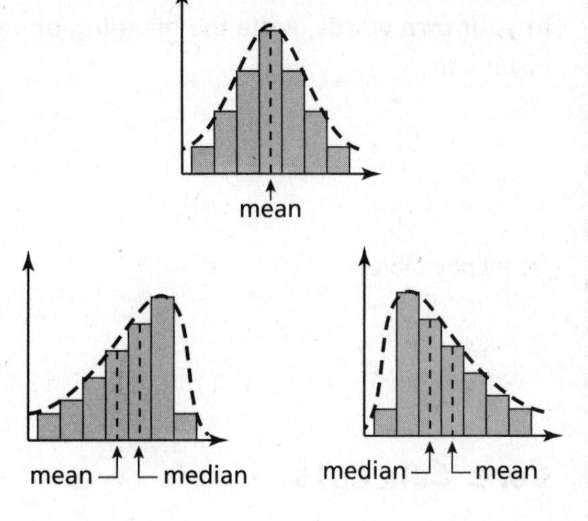

When a data distribution is symmetric,

- use the mean to describe the center and
- use the standard deviation to describe the variation.

When a data distribution is skewed,

- use the median to describe the center and
- use the five-number summary to describe the variation.

Notes:

Extra Practice

1. The table shows the average annual snowfall (in inches) of 26 cities.

   a. Display the data in a histogram using six intervals beginning with 15–28.

Average Annual Snowfall (inches)		
22	68	33
15	28	31
20	18	30
15	54	16
44	43	17
95	41	30
29	23	47
37	26	54
16	30	

   b. Which measures of center and variation best represent the data? Explain.

   c. A weather station lists the top 20 snowiest major cities. The city in 20th place had 51 inches of snow. How would you interpret the data?

11.3 Notetaking with Vocabulary (continued)

2. The double histogram shows the distributions of monthly precipitation for two towns over a 50-month period. Compare the distributions using their shapes and appropriate measures of center and variation.

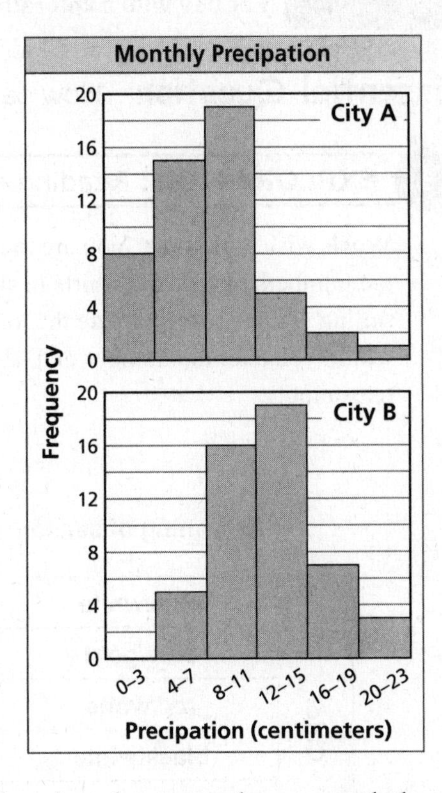

3. The table shows the results of a survey that asked high school students how many hours a week they listen to music.

a. Make a double box-and-whisker plot that represents the data. Describe the shape of each distribution.

	Females	Males
Survey size	50	58
Minimum	16	18
Maximum	40	52
1st Quartile	24	30
Median	28	38
3rd Quartile	32	46
Mean	28	30
Standard deviation	6	12

b. Compare the number of hours of music listened to by females to the number of hours of music listened to by males.

c. About how many females surveyed would you expect to listen to music between 22 and 34 hours per week?

d. If you survey 100 more females, about how many would you expect to listen to music between 16 and 40 hours per week?

Name ______________________________ Date __________

11.4 Two-Way Tables

For use with Exploration 11.4

Essential Question How can you read and make a two-way table?

1 EXPLORATION: Reading a Two-Way Table

Work with a partner. You are the manager of a sports shop. The two-way tables show the numbers of soccer T-shirts in stock at your shop at the beginning and end of the selling season. (a) Complete the totals for the rows and columns in each table. (b) How would you alter the number of T-shirts you order for next season? Explain your reasoning.

	Beginning of season	T-Shirt Size					
		S	M	L	XL	XXL	Total
Color	blue/white	5	6	7	6	5	
	blue/gold	5	6	7	6	5	
	red/white	5	6	7	6	5	
	black/white	5	6	7	6	5	
	black/gold	5	6	7	6	5	
	Total						145

	End of season	T-Shirt Size					
		S	M	L	XL	XXL	Total
Color	blue/white	5	4	1	0	2	
	blue/gold	3	6	5	2	0	
	red/white	4	2	4	1	3	
	black/white	3	4	1	2	1	
	black/gold	5	2	3	0	2	
	Total						

Name____ Date__________

2 EXPLORATION: Making a Two-Way Table

Work with a partner. The three-dimensional bar graph shows the numbers of hours students work at part-time jobs.

a. Make a two-way table showing the data. Use estimation to find the entries in your table.

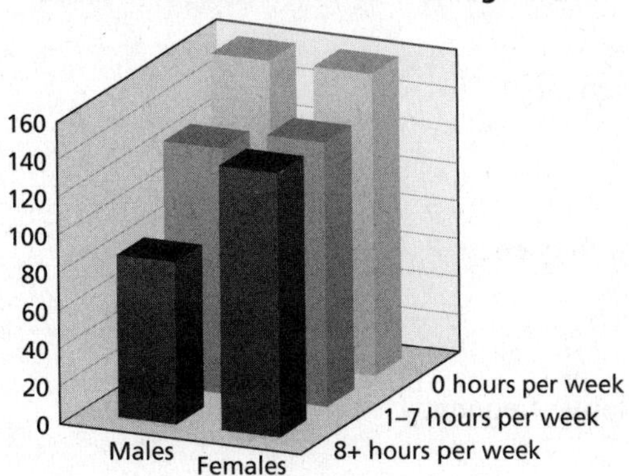

b. Write two observations that summarize the data in your table.

Communicate Your Answer

3. How can you read and make a two-way table?

Name ____________________ Date ________

11.4 Notetaking with Vocabulary
For use after Lesson 11.4

In your own words, write the meaning of each vocabulary term.

two-way table

joint frequency

marginal frequency

joint relative frequency

marginal relative frequency

conditional relative frequency

Core Concepts

Relative Frequencies

A **joint relative frequency** is the ratio of a frequency that is not in the "total" row or the "total" column to the total number of values or observations.

A **marginal relative frequency** is the sum of the joint relative frequencies in a row or column.

When finding relative frequencies in a two-way table, you can use the corresponding decimals or percents.

Notes:

Name______________________________ Date__________

11.4 Notetaking with Vocabulary (continued)

Conditional Relative Frequencies

A **conditional relative frequency** is the ratio of a joint relative frequency to the marginal relative frequency. You can find a conditional relative frequency using a row total or a column total of a two-way table.

Notes:

Extra Practice

In Exercises 1 and 2, find and interpret the marginal frequencies.

1.

		Attend College	
		Yes	**No**
Gender	**Male**	98	132
	Female	120	88

2.

		Own a Car	
		Yes	**No**
Gender	**Male**	54	136
	Female	45	137

3. You conduct a survey that asks 85 students in your school whether they are in Math Club or Chess Club. Thirty-five of the students are in Math Club, and 20 of those students are also in Chess Club. Thirty-eight of the students are not in Math Club or Chess Club. Organize the results in a two-way table. Include the marginal frequencies.

11.4 Notetaking with Vocabulary (continued)

4. Make a two-way table that shows the joint and marginal relative frequencies.

		Read *Catcher in the Rye*	
		Yes	**No**
Gender	**Male**	96	80
	Female	54	88

5. A company is organizing a baseball game for their employees. The employees are asked whether they prefer to attend a day game or a night game. They are also asked whether they prefer to sit in the upper deck or lower deck. The results are shown in a two-way table. Make a two-way table that shows the conditional relative frequencies based on the row totals. Given that an employee prefers to go to a day game, what is the conditional relative frequency that he or she prefers to sit in the lower deck?

		Seat	
		Upper	**Lower**
Game Time	**Day**	28	34
	Night	22	52

Name____ Date__________

11.5 Choosing a Data Display
For use with Exploration 11.5

Essential Question How can you display data in a way that helps you make decisions?

1 EXPLORATION: Displaying Data

Work with a partner. Analyze the data and then create a display that best represents the data. Explain your choice of data display.

a. A group of schools in New England participated in a 2-month study and reported 3962 animals found dead along roads.

birds: 307
mammals: 2746
amphibians: 145
reptiles: 75
unknown: 689

b. The data below show the numbers of black bears killed on a state's roads from 1993 to 2012.

1993: 30	2003: 74
1994: 37	2004: 88
1995: 46	2005: 82
1996: 33	2006: 109
1997: 43	2007: 99
1998: 35	2008: 129
1999: 43	2009: 111
2000: 47	2010: 127
2001: 49	2011: 141
2002: 61	2012: 135

c. A 1-week study along a 4-mile section of road found the following weights (in pounds) of raccoons that had been killed by vehicles.

13.4	14.8	17.0	12.9	21.3	21.5	16.8	14.8
15.2	18.7	18.6	17.2	18.5	9.4	19.4	15.7
14.5	9.5	25.4	21.5	17.3	19.1	11.0	12.4
20.4	13.6	17.5	18.5	21.5	14.0	13.9	19.0

Name ________________________________ Date __________

1 EXPLORATION: Displaying Data (continued)

d. A yearlong study by volunteers in California reported the following numbers of animals killed by motor vehicles.

raccoons: 1693

skunks: 1372

ground squirrels: 845

opossum: 763

deer: 761

gray squirrels: 715

cottontail rabbits: 629

barn owls: 486

jackrabbits: 466

gopher snakes: 363

Communicate Your Answer

2. How can you display data in a way that helps you make decisions?

3. Use the Internet or some other reference to find examples of the following types of data displays.

bar graph

circle graph

scatter plot

stem-and-leaf plot

pictograph

line graph

box-and-whisker plot

histogram

dot plot

Name____ Date__________

11.5 Notetaking with Vocabulary

For use after Lesson 11.5

In your own words, write the meaning of each vocabulary term.

qualitative (categorical) data

quantitative data

misleading graph

Core Concepts

Types of Data

Qualitative data, or **categorical data,** consist of labels or nonnumerical entries that can be separated into different categories. When using qualitative data, operations such as adding or finding a mean do not make sense.

Quantitative data consist of numbers that represent counts or measurements.

Notes:

Name ____ Date __________

Extra Practice

In Exercises 1–4, tell whether the data are *qualitative* or *quantitative*. Explain your reasoning.

1. bookmarks in your web browser

2. heights of players on a basketball team

3. the number of kilobytes in a downloaded file

4. FM radio station numbers

In Exercises 5 and 6, analyze the data and then create a display that best represents the data. Explain your reasoning.

5.

Home Runs Each Year											
Babe Ruth						Hank Aaron					
0	4	3	2	11	29	13	27	26	44	30	39
54	59	35	41	46	25	40	34	45	44	24	32
47	60	54	46	49	46	44	39	29	44	38	47
41	34	22	6			34	40	20	12	10	

Name____ Date__________

6.

Total Points Scored by a Basketball Team for Each Game					
48	56	49	52	40	65
30	47	62	40	59	37
45	41	44	33	44	30

In Exercises 7 and 8, describe how the graph is misleading. Then explain how someone might misinterpret the graph.

7.

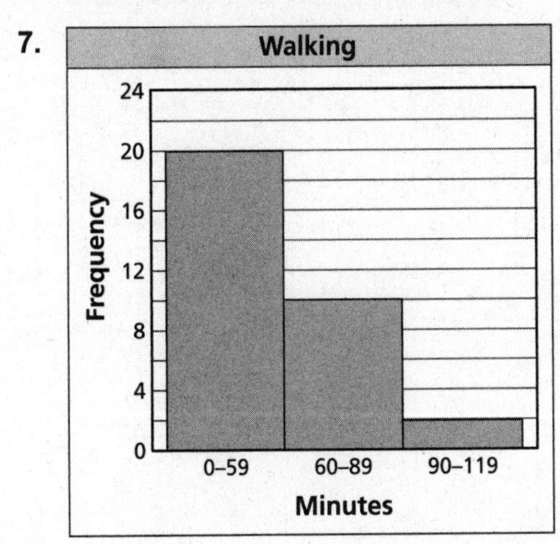

8.

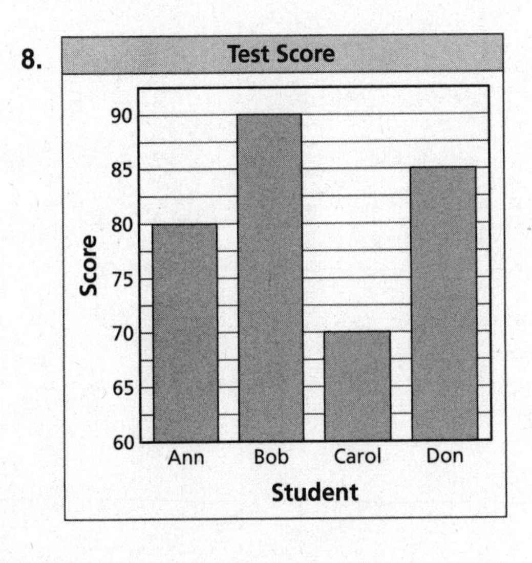